A FIRST COURSE IN LINEAR ALGEBRA

SECOND EDITION

A FIRST COURSE IN LINEAR ALGEBRA

SECOND EDITION

Daniel Zelinsky

Northwestern University

ACADEMIC PRESS New York San Francisco London

A Subsidiary of Harcourt Brace Jovanovich, Publishers

ACADEMIC PRESS, INC.
111 Fifth Avenue, New York, New York 10003

United Kingdom Edition published by
ACADEMIC PRESS, INC. (LONDON) LTD.
24/28 Oval Road, London NW1

LIBRARY OF CONGRESS CATALOG CARD NUMBER: 72-88368

PRINTED IN THE UNITED STATES OF AMERICA

CONTENTS

PREFACE

This book is meant as an introduction to the algebra and geometry of vectors, matrices, and linear transformations. The concept of function is used freely and the general level of sophistication assumed is roughly that of a student who has studied some calculus. However, calculus is not required for an understanding of the text. At Northwestern we have used the book as a text for the third quarter of the freshman year, preceded by two quarters of calculus of functions of one variable. During a ten-week quarter we have usually covered the first six chapters, assuming the students have no knowledge of vectors of any kind. By skimming over a few topics (for example: the cross product, which is used after Chapter 1 only as a convenient device for finding one vector perpendicular to two others; abstract vector spaces; and determinants), it should be possible to introduce part of Chapter 7 as well.

A much richer course than this would be desirable. One would like, for example, to deal more deeply with vector spaces of functions, to exhibit and exploit the concept of isomorphism of vector spaces, and to study eigenvalues and their properties much more thoroughly. However, we have found that at this level restraint is the better part of pedagogy. The student seriously interested in mathematics and its applications will deepen his understanding in later courses, especially when he begins to use this linear algebra for other things. The teacher who chooses to add this material to the course will find many opportunities and examples ready and waiting.

This text is designed as a background for second-year courses in: (1) calculus of several variables, which studies nonlinear functions on n-space by approximating such nonlinear functions by linear ones (the present text is the study of those linear ones); (2) differential equations, where the theory of linear differential equations parallels that of linear algebraic equations. Linear differential operators can be treated as linear functions, as in the present text, provided that a little work on function spaces is introduced. These ideas exactly parallel the suggestions of the Mathematical Association of America's Committee on the Undergraduate Program in Mathematics. No serious attempt is made to cover applications of linear algebra, but illustrations with a flavor of linear programming, input–output analysis, and Markov processes appear in the text and the problems.

If this book has a theme, it is that classical vector algebra and semiclassical linear algebra are not distinct subjects. The n in n-space is not necessarily a large or indeterminate number, but can profitably be thought of as being 2 or 3; mastery of these cases will make mastery of the general case an easy exercise. But if $n = 2$ or 3, there are pictures to draw; solving linear equations is the same as intersecting planes; diagonalizing matrices is the same as rotating axes to simplify the analytic description of a "stretching transformation," and so on. It is hoped that with these hooks on which to hang his intuition, the student will remember the parts of linear algebra that are needed in his later experiences, and that readers and teachers using this book will keep this spirit in mind even in places where the author has neglected to make it explicit.

I would like to offer my sincere thanks to my colleagues and to other readers and users of the first edition of this text, who have made comments and suggestions, many of which have been invaluable and much appreciated. And special thanks go to my wife and my son, Paul, who helped design the illustrations; and to my son, David, who caught more errors in the first edition than anyone else.

A FIRST COURSE
IN LINEAR ALGEBRA

SECOND EDITION

CHAPTER

1

Vectors

1. COORDINATE SYSTEMS

Just as a point in the plane can be described by a pair of numbers, so each point in space can be described by a triple of numbers. To do this we choose a *coordinate system*, namely, (1) a point, called the origin; (2) three lines, the *coordinate axes*, through the origin, each perpendicular to the other two (we also must choose which of these axes is to be called first, which second, and which third; these are usually referred to as the *x-axis*, *y-axis*, and *z-axis*, respectively); and (3) one other point besides the origin on each axis. This gives us two things, a measure of lengths on lines parallel to an axis (the distance from the origin to the other given point on the axis is the unit of length) and a *positive direction*, or positive sense, on each axis (direction from the origin to the other given point). Moreover, this specifies a coordinate system in the usual sense on each axis: a one-to-one correspondence between points on the axis and real numbers. Usually, but not always, the three given points on the three axes are all taken the same distance from the origin; in other words, the units of length are taken to be the same in all three directions. We shall always do this.

There are obviously infinitely many ways of making the choices (1), (2), and (3) in choosing a coordinate system. For convenience, we shall usually draw the positive z-axis pointing up, the positive y-axis pointing to the right, and (using perspective drawing) the positive x-axis pointing forward from the plane of the paper toward the reader. Half the possible coordinate systems can be rotated into this position; the other half, once you rotate the z-axis and the y-axis into this position, will have their positive x-axis pointing back instead of forward. The first half are referred to as right-handed systems (if you wrap the fingers of your right hand around the z-axis with the tips of your fingers pointing around from the positive x-axis to the positive y-axis, then your thumb points in the positive direction on the z-axis) (see Fig. 1.1); the others are left-handed.

Once the coordinate system is chosen we have also specified three *coordinate planes*: the xy-plane is the plane containing the x- and y-axes (or, equivalently, the plane through the origin perpendicular to the z-axis), the xz-plane contains the x- and z-axes (it is the plane through the origin perpendicular to the y-axis), and the yz-plane contains the y- and z-axes (it is the plane through the origin perpendicular to the

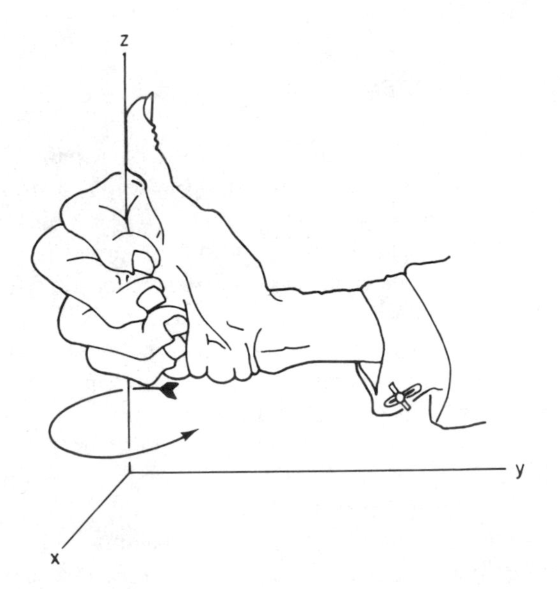

Figure 1.1

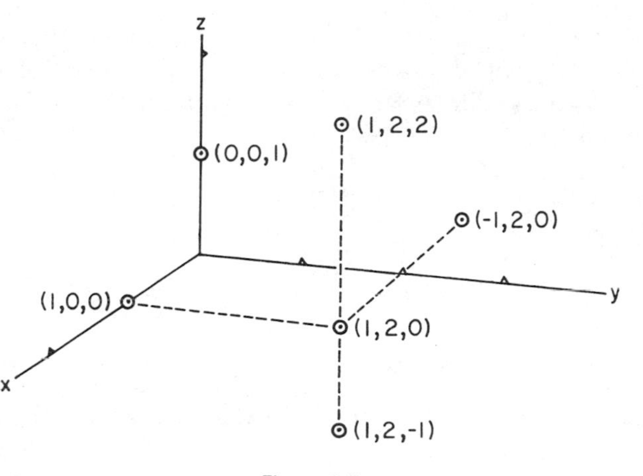

Figure 1.2

x-axis). These three planes cut all of space into eight *octants*, one above and one below each of the quadrants in the xy-plane. The *first octant* is the one above the first quadrant in the xy-plane. After we define coordinates of points, the first octant can be described as the set of all points that have all three coordinates positive. We do not bother to number the other octants.

Once the coordinate system is chosen, to each point we associate a triple of numbers (a, b, c) where a is the coordinate of (real number associated with) the projection of P on the x-axis, b is the coordinate of the projection of P on the y-axis, and c is the coordinate of the projection of P on the z-axis. It is probably easier to project first into the xy-plane, then onto the x- and y-axes. The three numbers a, b, c are, respectively, the x-coordinate, y-coordinate, and z-coordinate of P. Examples of points and their corresponding coordinates are shown in Fig. 1.2.

As you might expect, if we have an equation involving the letters x, y, and z, we define the *graph of the equation* as the set of all points whose coordinates, substituted for x, y, z in the equation, make the equation true. For example, the graph of the equation $x = z$ is the set of all points with equal x-coordinate and z-coordinate. Some sample points on the graph are $(1, 0, 1)$, $(1, 2, 1)$, $(1, 3, 1)$, $(2, 0, 2)$, ... You should convince yourself that the set of *all* points on the graph is a plane through the y-axis making a 45-degree angle with the xy-plane and the yz-plane.

Similarly, if we have several equations (we often speak of a system of equations), the graph of the system of equations is the set of all points whose coordinates satisfy all of the equations. For example, the graph of the pair of equations $x = 0$, $y = 0$ is the set of all points whose first two coordinates are zero; the graph is the z-axis.

The graph of the system

$$\begin{aligned} x + y + z &= 1 \\ x + y \quad &= 0 \\ x \quad &= 0 \\ z &= 0 \end{aligned}$$

is the set of all points satisfying all of the equations. It is the intersection of the graphs of the four individual equations. The reader should convince himself that the graph of this system of equations is the intersection of four planes that have no point in common. The graph of this system has no points at all; it is the empty set of points.

In reverse, we speak of an *equation of a geometrical figure*, meaning an equation whose graph is the given figure. For example, the equation of the xy-plane is $z = 0$, because the set of all points whose z-coordinate is zero is exactly the xy-plane. Similarly we speak of equations of a figure, meaning a system of equations whose graph is the figure. Equations of the z-axis are $x = 0$, $y = 0$.

PROBLEMS

1. Sketch a right-handed coordinate system and indicate the points whose coordinates are $(2, 0, 0)$, $(0, 0, -1)$, $(2, 0, 1)$, $(-2, -1, -1)$, $(\pi, \pi, 0)$, $(\sqrt{2}, -1, 0)$.

2. Where are all the points whose y-coordinates are zero? What is an equation of this set of points?

3. Where are all the points whose x- and y-coordinates add up to 1? What is an equation of this figure?

4. Find an equation of the plane that consists of all points one unit above the xy-plane.

5. What are the graphs of the following equations? Sketch them.

(a) $x = 0$. (b) $y = 0$. (c) $y = 1$.

(d) $x = -1$. (e) $2z + 1 = 0$. (f) $3z + \pi = 1$.

6. What are the graphs of the following equations? Sketch them.

(a) $x^2 + y^2 + z^2 = 1$. (b) $x^2 + y^2 = 1$.

(c) $x^2 + z^2 = 1$. (d) $y + z = 1$.

(e) $x + y + z = 1$. (f) $z = x^2 + y^2$.

(g) $z = x^2 - y^2$. (h) $x^2 + y^2 = 0$.

7. What would you mean by the graph of an inequality? What are the graphs of the following inequalities?

(a) $x \geq 0$. (b) $x + y \leq 1$. (c) $x^2 + y^2 + z^2 \leq 1$.

8. What are the general properties of the graph of an equation that has no z explicitly occurring (for example, Problem 5(a)–(d), Problem 6(b))?

9. What are the graphs of the following pairs? Sketch.

(a) $\{x = 0, z = 0\}$. (b) $\{x = 1, z = 0\}$.

(c) $\{x + y = 1, z = 0\}$. (d) $\{x + y + z = 1, x = 1\}$.

(e) $\{y + z = 0, x = 1\}$. (f) $\{x^2 + y^2 = 1, y = z\}$.

(g) $\{x \geq 0, x + y \leq 1\}$.

10. Later we shall show that the graph of every linear equation is a plane, and every plane is the graph of some linear equation. Assume this, and show that the graph of (almost) every pair of linear equations is a line and every line is the graph of some pair of linear equations.

2. VECTORS AND THEIR COMPONENTS

Given two points P and Q, the *vector* PQ is the line segment joining P and Q, directed from P to Q. We usually draw this as an arrow with tail at P and head at Q. In fact, we refer to P and Q, respectively, as the tail and the head of the vector PQ. One special case needs comment: If P and Q coincide, we still call PQ a vector, the *zero vector* at P, and denote it by $\mathbf{0}$.

We shall denote vectors either by a pair of points, as in the preceding paragraph, or by a boldface letter. Using different type fonts for vectors and numbers is very helpful, for it is always important *to be conscious of which letters denote vectors and which denote numbers,* especially after we introduce an arithmetic for vectors. Incidentally, in connection with vectors, a synonym for number is *scalar*. As with many questions of language, there is no sensible reason for such redundancy, but you have to learn it if you are going to understand the language.

It should then be clear what it means for two vectors to be *perpendicular*, to be *parallel*, to have the *same direction* (that is, not only parallel but arrows pointing the same way), to have *opposite directions*. We agree to consider the zero vectors as being parallel to and perpendicular to every vector. It should also be clear what the *length* of a vector means. Note that there is an infinitude of vectors all of the same length; the length does not determine the vector. Similarly, there is an infinitude of vectors with the same direction; the direction does not determine the vector.

Many physical quantities are appropriately represented by vectors. A force can be denoted by a vector whose length in inches equals the magnitude of the force in pounds and whose direction is the same as the direction of the force. Likewise displacements (movement of a particle from one point to another), velocities, and accelerations are effectively represented by arrows. See also the commments in Sections 5 and 8.

Given a fixed point O called the origin (the rest of the coordinate system is not needed here), we can make a one-to-one correspondence between vectors (with their tails at O) and points thus: to each point P we make correspond its *position vector*, which is the vector OP from the origin to the point. To each vector with its tail at O we make correspond the point that is at the head of the vector. We shall have frequent use for this correspondence. All sorts of statements about points will be translated into statements about their position vectors.

Given a coordinate system, each vector has three *components*, which are numbers. We shall give the definition in terms of some vectors that are called *projections*. First, given a point and a line, you know what the projection of the point on the line is: it is the point at the foot of a perpendicular from the given point to the given line. Now the projection of a vector $\mathbf{v}$ on a line is a vector whose head is the projection of the head of $\mathbf{v}$ and whose tail is the projection of the tail of $\mathbf{v}$. The projection of $\mathbf{v}$ on a given line is then a vector lying on the line. In Fig. 1.3 $P'Q'$ is the projection on the y-axis of the vector PQ.

Given any vector $\mathbf{v}$, then, we proceed to construct its projections on the x-, y-, and z-axes. This first projection, being a vector along the x-axis, is well described by a single number: plus or minus the

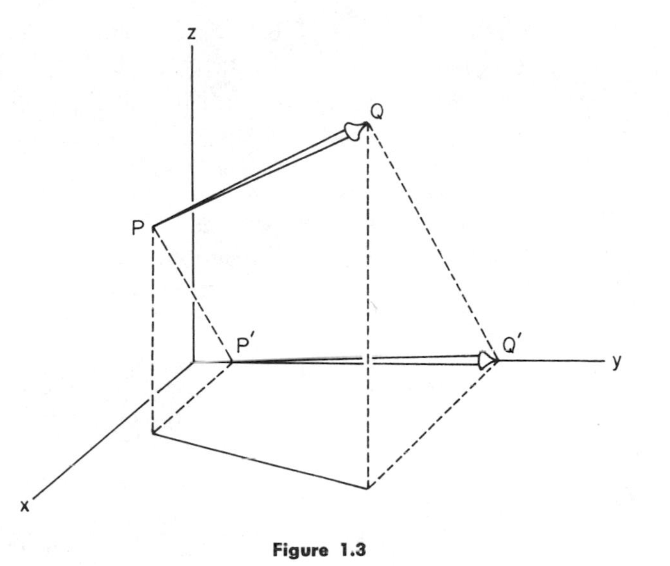

Figure 1.3

length of the projection, plus if the vector points in the direction of the positive x-axis, minus if the vector points in the direction of the negative x-axis. This number is the *x-component* of the original vector **v**. In similar fashion, the y-component of **v** is plus or minus the length of the projection of **v** on the y-axis, and the *z-component* of **v** is plus or minus the length of the projection of **v** on the z-axis. In all cases, the plus or minus sign is chosen according to the direction of the projection as just described for the x-component.

It is easy to see that if a point P has coordinates (x, y, z), then the projection of P on the x-axis has coordinates $(x, 0, 0)$. If another point Q has coordinates (x', y', z'), then its projection on the x-axis is the point $(x', 0, 0)$. The projection of the vector PQ on the x-axis is then the vector with tail at $(x, 0, 0)$ and head at $(x', 0, 0)$. The length of this projection is $|x' - x|$, so the x-component of PQ is $\pm |x' - x|$. We are to use the plus sign if the point $(x', 0, 0)$ is in front of $(x, 0, 0)$, that is, if $x' > x$, and the minus sign otherwise. In other words, the x-component of PQ is

$$\begin{array}{rl} |x' - x| & \text{if} \quad x' - x > 0, \\ -|x' - x| & \text{if} \quad x' - x \leq 0. \end{array}$$

From the definition of absolute values, we can write this in a single simple formula:

2.1 If the coordinates of the point P are (x, y, z) and the coordinates of the point Q are (x', y', z'), then the x-component of the vector PQ is $x' - x$; likewise, the y-component is $y' - y$, and the z-component is $z' - z$.

Note that when you subtract coordinates of one point from coordinates of the other, you must subtract in this order: head minus tail. The three components of a vector will usually be displayed between square brackets to remind you that they are components of a vector and not coordinates of a point. However,

2.2 If $\mathbf{v}$ is a vector with its tail at the origin, then the components of the vector $\mathbf{v}$ are the same as the coordinates of the point at the head of $\mathbf{v}$.

For the most part we shall be concerned with components only of position vectors, that is, of vectors with tails at the origin. *If nothing else is said, especially from Chapter 3 on, we shall always assume that the vectors under discussion have their tails at the origin.* In this case, the vectors will be completely determined by their components and we shall take the liberty of writing statements like $\mathbf{v} = [a, b, c]$, meaning that $\mathbf{v}$ has its tail at the origin and that the three components of $\mathbf{v}$ are the numbers a, b, and c. Clearly we can only do this if a definite coordinate system is agreed on in advance. If ever we rotate the coordinate system, the same vector will have different components in the new coordinate system, and our present convention will have to be abandoned.

Similarly, we indulge in the following very standard ellipses (the plural of ellipsis, not ellipse): If a point P has coordinates (a, b, c), we shall say $P = (a, b, c)$, provided we have agreed on a coordinate system; we shall also refer to (a, b, c) as a point. (Note that we may think of a triple as either a point or a vector. The point and the vector we get from one triple are related as in 2.2—the vector is the position vector of the point.) If an equation $f(x, y, z) = 0$ has a graph that is a figure (say, a plane), we refer to the figure itself as $f(x, y, z) = 0$. We also refer to the equation $f(x, y, z) = 0$ as a geometric figure. For example, we refer to the xy-plane as "the plane $z = 0$"; we talk of the "point $(2, -4, 1)$" or "the vector $[7, 0, 1]$," and so on.

Thus every vector has two aspects: It is a geometric object and (once a coordinate system is agreed on and the tail of the vector is at the origin) it "is" a triple of numbers. This dual personality will persist throughout our use of vectors. It is the key idea in all analytic geometry and provides twice as much power and understanding as either personality alone. Every statement about vectors will have two interpretations, one geometric and one analytical, that usually will look quite different from each other, though both interpretations say the same thing about vectors. Every operation on vectors will have to be learned in two guises, one geometric (and without coordinate systems), the other analytical or algebraic. You are advised to be conscious of this duality in all the sections that follow.

Before we end this section, we settle the question of exactly what it means for two vectors to have the same components. If the vectors both have their tails at the origin, 2.2 already tells us that the vectors are the same if their components are the same. In general,

2.3 If two vectors have their tails at the same point, then the two vectors have the same components if and only if they are equal (two vectors are equal, of course, if they have the same head and the same tail). If two vectors have their tails at two different points, then they have the same components if and only if they have the same length and the same direction; that is, one of the vectors can be moved without turning or changing length until it coincides with the other.

Proof. Suppose the two vectors have the common tail (x, y, z) and that one of them has the head (x', y', z') and the other has the head (x'', y'', z''). Then the first has components $[x' - x, y' - y, z' - z]$ and the second has components $[x'' - x, y'' - y, z'' - z]$. These components are the same (by this we mean $x' - x = x'' - x$ and $y' - y = y'' - y$ and $z' - z = z'' - z$) if and only if $x' = x''$ and $y' = y''$ and $z' = z''$. This last trio of conditions means that the two vectors have the same head; since they already have the same tail, this means they are equal.

For the case where the two tails do not coincide, let the two vectors be PQ and $P'Q'$. Draw PR parallel to the y-axis, RS parallel to the x-axis, and SQ parallel to the z-axis, and do the same for $P'R'$, $R'S'$, $S'Q'$ (Fig. 1.4). If PQ and $P'Q'$ have the same length and direction, these two figures will be congruent and the length and direction of PR will be the same as those of $P'R'$, so PQ and $P'Q'$ have the same y-component; similar arguments hold for RS, $R'S'$ and SQ, $S'Q'$. Conversely, if the components of PQ and $P'Q'$ are the same, then PR has the same length and direction as $P'R'$; so a motion without turning or stretching will move PR to coincide with $P'R'$. In the process RS will be carried to a vector still parallel to the x-axis; since RS and $R'S'$ are assumed to have the same length, S will be carried into

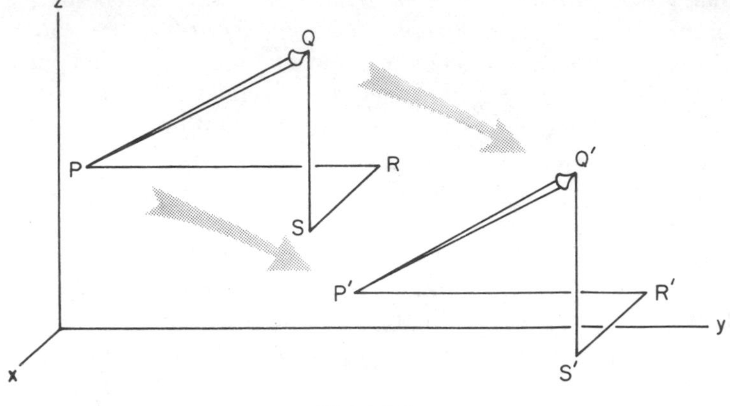

Figure 1.4

S'. A similar argument shows that Q is carried into Q', so PQ is carried into $P'Q'$.

2.4 Definition

Two vectors (like PQ and $P'Q'$ in the preceding proof) are called *equivalent* when they have the same length and direction; that is, when one can be carried into the other without turning or changing length.

Equivalent vectors are parallel, but the converse is not necessarily so, since parallel vectors can have different lengths and can even have opposite directions. There is an alternative approach to vectors that considers equivalent vectors as equal, so that you can move a vector to other locations and still have "the same vector." It is just a matter of taste which of these two approaches one uses in developing the theory and the calculations. We shall choose the one we have indicated: vectors in different places are equivalent at best, but not equal. When we restrict ourselves to vectors with tails at the origin, as we have agreed to do most of the time, the problems vanish: equivalent vectors with the same tail are equal.

3. LENGTH OF A VECTOR

We denote the length of PQ by $\| PQ \|$; the length of $\mathbf{v}$ is $\| \mathbf{v} \|$. It is a positive number unless $\mathbf{v}$ is a zero vector, in which case this length is zero. As promised in Section 2, we investigate the dual personality:

3.1 If a vector $\mathbf{v}$ has components $[a, b, c]$, then

$$\|\mathbf{v}\| = (a^2 + b^2 + c^2)^{1/2}.$$

Proof. Let $\mathbf{v} = PQ$ in Fig. 1.5 have components $[a, b, c]$. Draw PP' parallel to the y-axis, then $P'Q'$ parallel to the x-axis of such a length that $Q'Q$ will be parallel to the z-axis. Then $\|P'Q'\| = |a|$, $\|PP'\| = |b|$, $\|QQ'\| = |c|$. (Why?) Since $PQ'Q$ is a right triangle (in a vertical plane), Pythagoras says $\|PQ\|^2 = \|PQ'\|^2 + \|QQ'\|^2 = \|PQ'\|^2 + c^2$. (Note: $|c|^2 = c^2$. Why?) Since $PP'Q'$ is a right triangle (in a horizontal plane), Pythagoras says $\|PQ'\|^2 = \|PP'\|^2 + \|P'Q'\|^2 = a^2 + b^2$. Combining the two equations and taking square roots (why is $\|PQ\| = (\|PQ\|^2)^{1/2}$?), we get $\|PQ\| = (a^2 + b^2 + c^2)^{1/2}$.

3.2 If P is a point and has coordinates (x, y, z) and if Q is the point with coordinates (x', y', z'), then the distance from P to Q is

$$((x' - x)^2 + (y' - y)^2 + (z' - z)^2)^{1/2}.$$

Proof. The distance equals $\|PQ\|$ and PQ has components $[x' - x, y' - y, z' - z]$.

The notation $\|\mathbf{v}\|$ is adopted because this length function (associating a scalar to each vector) is formally analogous to the absolute value function of numbers. For example, the absolute value function has the following properties: $|a + b| \leq |a| + |b|$, $|ab| = |a||b|$,

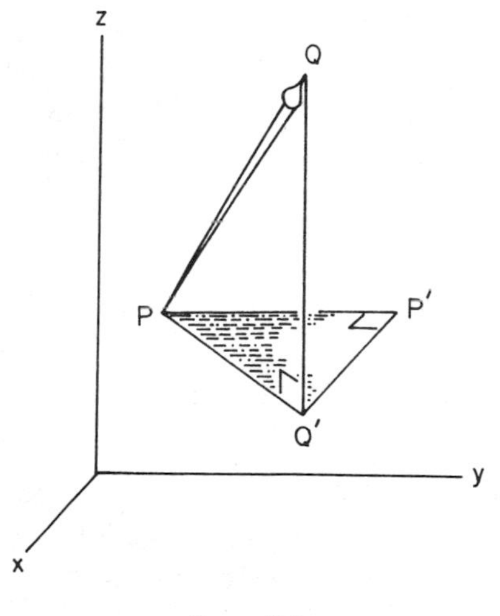

Figure 1.5

$|a| \geq 0$, and $|a| = 0$ if and only if $a = 0$. These same properties hold if $|\ |$ is replaced by $\|\ \|$. See 4.3(d), 5.2(c), and Problem 6 for the exact statements. See also Problems 10 and 14, which show how the absolute value of a number can be considered as a special case of the length of a vector.

PROBLEMS

1. If P is $(1, 3, 5)$ and Q is $(2, 2, 4)$ and O is the origin, sketch the vectors OP, OQ, PQ, and QP and find their components. Find their lengths.

2. Same question with $P = (-1, 0, -2)$ and $Q = (4, 1, -1)$.

3. If the vector $[2, 1, -7]$ has its tail at the origin, where is its head? If it has its tail at the point $(0, 0, 1)$, where is its head? If it has its tail at the point (a, b, c), where is its head?

4. If the tail of a vector is at (a, b, c) and the components of the vector are $[d, e, f]$, find the coordinates of the head.

5. (a) Find z so that the triangle with vertices $(2, 2, -5)$, $(-2, 4, -5)$, $(1, 5, z)$ is equilateral. Do the same for the points $(0, 3, 0)$, $(2, 4, 0)$, $(2, 3, z)$.

(b) Consider the triangle PQR where P is the point $(2, 3, 5)$, Q is $(-1, 0, -2)$, and R is $(-4, 1, z)$. Find z so the triangle is isosceles. There are five answers.

6. Prove $\mathbf{v}$ is a zero vector if and only if its components are all zero, and this in turn is true if and only if $\|\mathbf{v}\| = 0$. (Note: This is a trivial problem.)

7. Given three numbers a, b, c, show how to construct a vector having components $[a, b, c]$.

8. Sketch vectors with the following components:

(a) $[1, 0, 0]$; (b) $[2, -1, 1]$;

(c) $[\pi, 0, 1]$; (d) $[-1, -2, -3]$.

9. Show that if the components of a vector $\mathbf{v}$ are all multiplied by the same scalar a, then we get components of a new vector whose length is $|a|$ times the length of $\mathbf{v}$.

10. Show that if $\mathbf{v} = [a, 0, 0]$, then $\|\mathbf{v}\| = |a|$; do the same for vectors along the y-axis and for vectors along the z-axis.

11. Use 3.2 to write an equation of the sphere of radius r centered at the origin. Do the same for a sphere of radius r with center at the point (a, b, c).

12. Find the distance from the origin to the point (a, b, c); from $(1, 1, 1)$ to $(2, -1, 6)$.

13. Develop the theory of vectors in a plane rather than in 3-space. Specifically, show how to introduce components and exhibit a one-to-one correspondence between *pairs* of numbers $[a, b]$ and vectors in the plane with tails at a fixed point. Do what we shall not do in 3-space: Define the slope of a vector and write the slope in terms of the components. Show that two vectors are parallel if and only if their components are proportional, and that $[a, b]$ is perpendicular to $[a', b']$ if and only if $aa' + bb' = 0$. Show that $\|[a, b]\| = (a^2 + b^2)^{1/2}$. Finally, compute the cosines of the angles between the vector and each of the axes in terms of the components.

14. Develop the theory of vectors along a line. Specifically, exhibit a one-to-one correspondence between single numbers $[a]$ and vectors along the line with tails at a fixed point. There is not much else to do at present but prove that the length of a vector $[a]$ is $|a|$.

4. MULTIPLICATION OF VECTORS BY SCALARS

4.1 Definition

If $\mathbf{v}$ is a vector and s is a positive scalar, $s\mathbf{v}$ will denote the vector with the same tail as $\mathbf{v}$ in the same direction as $\mathbf{v}$, but s times as long; if s is a negative scalar, $s\mathbf{v}$ denotes the vector in the opposite direction from $\mathbf{v}$ and $|s|$ times as long, and $0\mathbf{v}$ is the zero vector at the tail of $\mathbf{v}$. In other words, $s\mathbf{v}$ is $|s|$ times as long as $\mathbf{v}$ and in the same direction as $\mathbf{v}$ if $s \geq 0$, opposite if $s < 0$. As promised in Section 2, this definition has a dual personality.

4.2 Each component of $s\mathbf{v}$ is s times the corresponding component of $\mathbf{v}$ for all scalars s and all vectors $\mathbf{v}$. When $\mathbf{v} = [a, b, c]$ (this notation is used for vectors with tail at the origin), then

$$s[a, b, c] = [sa, sb, sc].$$

Proof. Each projection of $s\mathbf{v}$ on a coordinate axis is $|\, s \,|$ times as long as the projection of $\mathbf{v}$ (argue via similar triangles). This projection of $s\mathbf{v}$ is in the same direction as the projection of $\mathbf{v}$ if $s \geq 0$ and in the opposite direction if s is negative. All this means that the component of $s\mathbf{v}$ is exactly s times the component of $\mathbf{v}$.

Example 1. (Geometric example): Figure 1.6.

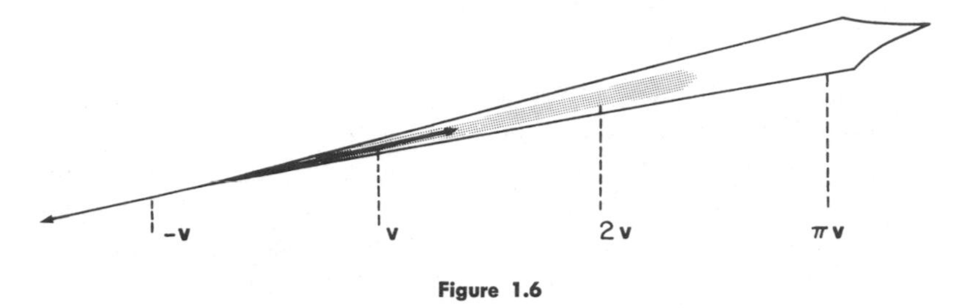

Figure 1.6

Example 2. (Analytic example): If $\mathbf{v}$ is [2, 5, 7], then $2\mathbf{v}$ is [4, 10, 14], $-\mathbf{v}$ is [−2, −5, −7], $\pi\mathbf{v}$ is $[2\pi, 5\pi, 7\pi]$.

Example 3. (A combination of the two personalities): If O is the origin and P is the point (2, 3, 7), we find the midpoint of the line segment OP: Its position vector is $\frac{1}{2}OP$ because of the geometric definition of scalar times vector. But $OP = [2, 3, 7]$, so $\frac{1}{2}OP = [1, \frac{3}{2}, \frac{7}{2}]$ by 4.2. Hence the coordinates of the midpoint are $(1, \frac{3}{2}, \frac{7}{2})$ by 2.2.

4.3 Some arithmetic of this multiplication

(a) $0\mathbf{v} = \mathbf{0}$ for every vector $\mathbf{v}$.

(b) $1\mathbf{v} = \mathbf{v}$ for every vector $\mathbf{v}$.

(c) $(st)\mathbf{v} = s(t\mathbf{v})$ for every vector $\mathbf{v}$ and all scalars s and t.

(d) $\| s\mathbf{v} \| = |\, s \,| \, \| \mathbf{v} \|$ for every vector $\mathbf{v}$ and every scalar s.

Proof. 4.3(a), (b), and (d) are direct consequences of the definition of multiplication by scalars. As for 4.3(c), the vectors on both sides of the equation have the same tail, namely, the point at the tail of $\mathbf{v}$; so we need only show that they have the same components (see 2.3). If the components of $\mathbf{v}$ are $[a, b, c]$, then the components of $(st)\mathbf{v}$ are $[(st)a, (st)b, (st)c]$ by 4.2. The components of $t\mathbf{v}$ are $[ta, tb, tc]$ and of $s(t\mathbf{v})$ are $[s(ta), s(tb), s(tc)]$; these last agree with the components of $(st)\mathbf{v}$ since we know the associative law for multiplication of numbers: $(st)a = s(ta)$, and so on.

4.4 Two vectors are parallel—that is, have the same or opposite directions—if and only if one of them is equivalent to a scalar times the other. Translated into a statement about the dual personality, two vectors are parallel if and only if the components of one can be obtained from the components of the other by multiplying by a common factor; two vectors are parallel if and only if their components are proportional.

Proof. Call the two vectors $\mathbf{v}$ and $\mathbf{w}$. Suppose first that they have the same tail and are parallel; then they are either in the same or opposite directions, and by multiplying one by a suitable positive or negative scalar we can get the other (the suitable scalar is plus or minus the ratio of their lengths). If the two vectors do not have the same tail, then $\mathbf{w}$ is equivalent to a vector $\mathbf{w}'$ that does have the same tail as $\mathbf{v}$ and, by the argument above, $\mathbf{v}$ is a scalar times $\mathbf{w}'$, which is what we were to prove. Conversely, if $\mathbf{v}$ is a scalar times $\mathbf{w}'$ and $\mathbf{w}'$ is equivalent to $\mathbf{w}$, then $\mathbf{v}$ is parallel to $\mathbf{w}'$ and $\mathbf{w}'$ is parallel to $\mathbf{w}$, so $\mathbf{v}$ is parallel to $\mathbf{w}$.

The translation into a statement about components is direct: If $\mathbf{w}$ and $\mathbf{w}'$ are equivalent, they have the same components by 2.3. The statement $\mathbf{v} = s\mathbf{w}'$ then means that the components of $\mathbf{v}$ are s times the components of $\mathbf{w}'$ (or of $\mathbf{w}$).

PROBLEMS

1. Sketch the vector [1, 1, 0]; then sketch $\sqrt{2}$[1, 1, 0] in two ways: first, using exactly the definition of multiplication by $\sqrt{2}$; then by computing the components of $\sqrt{2}$[1, 1, 0] by 4.2 and sketching the resulting vector. Do the same for 4[0, 1, 0] and (−1)[2, 1, 1].

2. Which of the following vectors are parallel, and why? (a) [−3.0, 1.5, 1.5]; (b) [2, 1, 1]; (c) [−2, 1, 1]; (d) [−1.0, −0.5, −0.5]; (e) the vector from the origin to (−4, 2, 2); (f) the vector from (2, 1, 1) to (−2, 1, 1); (g) $\frac{1}{2}$ times the vector in (e); (h) the vector from (3, 1, 2) to (1, 0, 1).

3. (a) Find a vector in the same direction as [2, 1, 3] but having length 1. Hint: Multiply by an unknown scalar; determine the scalar to make the final length 1.

(b) Find the vector in the same direction as $\mathbf{v}$ = [a, b, c] but having length 1. Write the components of the answer in terms of a, b, and c. Then write the answer in another form directly in terms of $\mathbf{v}$, without using components.

4. (a) Let O be the origin and P be the point $(2, 5, 6)$. Find the coordinates of the midpoint of the line segment joining O and P. Find the coordinates of the two trisection points of this segment.

(b) Do the same when the point P has coordinates (a, b, c).

5. Give a more geometric proof of 4.3(c) when s and t are positive scalars, using only definition 4.1 and not components as in 4.2. What modifications are necessary to give a proof when one of s and t is negative? When both are negative?

6. (a) Consider the triangle OPQ where O is the origin, P is the point $(1, 2, 0)$, and Q is $(1, 1, 3)$. Find the midpoints M and N of the line segments OP and OQ.

(b) Find the components of the vector MN joining these midpoints.

(c) Show that MN is equivalent to $\frac{1}{2}PQ$.

(d) Repeat parts (a), (b), and (c) when P is the point (x, y, z) and Q is the point (x', y', z').

(e) Show that part (d) proves the following theorem: In every triangle, the line joining the midpoints of two sides is parallel to the third side and half as long.

7. The *direction cosines* of a vector are the cosines of the three angles between the vector and the positive ends of the coordinate axes. Derive formulas for the direction cosines of the vector $[a, b, c]$ directly from this definition.

8. Derive the formulas in Problem 7 as follows:

(a) If two vectors have the same direction, their direction cosines are equal.

(b) If a vector has length 1, its direction cosines are its components.

(c) For any vector $\mathbf{v}$, the vector of length 1 in the same direction as $\mathbf{v}$ is $(1/\|\mathbf{v}\|)\mathbf{v}$ (Problem 3).

9. Compute the direction cosines of $[2, 1, 3]$, $[0, -1, 1]$, $[0, 0, 1]$, $[-2, -2, -2]$. Find the angle each vector makes with the x-axis.

10. (a) Prove that if λ, μ, ν are the direction cosines of a vector, then $\lambda^2 + \mu^2 + \nu^2 = 1$.

(b) As a consequence of part (a), once two direction angles are given, there are only two possibilities for the third. Check this geometrically.

(c) Suppose a vector is in the xy-plane. Then part (a) reduces to a very familiar trigonometric formula. What is it?

11. Show that the direction cosines of $(-1)\mathbf{v}$ are the negatives of the direction cosines of $\mathbf{v}$. How then does the angle between the x-axis and $\mathbf{v}$ compare to the angle between that axis and $(-1)\mathbf{v}$?

12. Find the angle between $[1, 2, 1]$ and $[1, 2, 0]$.

13. Find a formula for the angle between the vector $[a, b, c]$ and the xy-plane.

5. ADDITION OF VECTORS; LINEAR COMBINATIONS

Given two vectors $\mathbf{v}$ and $\mathbf{w}$ with the same tail, we define their sum $\mathbf{v} + \mathbf{w}$ as follows: Move $\mathbf{w}$ without stretching or turning until its tail is at the head of $\mathbf{v}$. That is, we are to draw a vector $\mathbf{w}'$ with tail at the head of $\mathbf{v}$ and equivalent to $\mathbf{w}$. Then $\mathbf{v} + \mathbf{w}$ is the vector from the tail of $\mathbf{v}$ to the head of $\mathbf{w}'$ (Fig. 1.7). This "addition," *a priori*, has nothing to do with addition of numbers. In particular, when you add vectors, you do *not* add their lengths (we shall see in a moment that you *do* add their components). In this sense it is unreasonable to use a plus sign for the operation, but it is so standard to do so, and the operation is so similar in its behavior to addition of numbers, that we soon get used to using + for both addition of numbers and addition of vectors. In view of this ambiguous use of +, we repeat the word of caution in Section 2: You must be conscious at all times of which letters represent vectors and which represent scalars.

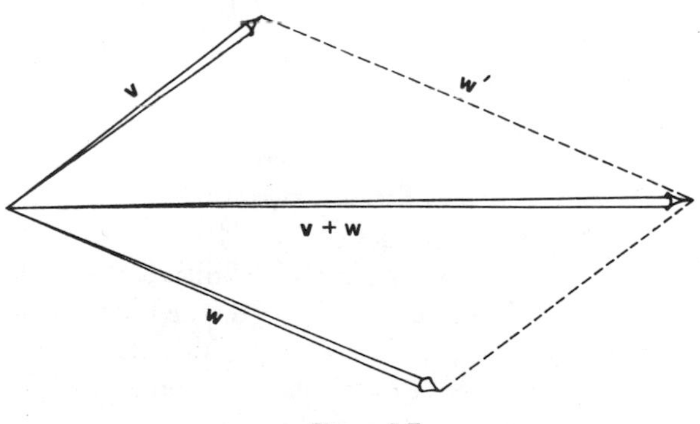

Figure 1.7

It is possible to define $\mathbf{v} + \mathbf{w}$ without moving $\mathbf{w}$: Draw the parallelogram two of whose edges are $\mathbf{v}$ and $\mathbf{w}$. The diagonal of this parallelogram (tail at the common tail of $\mathbf{v}$ and $\mathbf{w}$, head at the diagonally opposite corner) is $\mathbf{v} + \mathbf{w}$. This construction gives rise to the name *parallelogram addition* for this addition of vectors.

This definition of addition is reasonable enough in itself, but its primary motivation is from the applications. For example, if a particle is subjected to two forces and if each force is represented by a vector, then the particle acts just as if it were acted on by a single force, which can be correctly represented by the sum of the two actual force vectors. The same remarks apply to velocities: for example, a boat in a river has two velocities, one caused by its motor, one by the current; the actual velocity is represented by a vector that is the sum of these two velocity vectors. Perhaps interpreting the vectors as displacements is easiest of all: a displacement $\mathbf{v}$ followed by a displacement of the same magnitude and direction as $\mathbf{w}$ will clearly result in a net displacement $\mathbf{v} + \mathbf{w}$.

We now display the analytical personality of vector addition.

5.1 Each component of $\mathbf{v} + \mathbf{w}$ is the sum of the corresponding components of $\mathbf{v}$ and $\mathbf{w}$. For vectors with their tails at the origin, this says

$$[a, b, c] + [a', b', c'] = [a + a', b + b', c + c'].$$

Proof. Let us assume at first that the vectors do have their tails at the origin and that the two vectors are $[a, b, c]$ and $[a', b', c']$. The head of $[a, b, c]$ is the point (a, b, c), by 2.2. Let $\mathbf{w}''$ be the vector equivalent to $[a', b', c']$ whose tail is at this point; then the head of $\mathbf{w}''$ is at $(a + a', b + b', c + c')$ (cf. Problem 4, Section 3). By the definition of addition, $[a, b, c] + [a', b', c']$ is the vector from the origin to this latter point. Again by 2.2 the vector sum has components $[a + a', b + b', c + c']$, proving the second half of 5.1.

If we are to add vectors $\mathbf{v}$ and $\mathbf{w}$ whose tails are not at the origin, we move the vectors until their tails are at the origin. We can consider this as a motion of all of space, without turning or stretching, that moves the common tail of $\mathbf{v}$ and $\mathbf{w}$ to the origin. This same motion will carry $\mathbf{v}$ into an equivalent vector $[a, b, c]$, $\mathbf{w}$ into a vector $[a', b', c']$ equivalent to $\mathbf{w}$, $\mathbf{w}'$ (as in the definition of $\mathbf{v} + \mathbf{w}$) into the vector $\mathbf{w}''$ just defined in the addition of $[a, b, c]$ and $[a', b', c']$, and hence $\mathbf{v} + \mathbf{w}$ into $[a, b, c] + [a', b', c']$. Thus these last two sums are equivalent. We have computed the components of $[a, b, c] + [a', b', c']$ above. Since equivalent vectors have the same components, the components of $\mathbf{v} + \mathbf{w}$ are the same, namely, $a + a', b + b', c + c'$; the same equivalence argument shows that the components of $\mathbf{v}$ are a, b, c and the components of $\mathbf{w}$ are a', b', c'. This proves the first statement in 5.1.

5.2 The following are the basic properties of addition and its connection with length and with multiplication by scalars: For all vectors $\mathbf{u}, \mathbf{v}, \mathbf{w}$ with a common tail and all scalars s, t,

(a) $\mathbf{v} + \mathbf{w} = \mathbf{w} + \mathbf{v}$,

(b) $\mathbf{u} + (\mathbf{v} + \mathbf{w}) = (\mathbf{u} + \mathbf{v}) + \mathbf{w}$,

(c) $\|\mathbf{v} + \mathbf{w}\| \leq \|\mathbf{v}\| + \|\mathbf{w}\|$ (triangle inequality),

(d) $s(\mathbf{v} + \mathbf{w}) = s\mathbf{v} + s\mathbf{w}$,

(e) $(s + t)\mathbf{v} = s\mathbf{v} + t\mathbf{v}$,

(f) $n\mathbf{v} = \mathbf{v} + \mathbf{v} + \cdots + \mathbf{v}$ (n terms, for every positive integer n),

(g) $\mathbf{v} + (-1)\mathbf{v} = \mathbf{0}$,

(h) $\mathbf{v} + \mathbf{0} = \mathbf{v}$.

The proof of 5.2 may be carried out in either of two ways: One way is to translate everything into components, using 4.2 and 5.1 (and 3.1 for 5.2(c)) and prove the corresponding statements for triples. This will finish the proof of 5.2(c). For each of the other parts, it will prove that the vectors on the two sides of the equality have the same components, hence are equivalent. But in each case both vectors have the same tail, too, so they are equal by 2.3.

The other way is to prove the statements geometrically, using the (geometric) definitions of the operations directly. We give one sample of each style of proof, and leave the rest to the reader.

Geometric proof of 5.2(a). Draw a parallelogram two of whose sides are $\mathbf{v}$ and $\mathbf{w}$. Then the diagonal is both $\mathbf{v} + \mathbf{w}$ and $\mathbf{w} + \mathbf{v}$.

Analytic proof of 5.2(e). Let $\mathbf{v}$ have components $[a, b, c]$ (that is, $\mathbf{v}$ is equivalent to the vector $[a, b, c]$ with tail at the origin). Then $s\mathbf{v}$ has components $[sa, sb, sc]$ by 4.2. Similarly, $t\mathbf{v}$ has components $[ta, tb, tc]$, so that $s\mathbf{v} + t\mathbf{v}$ has components $[sa + ta, sb + tb, sc + tc]$ by 5.1. On the other hand, $(s + t)\mathbf{v}$ has components $[(s + t)a, (s + t)b, (s + t)c]$, which are the same as the components of $s\mathbf{v} + t\mathbf{v}$ since we know the distributive law for numbers: $(s + t)a = sa + ta$, and so on. Thus $(s + t)\mathbf{v}$ and $s\mathbf{v} + t\mathbf{v}$ are equivalent; since they also have the same tail, they are equal.

Note that some of these proofs will be much easier geometrically than analytically, especially 5.2(c), whereas others are much easier analytically, especially that of 5.2(e), where the geometric definition would require investigation of many special cases, depending on the signs of s, t, and $s + t$. However, it is instructive to try them all both ways.

To continue the analogy between addition of vectors and addition of numbers, for each vector $\mathbf{v}$ we define $-\mathbf{v}$ to be $(-1)\mathbf{v}$, that is, the vector of the same length as $\mathbf{v}$, in the opposite direction. Then 5.2(g) says that $-\mathbf{v}$ is the "additive inverse" of $\mathbf{v}$, that is, $\mathbf{v} + (-\mathbf{v}) = \mathbf{0}$. We also define $\mathbf{v} - \mathbf{w}$ ($\mathbf{v}$ and $\mathbf{w}$ are again vectors with the same tail) to be $\mathbf{v} + (-1)\mathbf{w}$ or $\mathbf{v} + (-\mathbf{w})$. Using 5.2, it is easy to check that $\mathbf{v} - \mathbf{w}$ is also the one and only vector that, added to $\mathbf{w}$, gives $\mathbf{v}$ as a sum. In principle we should not have to say more about this subtraction of vectors. However, the operation comes up frequently enough that you should recognize the geometric construction: Put the tails of $\mathbf{v}$ and $\mathbf{w}$ together; draw the vector joining the heads of $\mathbf{v}$ and $\mathbf{w}$, its head at the head of $\mathbf{v}$ (the refrain "head minus tail" can prevent you from confusing $\mathbf{v} - \mathbf{w}$ with its opposite, $\mathbf{w} - \mathbf{v}$). Then $\mathbf{v} - \mathbf{w}$ is the result of moving this vector until its tail is at the tail of $\mathbf{v}$ (the vector joining the heads of $\mathbf{v}$ and $\mathbf{w}$ is equivalent to $\mathbf{v} - \mathbf{w}$ in the language of 2.4). That this construction is correct is checked instantaneously by noticing that it constructs the vector that, when added to $\mathbf{w}$, gives $\mathbf{v}$ (Fig. 1.8).

For example, if P is the point (x, y, z) and Q is the point (x', y', z') and O is the origin, then PQ is equivalent to $OQ - OP$. Converting to components, we get $OQ - OP = [x', y', z'] - [x, y, z] = [x' - x, y' - y, z' - z]$, retrieving the formula for the components of PQ in 2.1.

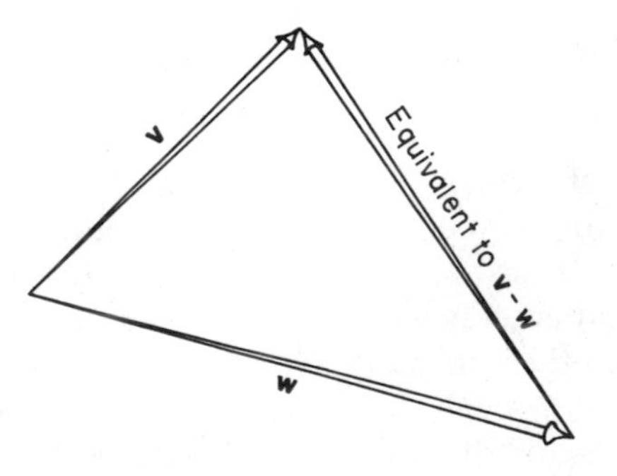

Figure 1.8

The results of 5.2 can be summarized roughly by saying that all the ordinary manipulations with addition and multiplication that can be performed with numbers work as well for vectors. For example,

$$\mathbf{u} - (\mathbf{v} - \mathbf{w}) = \mathbf{u} - \mathbf{v} + \mathbf{w}, \qquad \mathbf{u} - 2(\mathbf{v} - 3\mathbf{w}) = \mathbf{u} - 2\mathbf{v} + 6\mathbf{w}.$$

Why?

Combining this arithmetic with the geometry can sometimes give interesting results, for example, 5.3 and 5.4:

5.3 The line segment joining the midpoints of two sides of a triangle is parallel to the third side and half as long.

Proof. Let the triangle have vertices P, Q, R. If M and N denote the midpoints of PQ and PR, respectively, all we need to show is that the vector MN is equivalent to $\frac{1}{2}QR$. (See Fig. 1.9.) By the geometric version of subtraction just given, MN is equivalent to $PN - PM$ and QR is equivalent to $PR - PQ$. But $PN = \frac{1}{2}PR$, $PM = \frac{1}{2}PQ$, so MN is equivalent to $PN - PM = \frac{1}{2}PR - \frac{1}{2}PQ = \frac{1}{2}(PR - PQ)$, which is equivalent to $\frac{1}{2}QR$. See Problem 6, Section 4 for the same theorem proved using components.

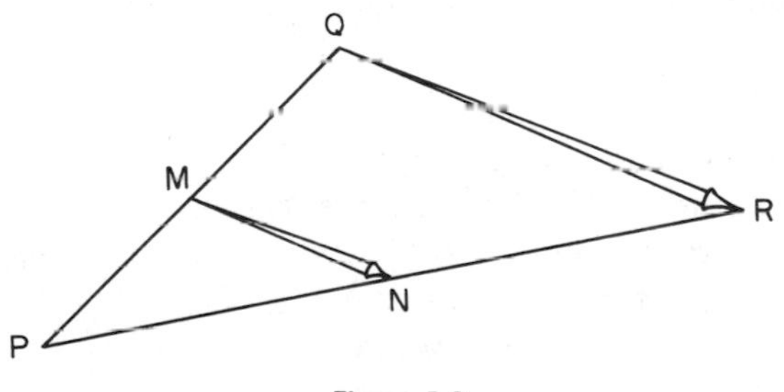

Figure 1.9

5.4 The coordinates of the midpoint of a line segment are the averages of the coordinates of the endpoints.

If $\mathbf{v}$ and $\mathbf{w}$ are vectors with the same tail O and with heads at P and Q, respectively, then the vector OM from the tails of $\mathbf{v}$ and $\mathbf{w}$ to the midpoint of the line segment PQ is $\frac{1}{2}(\mathbf{v} + \mathbf{w})$.

Proof. First prove the second statement: $OM = OP +$ "PM" where "PM" denotes a vector with tail at O (so it can be added to OP) equivalent to PM. $PM = \frac{1}{2}PQ$ and PQ is equivalent to $\mathbf{w} - \mathbf{v}$, so PM and "PM" are equivalent to $\frac{1}{2}(\mathbf{w} - \mathbf{v})$. But the last two of these three vectors have their tails at O, so they are equal vectors. Therefore, $OM = OP +$ "PM" $= \mathbf{v} + \frac{1}{2}(\mathbf{w} - \mathbf{v}) = \mathbf{v} + \frac{1}{2}\mathbf{w} - \frac{1}{2}\mathbf{v} = \frac{1}{2}\mathbf{v} + \frac{1}{2}\mathbf{w} = \frac{1}{2}(\mathbf{v} + \mathbf{w})$.

The first half of 5.4 follows immediately. Let the end points of the line segment in question be P and Q and let $\mathbf{v} = [a, b, c]$ and $\mathbf{w} = [a', b', c']$ be the position vectors of P and Q, respectively. We have exactly the same picture as in the second half of 5.4, with O the origin. The coordinates of the midpoint M will be the components of $OM = \frac{1}{2}(\mathbf{v} + \mathbf{w}) = \frac{1}{2}([a, b, c] + [a', b', c']) = [(a + a')/2, (b + b')/2, (c + c')/2]$.

5.5 Definition

Once our coordinate system is fixed, we let $\mathbf{i}$ denote the vector of length 1 in the positive x-direction, $\mathbf{j}$ denote the vector of length 1 in the positive y-direction, and $\mathbf{k}$ the vector of length 1 in the positive z-direction. In other words, $\mathbf{i} = [1, 0, 0]$, $\mathbf{j} = [0, 1, 0]$, and $\mathbf{k} = [0, 0, 1]$.

Then, using vector addition and multiplication by scalars, every vector is uniquely expressible in the form $a\mathbf{i} + b\mathbf{j} + c\mathbf{k}$. In fact (but check this),

$$a\mathbf{i} + b\mathbf{j} + c\mathbf{k} = [a, b, c],$$

which says that if a vector is expressed as $a\mathbf{i} + b\mathbf{j} + c\mathbf{k}$, then a, b, and c are the components of the vector.

5.6 Definition

A *linear combination* of two vectors $\mathbf{u}$ and $\mathbf{v}$ with a common tail is any vector that is equal to $a\mathbf{u} + b\mathbf{v}$ for some scalars a and b. A linear combination of several vectors $\mathbf{v}_1, \mathbf{v}_2, \ldots, \mathbf{v}_n$ is a vector equal to $a_1\mathbf{v}_1 + a_2\mathbf{v}_2 + \cdots + a_n\mathbf{v}_n$ for some scalars $a_1, a_2, \ldots, a_n$. These scalars are called *coefficients of the linear combination.*

5.7 Every vector with tail at the origin is a linear combination of $\mathbf{i}$, $\mathbf{j}$, and $\mathbf{k}$. Moreover, the coefficients of the linear combination are the components of the vector; $a\mathbf{i} + b\mathbf{j} + c\mathbf{k}$ equals $a'\mathbf{i} + b'\mathbf{j} + c'\mathbf{k}$ if and only if $a = a'$, $b = b'$, and $c = c'$.

PROBLEMS

1. Compute the following vector sums and differences both analytically (using 5.2) and geometrically.

$$[1, 1, 0] + [4, -1, 0] \qquad (2\mathbf{i} - \mathbf{j}) + (\mathbf{i} + \mathbf{j})$$

$$2\mathbf{i} + \mathbf{j} \qquad (2\mathbf{i} - \mathbf{j}) - (\mathbf{i} - \mathbf{j})$$

$$[1, 0, 1] - [0, 1, 1]$$

2. Write the specified vectors as linear combinations of **i**, **j**, **k**:

(a) [2, 1, 3], [2, 0, 3], [−1, 0, 1], [1, 0, 0], [0, 1, 0], [0, 0, 1], [1, $\sqrt{3}$, 0]. Check by performing the linear combinations (using the geometrical definitions of addition and of multiplication by scalars).

(b) The vector from the origin to (2, −1, 1); the vector with tail at the origin and equivalent to the vector from (1, 2, −1) to (−1, 1, 1); the vector from the origin to (a_1, a_2, a_3); the vector with tail at the origin and equivalent to the vector from (a_1, a_2, a_3) to (b_1, b_2, b_3).

3. Find $\| 2\mathbf{i} + 3\mathbf{j} - \mathbf{k} \|$, $\| \mathbf{i} + 2\mathbf{k} \|$, $\| a\mathbf{i} + b\mathbf{j} + c\mathbf{k} \|$.

4. (a) A certain linear combination of $2\mathbf{i} + \mathbf{j}$ and $\mathbf{i} - \mathbf{j}$ equals **i**. Find the coefficients of the linear combination.

(b) Find the coefficients of another linear combination of $2\mathbf{i} + \mathbf{j}$ and $\mathbf{i} - \mathbf{j}$ that will equal **0**; that will equal $a\mathbf{i} + b\mathbf{j}$ (a and b are arbitrary, unspecified scalars).

5. A certain linear combination of $2\mathbf{i} + \mathbf{j}$, $\mathbf{i} - \mathbf{j} + \mathbf{k}$ and $6\mathbf{i} + 2\mathbf{k}$ is **0**. Find the coefficients of the linear combination. The answer is not unique.

6. (a) Find a number x such that $x[2, 1, 1] + [1, 0, 2] = [-1, -1, 1]$.

(b) Find x such that $x[2, 1, 1] + [1, 0, 2] = [-1, 0, 0]$.

7. Find scalars x, y, and z if

$$x(5\mathbf{i} + \mathbf{j}) + y(\mathbf{j} + \mathbf{k}) + z\mathbf{k} = 5\mathbf{i} + 3\mathbf{j} + \mathbf{k}.$$

8. In each of the following parts, find x and y.

(a) $x(5\mathbf{i} + \mathbf{j}) + y(\mathbf{i} - \mathbf{j}) = \mathbf{i} + 5\mathbf{j}$.

(b) $x(5\mathbf{i} + \mathbf{j}) + y(\mathbf{i} - \mathbf{j}) = \mathbf{i}$.

(c) $x(5\mathbf{i} + \mathbf{j}) + y(\mathbf{i} - \mathbf{j}) = \mathbf{0}$.

(d) $x(5\mathbf{i} + \mathbf{j} + \mathbf{k}) + y(\mathbf{i} - \mathbf{j} - \mathbf{k}) = \mathbf{i} + 2\mathbf{j} + 2\mathbf{k}$.

(e) $x(5\mathbf{i} + \mathbf{j} + \mathbf{k}) + y(\mathbf{i} - \mathbf{j} + 3\mathbf{k}) = \mathbf{i}$.

(f) $[2x + 3y - 1, x - y + 1, 0] = [x, y, 0]$.

9. Let P be the point (3, 4, −1), Q be (0, 1, 4), and M be the midpoint of the segment joining P to Q. Find the coordinates of M. Sketch to see that your answer is reasonable. Find the vectors OM,

PM, PQ. Check again that $PM = \frac{1}{2}PQ$. Verify this further by computing the distance from P to M, from M to Q, and from P to Q, and then showing that the first two distances are half the last one.

10. (a) Using the notations of 5.4, find a formula for the position vector of the point that lies on the same line segment as P and Q, one third of the way from P to Q (this is called a trisection point of the segment); one nth of the way. Answers: $\frac{2}{3}\mathbf{v} + \frac{1}{3}\mathbf{w}$; $[(n-1)/n]\mathbf{v} + (1/n)\mathbf{w}$. Note the "weighted average."

(b) Using part (a), find the coordinates of the trisection point.

(c) Find the trisection point T if P and Q are as in Problem 9. Sketch. Check that $PT = \frac{1}{3}PQ$ and $QT = \frac{2}{3}QP$.

(d) Use part (a) as a guide to show that if P and Q are distinct points, then as t ranges through all real numbers, the vectors $tOP + (1 - t)OQ$ comprise the position vectors of all the points on the line through P and Q.

11. Let P be the point $(2, 3, 1)$ and Q be the point $(1, -1, 2)$. Use the result of Problem 10(d) to write the coordinates of every point on the line through P and Q (the coordinates will be expressed as functions of t). Then find the value of t that gives the point closest to the origin (use calculus). Find the point on this line that is closest to the origin.

12. Show that the diagonals of a parallelogram bisect each other. Partial solution: Let the origin be at one corner and R be the opposite corner; let P and Q denote the other two corners. Then the position vector of the bisection point of the diagonal OR is $\frac{1}{2}OR$. Use 5.4 to write the position vector of the bisection point of the other diagonal PQ. By using vector arithmetic, you should be able to show that these two position vectors are equal, which shows that the two bisection points coincide. Further suggestion: Write all the relevant vectors in terms of the position vectors OP and OQ.

13. Let P, Q, and R be the vertices of a triangle. Let M be the midpoint of the line segment QR.

Write the position vector OM as a linear combination of OQ and OR.

Write PM as a linear combination of OP, OQ, and OR. The line segment PM is a median of the triangle.

Let T be the trisection point of PM that is closer to M. Write OT as a linear combination of OP, OQ, and OR.

Find the position vectors of the trisection points of the other two medians of the triangle and show that they are equal to OT.

Conclude that the medians of every triangle meet at a point that is a trisection point of each median.

The foregoing argument would be equally correct if we chose the origin of coordinates to be at one of the vertices. What do the calculations look like if you take $P = O$?

14. Show that for any quadrilateral, the lines joining the midpoints of opposite sides bisect each other. (The position vector of the midpoint of one of these lines equals the position vector of the midpoint of the other; it is easier in this problem not to take the origin at one of the vertices, but rather at some arbitrary point outside the quadrilateral.)

15. Consider a cube with one corner at the origin. Let $\mathbf{u}$, $\mathbf{v}$, and $\mathbf{w}$ be vectors with tails at the origin lying along three edges of the cube. Show geometrically that $\mathbf{u} + \mathbf{v} + \mathbf{w}$ is the vector along the diagonal of the cube. Find the length of the diagonal.

16. If $\mathbf{u} = \mathbf{i} + \mathbf{k}$ and $\mathbf{v} = \mathbf{j} + \mathbf{k}$, express $2\mathbf{u} + 3\mathbf{v}$ as a linear combination of $\mathbf{i}, \mathbf{j}, \mathbf{k}$. Express $x\mathbf{u} + y\mathbf{v}$ likewise. Which of the following vectors are linear combinations of $\mathbf{u}$ and $\mathbf{v}$ (that is, expressible as $x\mathbf{u} + y\mathbf{v}$ for suitable x and y):

$$\mathbf{i} + 2\mathbf{j} + \mathbf{k}; \qquad \mathbf{i} + \mathbf{k}; \qquad 2\mathbf{i} + 5\mathbf{j} + 3\mathbf{k}; \qquad \mathbf{i} - \mathbf{j}; \qquad \mathbf{i}.$$

17. (a) Let $\mathbf{v}$ and $\mathbf{w}$ be vectors with tails at the origin and which are not parallel to each other. Describe the set of all linear combinations of $\mathbf{v}$ and $\mathbf{w}$ (all $a\mathbf{v} + b\mathbf{w}$ as a and b vary through all numbers). Answer: The set of all vectors in the plane that contains $\mathbf{v}$ and $\mathbf{w}$.

(b) What is the set of all linear combinations of $\mathbf{v}$ and $\mathbf{w}$ if $\mathbf{v}$ and $\mathbf{w}$ are parallel?

18. (a) What is the set of all linear combinations of a single nonzero vector $\mathbf{v}$ with tail at the origin? Answer: All vectors on the line through $\mathbf{v}$ (a line through the origin).

(b) What is the answer in part (a) if $\mathbf{v}$ is $\mathbf{0}$?

19. What is the set of all linear combinations of three vectors $\mathbf{v}_1, \mathbf{v}_2, \mathbf{v}_3$ that have their tails at the origin but do not lie in a plane? What happens if $\mathbf{v}_1, \mathbf{v}_2, \mathbf{v}_3$ all lie in a plane?

20. What is the set of all linear combinations of $\mathbf{v}_1, \mathbf{v}_2, \mathbf{v}_3, \mathbf{v}_4$ in general? When will the general answer fail?

21. The set of all linear combinations of one nonzero position vector $\mathbf{v}$ (that is, the set of all scalar multiples of $\mathbf{v}$) is the set of all position vectors of points on a line through the origin. If $\mathbf{u}$ is another vector, the set of all $\mathbf{u} + t\mathbf{v}$ as t ranges over all scalars is the set of position vectors of points on a line through the head of $\mathbf{u}$. Show that the first of these statements (which is the same as Problem 18) is a consequence of the second, and reconcile the second with Problem 10(d).

22. Show that the set of all vectors of the form

$$[2x,\ x + y,\ x - y]$$

is the set of all linear combinations of $[2, 1, 1]$ and $[0, 1, -1]$.

23. Express the set of all $[x + y + z,\ 2x + y + 2z,\ x + y]$ as the set of all linear combinations of three vectors.

24. Express the set of all $(y - z)\mathbf{i} + (x - y)\mathbf{j} + z\mathbf{k}$ as the set of all linear combinations of some vectors.

25. Show that the set of vectors in Problem 24 is the set of all (position) vectors in 3-space. Do the same for the set in Problem 23. How about Problem 22?

26. Here is an example of a linear combination where the coefficients are not unique. Let

$$\mathbf{u} = \mathbf{i} + 2\mathbf{j} + 8\mathbf{k}, \qquad \mathbf{v} = -\mathbf{i} + 2\mathbf{j} + 4\mathbf{k}, \qquad \mathbf{w} = \mathbf{i} - \mathbf{j} - \mathbf{k},$$

and show that

$$3\mathbf{u} + 4\mathbf{v} + 5\mathbf{w} = 4\mathbf{u} + \mathbf{v} + \mathbf{w}.$$

"Solve" for $\mathbf{u}$ to show that

$$\mathbf{u} = 3\mathbf{v} + 4\mathbf{w}$$

and hence any linear combination of $\mathbf{u}$, $\mathbf{v}$, and $\mathbf{w}$ is a linear combination of $\mathbf{v}$ and $\mathbf{w}$. Show that the geometric meaning of this is that $\mathbf{u}$, $\mathbf{v}$, and $\mathbf{w}$ lie in the same plane.

27. Show that whenever coefficients of some linear combination of vectors $\mathbf{v}_1, \ldots, \mathbf{v}_n$ are not unique, then one of the $\mathbf{v}_p$ is a linear combination of the others (say this one is $\mathbf{v}_1$) and every linear combination of $\mathbf{v}_1, \ldots, \mathbf{v}_n$ is a linear combination of $\mathbf{v}_2, \ldots, \mathbf{v}_n$. If $n = 3$, this means $\mathbf{v}_1$, $\mathbf{v}_2$, $\mathbf{v}_3$ are coplanar. If $n = 2$, this means $\mathbf{v}_1$ and $\mathbf{v}_2$ are collinear (parallel). Compare Problems 17 and 19.

28. The set of all linear combinations of a fixed set of vectors $\mathbf{v}_1, \ldots, \mathbf{v}_n$ has the following properties.

If two vectors $\mathbf{v}$, $\mathbf{w}$ are in the set, then so is $\mathbf{v} + \mathbf{w}$.

If a vector $\mathbf{v}$ is in the set, then so is every scalar times $\mathbf{v}$.

If two vectors $\mathbf{v}$, $\mathbf{w}$ are in the set, then so is $\mathbf{v} - \mathbf{w}$.

If two vectors $\mathbf{v}$, $\mathbf{w}$ are in the set, then so is every linear combination of $\mathbf{v}$ and $\mathbf{w}$.

If m vectors $\mathbf{w}_1, \ldots, \mathbf{w}_m$ are in the set, then so is every linear combination of $\mathbf{w}_1, \ldots, \mathbf{w}_m$.

Paraphrase: Every linear combination of linear combinations of $\mathbf{v}_1, \ldots, \mathbf{v}_n$ is a linear combination of $\mathbf{v}_1, \ldots, \mathbf{v}_n$.

6. DOT PRODUCTS

We cannot multiply vectors to get vectors and still retain all the arithmetic properties of multiplication that we expect from our experience with multiplication of numbers. The proof of this fact is not easy, but is not impossible at your level; you might want to experiment with multiplications to show that it is impossible to have a multiplication of vectors in 3-space which is commutative, associative, distributive over addition, and for which division except by $\mathbf{0}$ is always possible.† There is, however, one important product that has many of these properties, except that the product of two vectors is not a vector but a scalar.

6.1 Definition

If $\mathbf{v}$ and $\mathbf{w}$ are vectors, the *dot product* $\mathbf{v} \cdot \mathbf{w}$ is the number $\|\mathbf{v}\| \|\mathbf{w}\| \cos\theta$, where θ is the angle between the vectors $\mathbf{v}$ and $\mathbf{w}$. If $\mathbf{v}$ and $\mathbf{w}$ do not have the same tail, then θ is to be the angle between $\mathbf{v}$ and a vector equivalent to $\mathbf{w}$ that does have the same tail as $\mathbf{v}$. We always take θ to be between 0 and π, which makes "the angle" unambiguous.

Because of this second agreement on how to measure θ, we see that the dot product does not change if we change the factors into equivalent vectors: if $\mathbf{v}$ is equivalent to $\mathbf{v}'$ and $\mathbf{w}$ is equivalent to $\mathbf{w}'$, then $\mathbf{v} \cdot \mathbf{w} = \mathbf{v}' \cdot \mathbf{w}'$. Thus in computing dot products, we can always reduce to dot products of vectors with their tails at the origin.

† See K. May, *Amer. Math. Monthly* **73,** 289–291 (1966).

Example 1.

$\mathbf{i} \cdot \mathbf{j} = 0$ because $\| \mathbf{i} \| = 1$, $\| \mathbf{j} \| = 1$, $\theta = \pi/2$, $\cos\theta = 0$;

$\mathbf{i} \cdot \mathbf{i} = 1$ because $\| \mathbf{i} \| = 1$, $\theta = 0$, $\cos\theta = 1$;

$\mathbf{i} \cdot 3\mathbf{i} = 3$;

$\mathbf{i} \cdot (\mathbf{i} + \mathbf{j}) = (1)(\sqrt{2}) \cos(\pi/4) = 1$ (draw $\mathbf{i} + \mathbf{j}$ to check this);

$\mathbf{v} \cdot (-\mathbf{v}) = -\| \mathbf{v} \|^2$ because $\theta = \pi$, $\cos\theta = -1$;

$\mathbf{v} \cdot \mathbf{w} \geq 0$ if and only if $\mathbf{v}$ and $\mathbf{w}$ make an acute angle with each other;

$\mathbf{v} \cdot \mathbf{w} < 0$ if the angle is obtuse (since the factors $\| \mathbf{v} \|$, $\| \mathbf{w} \|$ are positive the sign of $\mathbf{v} \cdot \mathbf{w}$ depends only on the sign of $\cos\theta$). Especially useful are the following two computations.

6.2 $\mathbf{v} \cdot \mathbf{w} = 0$ if and only if $\mathbf{v}$ and $\mathbf{w}$ are perpendicular.

Proof. $\| \mathbf{v} \| \, \| \mathbf{w} \| \cos\theta = 0$ if and only if either $\| \mathbf{v} \| = 0$ (so $\mathbf{v} = \mathbf{0}$ and $\mathbf{v}$ is perpendicular to $\mathbf{w}$ because we agreed to say the zero vector is parallel to and perpendicular to every vector) or $\| \mathbf{w} \| = 0$ (so $\mathbf{w} = \mathbf{0}$ and $\mathbf{w}$ is perpendicular to $\mathbf{v}$) or $\cos\theta = 0$ (so, since $0 \leq \theta \leq \pi$, $\theta = \pi/2$ and $\mathbf{v}$ and $\mathbf{w}$ are perpendicular).

6.3 $\mathbf{v} \cdot \mathbf{v} = \| \mathbf{v} \|^2$ or $\| \mathbf{v} \| = (\mathbf{v} \cdot \mathbf{v})^{1/2}$.

Proof. $\theta = 0$, $\cos\theta = 1$, $\mathbf{v} \cdot \mathbf{v} = \| \mathbf{v} \| \, \| \mathbf{v} \|$.

Before we compute the dual personality of the dot product ($[a, b, c] \cdot [a', b', c'] = ?$), we need the following:

6.4 The ordinary arithmetic properties of products hold for dot products:

(a) $\mathbf{v} \cdot \mathbf{w} = \mathbf{w} \cdot \mathbf{v}$.

(b) $\mathbf{v} \cdot (\mathbf{w} + \mathbf{w}') = \mathbf{v} \cdot \mathbf{w} + \mathbf{v} \cdot \mathbf{w}'$.

(c) $(\mathbf{v} + \mathbf{v}') \cdot \mathbf{w} = \mathbf{v} \cdot \mathbf{w} + \mathbf{v}' \cdot \mathbf{w}$.

(d) $\mathbf{v} \cdot (\mathbf{w} + \mathbf{w}' + \mathbf{w}'') = \mathbf{v} \cdot \mathbf{w} + \mathbf{v} \cdot \mathbf{w}' + \mathbf{v} \cdot \mathbf{w}''$.

(e) $(\mathbf{v} + \mathbf{v}' + \mathbf{v}'') \cdot \mathbf{w} = \mathbf{v} \cdot \mathbf{w} + \mathbf{v}' \cdot \mathbf{w} + \mathbf{v}'' \cdot \mathbf{w}$, etc.

Note that the associative law $(\mathbf{u} \cdot \mathbf{v}) \cdot \mathbf{w} = \mathbf{u} \cdot (\mathbf{v} \cdot \mathbf{w})$ cannot hold. It does not even make sense since $\mathbf{u} \cdot \mathbf{v}$ is a scalar and you cannot take a dot product of a scalar and a vector $\mathbf{w}$. It is also false in general that $(\mathbf{u} \cdot \mathbf{v})\mathbf{w} = \mathbf{u}(\mathbf{v} \cdot \mathbf{w})$, where on each side of the equation you have one dot product and one multiplication of a vector by a scalar; the left-hand side is a vector parallel to $\mathbf{w}$ and the right-hand side is a vector parallel to $\mathbf{u}$, so the two cannot be equal unless $\mathbf{u}$ and $\mathbf{w}$ are parallel.

Proof. (a) Interchanging $\mathbf{v}$ and $\mathbf{w}$ in the definition of dot product does not change θ and only interchanges the order of multiplying the scalars $\|\mathbf{v}\|$ and $\|\mathbf{w}\|$. But multiplication of scalars is commutative.

(b) We define the *component of* $\mathbf{w}$ *along* $\mathbf{v}$ to be plus or minus the length of the projection of $\mathbf{w}$ on the line of $\mathbf{v}$; plus if the projection has the same direction as $\mathbf{v}$ and minus if opposite. In other words, the component of $\mathbf{w}$ along $\mathbf{v}$ is $\|\mathbf{w}\| \cos\theta$ (see Fig. 1.10). Thus

6.5 $\mathbf{v} \cdot \mathbf{w} = \|\mathbf{v}\|$ times the component of $\mathbf{w}$ along $\mathbf{v}$.

This is a more convenient formula for the dot product in proving 6.4(b), because the vector sum (projection of $\mathbf{w}$) + (projection of $\mathbf{w}'$) equals the projection of $(\mathbf{w} + \mathbf{w}')$, and so the scalar sum (component of $\mathbf{w}$ along $\mathbf{v}$) + (component of $\mathbf{w}'$ along $\mathbf{v}$) equals the component of $(\mathbf{w} + \mathbf{w}')$ along $\mathbf{v}$. Multiplying this last equation by $\|\mathbf{v}\|$, we get

$$\mathbf{v} \cdot \mathbf{w} + \mathbf{v} \cdot \mathbf{w}' = \mathbf{v} \cdot (\mathbf{w} + \mathbf{w}').$$

The proof of (c) is exactly similar, or we can give another proof using 6.4(a), 6.4(b), and 6.4(a), thus:

$$(\mathbf{v} + \mathbf{v}') \cdot \mathbf{w} = \mathbf{w} \cdot (\mathbf{v} + \mathbf{v}') = \mathbf{w} \cdot \mathbf{v} + \mathbf{w} \cdot \mathbf{v}' = \mathbf{v} \cdot \mathbf{w} + \mathbf{v}' \cdot \mathbf{w}.$$

(d) $$\mathbf{v} \cdot ((\mathbf{w} + \mathbf{w}') + \mathbf{w}'') = \mathbf{v} \cdot (\mathbf{w} + \mathbf{w}') + \mathbf{v} \cdot \mathbf{w}''$$
$$= \mathbf{v} \cdot \mathbf{w} + \mathbf{v} \cdot \mathbf{w}' + \mathbf{v} \cdot \mathbf{w}'',$$

and similarly for (e).

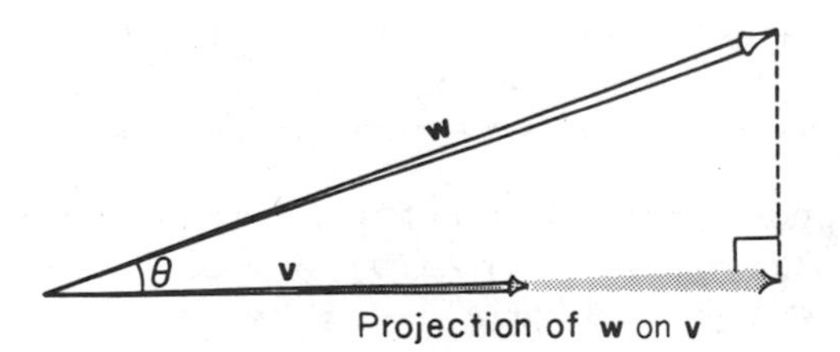

Figure 1.10

6.6 Theorem

$$[a, b, c] \cdot [a', b', c'] = aa' + bb' + cc',$$

$$(a\mathbf{i} + b\mathbf{j} + c\mathbf{k}) \cdot (a'\mathbf{i} + b'\mathbf{j} + c'\mathbf{k}) = aa' + bb' + cc'.$$

Proof. By 6.4(d), with $\mathbf{v} = a\mathbf{i} + b\mathbf{j} + c\mathbf{k}$, we get

$$(a\mathbf{i} + b\mathbf{j} + c\mathbf{k}) \cdot (a'\mathbf{i} + b'\mathbf{j} + c'\mathbf{k})$$

$$= \mathbf{v} \cdot (a'\mathbf{i}) + \mathbf{v} \cdot (b'\mathbf{j}) + \mathbf{v} \cdot (c'\mathbf{k}).$$

Use 6.4(e) to get $\mathbf{v} \cdot (a'\mathbf{i}) = a\mathbf{i} \cdot a'\mathbf{i} + b\mathbf{j} \cdot a'\mathbf{i} + c\mathbf{k} \cdot a'\mathbf{i}$ and do the same with the other two dot products, finally getting the whole dot product expressed as a sum of nine terms

$$(a\mathbf{i}) \cdot (a'\mathbf{i}) + (b\mathbf{j}) \cdot (a'\mathbf{i}) + (c\mathbf{k}) \cdot (a'\mathbf{i})$$

$$+ (a\mathbf{i}) \cdot (b'\mathbf{j}) + (b\mathbf{j}) \cdot (b'\mathbf{j}) + (c\mathbf{k}) \cdot (b'\mathbf{j})$$

$$+ (a\mathbf{i}) \cdot (c'\mathbf{k}) + (b\mathbf{j}) \cdot (c'\mathbf{k}) + (c\mathbf{k}) \cdot (c'\mathbf{k}).$$

Each of these terms can be computed directly from the definition: the cross-product terms are all zero since each is a product of two perpendicular vectors (see 6.2). The only ones of the nine products that are not zero are $(a\mathbf{i}) \cdot (a'\mathbf{i})$, $(b\mathbf{j}) \cdot (b'\mathbf{j})$, and $(c\mathbf{k}) \cdot (c'\mathbf{k})$. But

$$(a\mathbf{i}) \cdot (a'\mathbf{i}) = \| a\mathbf{i} \| \, \| a'\mathbf{i} \| \cos \theta = | a | \, | a' | \cos \theta$$

where θ is either 0 or π, depending on whether $a\mathbf{i}$ and $a'\mathbf{i}$ have the same or opposite directions, that is, whether a and a' have the same or opposite sign. If a and a' have the same sign, we get $| a | \, | a' | \cos \theta = | a | \, | a' | = aa'$; and if a and a' have opposite signs, we get $| a | \, | a' | \cos \pi = -| a | \, | a' | = aa'$ again. Thus in both cases $a\mathbf{i} \cdot a'\mathbf{i} = aa'$. Similarly,

$$(b\mathbf{j}) \cdot (b'\mathbf{j}) = bb', \qquad (c\mathbf{k}) \cdot (c'\mathbf{k}) = cc'$$

and so

$$(a\mathbf{i} + b\mathbf{j} + c\mathbf{k}) \cdot (a'\mathbf{i} + b'\mathbf{j} + c'\mathbf{k}) = aa' + bb' + cc'.$$

6.7 $(s\mathbf{v}) \cdot (t\mathbf{w}) = (st)(\mathbf{v} \cdot \mathbf{w})$.

Proof. Suppose $\mathbf{v}$ has components $[a, b, c]$ and $\mathbf{w}$ has components $[a', b', c']$. Then by 6.6, $(s\mathbf{v}) \cdot (t\mathbf{w}) = (sa)(ta') + (sb)(tb') + (sc)(tc') = (st)(aa' + bb' + cc') = (st)(\mathbf{v} \cdot \mathbf{w})$.

6.8 $\mathbf{u} \cdot (a\mathbf{v} + b\mathbf{w}) = a(\mathbf{u} \cdot \mathbf{v}) + b(\mathbf{u} \cdot \mathbf{w})$.

Example 1. $[5, 0, 3] \cdot [2, -1, -4] = 5(2) + 0(-1) + 3(-4) = 10 + 0 - 12 = -2$. Thus $[5, 0, 3]$ and $[2, -1, -4]$ make an obtuse angle with each other.

Example 2. $(2\mathbf{i} + 3\mathbf{j}) \cdot (5\mathbf{i} - 6\mathbf{k}) = 2(5) + 3(0) + 0(-6) = 10$.

Example 3. $[4, 1, 1] \cdot [1, -2, -2] = 4 - 2 - 2 = 0$. Thus these two vectors are perpendicular, by 6.2.

Example 4. Find the angle between the vectors $\mathbf{v} = [5, 1, 1]$ and $\mathbf{w} = [-1, 2, 4]$:

$$\|\mathbf{v}\|\,\|\mathbf{w}\| \cos\theta = \mathbf{v} \cdot \mathbf{w} = -5 + 2 + 4 = 1.$$

But

$$\|\mathbf{v}\| = (25 + 1 + 1)^{1/2} = (27)^{1/2}$$

and

$$\|\mathbf{w}\| = (1 + 4 + 16)^{1/2} = (21)^{1/2}.$$

Thus $\cos\theta = [(27)^{1/2}(21)^{1/2}]^{-1} = 0.042$; θ is approximately $[(\pi/2) - 0.042]$ radians $= 87.5$ degrees.

Example 5. The cosine of the angle α between $\mathbf{v}$ and the positive x-axis is $(1/\|\mathbf{v}\|)(\mathbf{v} \cdot \mathbf{i})$ because by definition of dot products, $\mathbf{v} \cdot \mathbf{i} = \|\mathbf{v}\|\,\|\mathbf{i}\| \cos\alpha$. If $\mathbf{v} = a\mathbf{i} + b\mathbf{j} + c\mathbf{k}$, this direction cosine is $(a^2 + b^2 + c^2)^{-1/2}(a)$. Compare Problems 9 and 10, Section 4.

Example 6. Suppose we have three vectors $\mathbf{v}_1, \mathbf{v}_2, \mathbf{v}_3$ (not necessarily $\mathbf{i}$, $\mathbf{j}$, and $\mathbf{k}$) each of length 1 and each perpendicular to the other two. Then any vector that is a linear combination of $\mathbf{v}_1, \mathbf{v}_2, \mathbf{v}_3$, say, obtained as $a_1\mathbf{v}_1 + a_2\mathbf{v}_2 + a_3\mathbf{v}_3$ (with some scalars a_1, a_2, a_3), has length given by the formula

$$(a_1^2 + a_2^2 + a_3^2)^{1/2}.$$

We could see this by essentially repeating the proof of the length formula in Section 3 with the directions of $\mathbf{v}_1, \mathbf{v}_2, \mathbf{v}_3$ replacing the coordinate axes used there, or we can proceed as follows:

If $\mathbf{v} = a_1\mathbf{v}_1 + a_2\mathbf{v}_2 + a_3\mathbf{v}_3$, then $\|\mathbf{v}\|^2 = \mathbf{v} \cdot \mathbf{v}$ by 6.3, but

$$\|\mathbf{v}\|^2 = a_1\mathbf{v}_1 \cdot a_1\mathbf{v}_1 + a_1\mathbf{v}_1 \cdot a_2\mathbf{v}_2 + \cdots \text{ to nine terms} \qquad \text{(by 6.4(d), (e))}$$

$$= a_1^2\mathbf{v}_1 \cdot \mathbf{v}_1 + a_2^2\mathbf{v}_2 \cdot \mathbf{v}_2 + a_3^2\mathbf{v}_3 \cdot \mathbf{v}_3$$

$$+ \text{ six cross-product terms} \qquad \text{(by 6.7)}$$

$$= a_1^2 + a_2^2 + a_3^2$$

because $\mathbf{v}_p \cdot \mathbf{v}_p = \|\mathbf{v}_p\|^2 = 1$ and, if $p \neq q$, $\mathbf{v}_p \cdot \mathbf{v}_q = 0$, by 6.2.

Example 7. (Law of cosines, but see also Problem 16):

$$\| \mathbf{u} + \mathbf{v} \|^2 = \| \mathbf{u} \|^2 + \| \mathbf{v} \|^2 + 2 \| \mathbf{u} \| \| \mathbf{v} \| \cos \theta$$

because

$$\begin{aligned} \| \mathbf{u} + \mathbf{v} \|^2 &= (\mathbf{u} + \mathbf{v}) \cdot (\mathbf{u} + \mathbf{v}) && \text{(by 6.3)} \\ &= \mathbf{u} \cdot \mathbf{u} + \mathbf{u} \cdot \mathbf{v} + \mathbf{v} \cdot \mathbf{u} + \mathbf{v} \cdot \mathbf{v} && \text{(by 6.4(b), (c))} \\ &= \| \mathbf{u} \|^2 + 2\mathbf{u} \cdot \mathbf{v} + \| \mathbf{v} \|^2 && \text{(by 6.3 and 6.4(a))} \end{aligned}$$

6.9 The components of every vector $\mathbf{v}$ (which were defined geometrically in Section 2, and which are the coefficients when $\mathbf{v}$ is expressed as (equivalent to) a linear combination of $\mathbf{i}$, $\mathbf{j}$, and $\mathbf{k}$) are also given by the formulas $\mathbf{v} \cdot \mathbf{i}$, $\mathbf{v} \cdot \mathbf{j}$, $\mathbf{v} \cdot \mathbf{k}$.

Proof. If $\mathbf{v} = a\mathbf{i} + b\mathbf{j} + c\mathbf{k}$ so that the components of $\mathbf{v}$ are a, b, c, then $\mathbf{v} \cdot \mathbf{i} = (a)(1) + 0 + 0 = a$ and similarly $\mathbf{v} \cdot \mathbf{j} = b$, $\mathbf{v} \cdot \mathbf{k} = c$ by 6.6.

PROBLEMS

1. Compute the following dot products: $[2, 1, 2] \cdot [3, 4, -7]$, $[1, 0, 3] \cdot [0, 1, 4]$, $[1, \sqrt{2}, -1] \cdot [\sqrt{2}, 4, 0]$, $(2\mathbf{i} - \mathbf{j} + \mathbf{k}) \cdot (\mathbf{i} + 2\mathbf{j} - \mathbf{k})$, $(2\mathbf{i} - \mathbf{j}) \cdot (\mathbf{j} + \mathbf{k})$.

2. Find the angle between each of the pairs of vectors in Problem 1.

3. Which of the following vectors are perpendicular to which: $2\mathbf{i} - \mathbf{j}$; $\mathbf{i}$; $\mathbf{i} + 2\mathbf{j}$; $\mathbf{k}$; $\mathbf{i} + 2\mathbf{j} + \mathbf{k}$?

4. Find the angle between one edge of a cube and a diagonal of the cube. (The edge and the diagonal should both issue from one vertex of the cube. The diagonal is the line from this vertex to the farthest of the other vertices; it is not the diagonal of one of the square faces.)

5. If $\mathbf{v}$ and $\mathbf{w}$ are perpendicular to $\mathbf{u}$, then every linear combination of $\mathbf{v}$ and $\mathbf{w}$ is perpendicular to $\mathbf{u}$. Prove this purely geometrically using the definition of $a\mathbf{v} + b\mathbf{w}$, and then prove it again using 6.2 and the arithmetic of dot products.

6. Find one vector perpendicular to both $2\mathbf{i} + 3\mathbf{j} + 4\mathbf{k}$ and $\mathbf{i} - \mathbf{j}$. Note from the geometry that there is an infinitude of answers: If one vector works, any scalar multiple of it will also work. Hence, by suitably adjusting this scalar, you can assume, for example, that your answer will have x-component equal to 1. Use 6.6, then, to find y and z so that $\mathbf{i} + y\mathbf{j} + z\mathbf{k}$ has the required properties. When you have finished, discuss this normalization of x-component more carefully; are there conceivably some similar problems where you could not have assumed that one answer could be found with x-component 1?

7. Show that a vector perpendicular to both $a_1\mathbf{i} + a_2\mathbf{j} + a_3\mathbf{k}$ and $b_1\mathbf{i} + b_2\mathbf{j} + b_3\mathbf{k}$ is

$$\begin{vmatrix} a_2 & a_3 \\ b_2 & b_3 \end{vmatrix} \mathbf{i} + \begin{vmatrix} a_3 & a_1 \\ b_3 & b_1 \end{vmatrix} \mathbf{j} + \begin{vmatrix} a_1 & a_2 \\ b_1 & b_2 \end{vmatrix} \mathbf{k},$$

where

$$\begin{vmatrix} a_2 & a_3 \\ b_2 & b_3 \end{vmatrix}$$

is the determinant $a_2b_3 - a_3b_2$, etc.

8. Prove Schwarz's inequality: For every six numbers a_1, a_2, a_3, b_1, b_2, b_3, we have $(a_1^2 + a_2^2 + a_3^2)(b_1^2 + b_2^2 + b_3^2) \geq (a_1b_1 + a_2b_2 + a_3b_3)^2$. This one is difficult analytically, but easy geometrically: Consider the vectors $\mathbf{a} = [a_1, a_2, a_3]$ and $\mathbf{b} = [b_1, b_2, b_3]$ and interpret both sides of the inequality in these terms (for example, $a_1^2 + a_2^2 + a_3^2 = \|\mathbf{a}\|^2$). The analogous inequality for $2n$ numbers $a_1, \ldots, a_n$, $b_1, \ldots, b_n$ is also true; this general inequality is what is properly called Schwarz's inequality; it cannot very well be given a geometrical proof, though (see Section 8).

9. Let

$$\mathbf{v}_1 = 3^{-1}(2\mathbf{i} + 2\mathbf{j} + \mathbf{k}), \qquad \mathbf{v}_2 = (\sqrt{2})^{-1}(\mathbf{i} - \mathbf{j}),$$

$$\mathbf{v}_3 = (3\sqrt{2})^{-1}(\mathbf{i} + \mathbf{j} - 4\mathbf{k}), \qquad \mathbf{v}_4 = 2^{-1}(2\mathbf{i} - \mathbf{j} - \mathbf{k}).$$

(a) Compute the dot products $\mathbf{v}_p \cdot \mathbf{v}_q$ for p and $q = 1, 2, 3, 4$.

(b) If $\mathbf{v}_4 = a_1\mathbf{v}_1 + a_2\mathbf{v}_2 + a_3\mathbf{v}_3$, show that the only possibilities for a_1, a_2, a_3 are $a_1 = \frac{1}{6}$, $a_2 = 3/(2\sqrt{2})$, $a_3 = 5/(6\sqrt{2})$. Hint: Dot the equation with $\mathbf{v}_1$, then with $\mathbf{v}_2$, then with $\mathbf{v}_3$.

(c) Conversely, show that in fact

$$\mathbf{v}_4 = \tfrac{1}{6}\mathbf{v}_1 + (3/2\sqrt{2})\mathbf{v}_2 + (5/6\sqrt{2})\mathbf{v}_3.$$

(d) Show that every vector in 3-space is equivalent to a unique $b_1\mathbf{v}_1 + b_2\mathbf{v}_2 + b_3\mathbf{v}_3$ for suitable choice of the scalars b_1, b_2, b_3.

10. Show that the components of $\mathbf{v}$ are $\mathbf{v} \cdot \mathbf{i}$, $\mathbf{v} \cdot \mathbf{j}$, and $\mathbf{v} \cdot \mathbf{k}$. Do this geometrically from the definition of the dot product. Do not use 6.9.

11. Suppose we have three vectors $\mathbf{v}_1, \mathbf{v}_2, \mathbf{v}_3$ with a common tail, each of length 1, and each one perpendicular to the other two (they could be $\mathbf{i}, \mathbf{j}, \mathbf{k}$ but need not be).

(a) Show that every vector in 3-space is equivalent to a linear combination of $\mathbf{v}_1, \mathbf{v}_2, \mathbf{v}_3$; that is, every vector $\mathbf{v}$ is equivalent to $a_1\mathbf{v}_1 + a_2\mathbf{v}_2 + a_3\mathbf{v}_3$ for some scalars a_1, a_2, a_3.

(b) Show that, in fact, $a_p = \mathbf{v} \cdot \mathbf{v}_p$ for $p = 1, 2, 3$.

(c) Show that $a_p =$ the component of $\mathbf{v}$ along $\mathbf{v}_p$, as defined in the proof of 6.4.

12. If $\mathbf{v}_1$ is perpendicular to $\mathbf{v}_2$, if $\| \mathbf{v}_1 \| = 1$ and $\| \mathbf{v}_2 \| = 1$ and $\mathbf{v} = a_1\mathbf{v}_1 + a_2\mathbf{v}_2$, then $\| \mathbf{v} \| = (a_1^2 + a_2^2)^{1/2}$. Prove this in two ways, as suggested in Example 6.

13. Given $\mathbf{v}_1, \mathbf{v}_2, \mathbf{v}_3$ each of length 1 and each perpendicular to the other two, and given two vectors $\mathbf{v} = a_1\mathbf{v}_1 + a_2\mathbf{v}_2 + a_3\mathbf{v}_3$, $\mathbf{w} = b_1\mathbf{v}_1 + b_2\mathbf{v}_2 + b_3\mathbf{v}_3$, show that $\mathbf{v} \cdot \mathbf{w} = a_1b_1 + a_2b_2 + a_3b_3$ and the angle between $\mathbf{v}$ and $\mathbf{w}$ is

$$\cos^{-1}\left[\frac{a_1b_1 + a_2b_2 + a_3b_3}{(a_1^2 + a_2^2 + a_3^2)^{1/2}(b_1^2 + b_2^2 + b_3^2)^{1/2}}\right].$$

14. Prove that the diagonals of a rhombus are perpendicular. (A rhombus is a parallelogram all of whose sides have equal length.)

15. 6.3 expresses $\| \mathbf{v} \|$ in terms of dot products. Show that dot products can be computed using only lengths and $+$ thus:

$$\mathbf{v} \cdot \mathbf{w} = (\tfrac{1}{2})(\| \mathbf{v} + \mathbf{w} \|^2 - \| \mathbf{v} \|^2 - \| \mathbf{w} \|^2).$$

16. The ordinary law of cosines in trigonometry is not quite the same as Example 7; it states that $c^2 = a^2 + b^2 - 2ab \cos \theta$ if a, b, c are the lengths of the sides of a triangle and θ is the angle between the sides of lengths a and b. Prove this by letting the origin be the vertex of the angle θ; $\mathbf{u}$ and $\mathbf{v}$ the vectors along the two sides of lengths a and b.

17. Prove that for every two vectors $\mathbf{u}$ and $\mathbf{v}$,

$$\| \mathbf{u} + \mathbf{v} \|^2 + \| \mathbf{u} - \mathbf{v} \|^2 = 2(\| \mathbf{u} \|^2 + \| \mathbf{v} \|^2).$$

Hint: Write each length as a dot product by 6.3, and use 6.4.

18. Use Problem 17 to show that the sum of the squares of the lengths of the four sides of any parallelogram equals the sum of the squares of the lengths of the diagonals.

7. CROSS PRODUCTS

Given two vectors **u** and **v**, the vector computed in Problem 7, Section 6 is called the cross product of **u** and **v**:

7.1 Definition

The *cross product* or vector product of **u** and **v**, written $\mathbf{u} \times \mathbf{v}$, is $(a_2b_3 - a_3b_2)\mathbf{i} + (a_3b_1 - a_1b_3)\mathbf{j} + (a_1b_2 - a_2b_1)\mathbf{k}$ where the a's are the components of **u** and the b's are the components of **v**.

Admittedly this definition arrives at a reasonably important product through the back door. It would have been more in the spirit of our previous definitions if we had defined the cross product by specifying its geometrical properties, independent of components or coordinate systems. We could have done this by specifying the length of the cross product as in 7.10 or 7.11 and its direction as in 7.2 plus the additional information that **u**, **v**, and $\mathbf{u} \times \mathbf{v}$ form a right-handed system in the sense of Section 1. We could then have derived all the other properties, including the formula 7.1. We adopt the present approach because it is slightly easier to implement, and we are trying to avoid undue length or emphasis on this section because, although the cross product is a useful device for finding a vector perpendicular to two others, it is restricted to 3-space. Later, when we deal predominantly with n-space for various n's, no cross product will be available. (As a matter of cultural interest, perhaps we should remark that there is an "exterior product" in n-space that is essentially the cross product when $n = 3$; but the exterior product of two vectors is a kind of tensor, not another vector.)

We emphasize that $\mathbf{u} \times \mathbf{v}$ is a vector, where $\mathbf{u} \cdot \mathbf{v}$ is a scalar. In all the following manipulations you should be conscious of this explicitly.

From the way it was found we know that

7.2 $\mathbf{u} \times \mathbf{v}$ is perpendicular to both **u** and **v**. (Or, you can check this by computing that $\mathbf{u} \cdot (\mathbf{u} \times \mathbf{v}) = 0$ and $\mathbf{v} \cdot (\mathbf{u} \times \mathbf{v}) = 0$).

It is also immediate that

7.3 $\mathbf{u} \times \mathbf{v} = -\mathbf{v} \times \mathbf{u}$ since $\mathbf{v} \times \mathbf{u}$ is computed from 7.1 by interchanging the roles of the a's and the b's, and this changes the sign of the three components.

Similarly, you can check that, for all vectors $\mathbf{u}$, $\mathbf{v}$, $\mathbf{w}$ and all scalars a,

7.4 $\mathbf{u} \times (\mathbf{v} + \mathbf{w}) = \mathbf{u} \times \mathbf{v} + \mathbf{u} \times \mathbf{w};$
$(\mathbf{u} + \mathbf{v}) \times \mathbf{w} = \mathbf{u} \times \mathbf{w} + \mathbf{v} \times \mathbf{w};$

7.5 $\mathbf{u} \times a\mathbf{v} = (a\mathbf{u}) \times \mathbf{v} = a(\mathbf{u} \times \mathbf{v});$

7.6 $\mathbf{u} \times \mathbf{0} = \mathbf{0} \times \mathbf{u} = \mathbf{0};$

7.7 $\mathbf{u} \times \mathbf{u} = \mathbf{0}.$

Thus, much of the ordinary arithmetic you are familiar with works for cross products, but the commutative law fails (by a minus sign) and the associative law is also false. For example, $\mathbf{i} \times (\mathbf{i} \times \mathbf{j}) = \mathbf{i} \times \mathbf{k} = -\mathbf{j}$, while $(\mathbf{i} \times \mathbf{i}) \times \mathbf{j} = \mathbf{0} \times \mathbf{j} = \mathbf{0}$. There are two substitutes for this associative law. One is Problem 10; the other is

7.8 $\mathbf{u} \times (\mathbf{v} \times \mathbf{w}) = (\mathbf{u} \cdot \mathbf{w})\mathbf{v} - (\mathbf{u} \cdot \mathbf{v})\mathbf{w}.$

To prove this, check it first for $\mathbf{w} = \mathbf{i}$:

$$\mathbf{v} \times \mathbf{i} = \begin{vmatrix} b_2 & b_3 \\ 0 & 0 \end{vmatrix} \mathbf{i} + \begin{vmatrix} b_3 & b_1 \\ 0 & 1 \end{vmatrix} \mathbf{j} + \begin{vmatrix} b_1 & b_2 \\ 1 & 0 \end{vmatrix} \mathbf{k} = b_3\mathbf{j} - b_2\mathbf{k};$$

$$\mathbf{u} \times (\mathbf{v} \times \mathbf{i}) = (-a_2b_2 - a_3b_3)\mathbf{i} + a_1b_2\mathbf{j} + a_1b_3\mathbf{k};$$

$$\begin{aligned}(\mathbf{u} \cdot \mathbf{i})\mathbf{v} - (\mathbf{u} \cdot \mathbf{v})\mathbf{i} &= a_1(b_1\mathbf{i} + b_2\mathbf{j} + b_3\mathbf{k}) \\ &\quad - (a_1b_1 + a_2b_2 + a_3b_3)\mathbf{i} \\ &= (-a_2b_2 - a_3b_3)\mathbf{i} + a_1b_2\mathbf{j} + a_1b_3\mathbf{k}.\end{aligned}$$

Next, check 7.8 when $\mathbf{w} = \mathbf{j}$ and when $\mathbf{w} = \mathbf{k}$. Finally, for arbitrary $\mathbf{w} = c_1\mathbf{i} + c_2\mathbf{j} + c_3\mathbf{k}$, both sides of 7.8 come apart into terms, each involving only one of **i**, **j**, **k** instead of **w**, thus:

$$\begin{aligned}
\mathbf{u} \times (\mathbf{v} \times \mathbf{w}) &= \mathbf{u} \times (\mathbf{v} \times c_1\mathbf{i} + \mathbf{v} \times c_2\mathbf{j} + \mathbf{v} \times c_3\mathbf{k}) \\
&= \mathbf{u} \times c_1(\mathbf{v} \times \mathbf{i}) + \mathbf{u} \times c_2(\mathbf{v} \times \mathbf{j}) \\
&\quad + \mathbf{u} \times c_3(\mathbf{v} \times \mathbf{k}) \\
&= c_1[\mathbf{u} \times (\mathbf{v} \times \mathbf{i})] + c_2[\mathbf{u} \times (\mathbf{v} \times \mathbf{j})] \\
&\quad + c_3[\mathbf{u} \times (\mathbf{v} \times \mathbf{k})] \\
&= c_1[(\mathbf{u} \cdot \mathbf{i})\mathbf{v} - (\mathbf{u} \cdot \mathbf{v})\mathbf{i}] + c_2[(\mathbf{u} \cdot \mathbf{j})\mathbf{v} - \text{etc.}] \\
&= [(\mathbf{u} \cdot c_1\mathbf{i})\mathbf{v} - (\mathbf{u} \cdot \mathbf{v})c_1\mathbf{i}] + [(\mathbf{u} \cdot c_2\mathbf{j})\mathbf{v} - \text{etc.}] \\
&= (\mathbf{u} \cdot c_1\mathbf{i})\mathbf{v} + (\mathbf{u} \cdot c_2\mathbf{j})\mathbf{v} + (\mathbf{u} \cdot c_3\mathbf{k})\mathbf{v} \\
&\quad - (\mathbf{u} \cdot \mathbf{v})c_1\mathbf{i} - (\mathbf{u} \cdot \mathbf{v})c_2\mathbf{j} - (\mathbf{u} \cdot \mathbf{v})c_3\mathbf{k} \\
&= (\mathbf{u} \cdot \mathbf{w})\mathbf{v} - (\mathbf{u} \cdot \mathbf{v})\mathbf{w}.
\end{aligned}$$

The cross product and the dot product behave well together:

7.9 $(\mathbf{u} \times \mathbf{v}) \cdot \mathbf{w} = \mathbf{u} \cdot (\mathbf{v} \times \mathbf{w})$.

This may be proved much as 7.8 was, by checking it first for $\mathbf{w} = \mathbf{i}$, $\mathbf{w} = \mathbf{j}$, and $\mathbf{w} = \mathbf{k}$, and then showing that the general case $\mathbf{w} = c_1\mathbf{i} + c_2\mathbf{j} + c_3\mathbf{k}$ reduces to these three special cases if you use the arithmetic of dot and cross products: see 7.4, 7.5, and 6.4.

Finally

7.10 $\|\mathbf{u} \times \mathbf{v}\| = \|\mathbf{u}\| \, \|\mathbf{v}\| \sin\theta$, where θ is the angle between **u** and **v**.

To prove this,

$$\begin{aligned}
\|\mathbf{u} \times \mathbf{v}\|^2 &= (\mathbf{u} \times \mathbf{v}) \cdot (\mathbf{u} \times \mathbf{v}) && \text{(by 6.3)} \\
&= \mathbf{u} \cdot (\mathbf{v} \times (\mathbf{u} \times \mathbf{v})) && \text{(by 7.9)} \\
&= \mathbf{u} \cdot ((\mathbf{v} \cdot \mathbf{v})\mathbf{u} - (\mathbf{v} \cdot \mathbf{u})\mathbf{v}) && \text{(by 7.8)} \\
&= (\mathbf{v} \cdot \mathbf{v})(\mathbf{u} \cdot \mathbf{u}) - (\mathbf{v} \cdot \mathbf{u})(\mathbf{u} \cdot \mathbf{v}) \\
& && \text{(by 6.4(b) and 6.7)} \\
&= \|\mathbf{v}\|^2 \|\mathbf{u}\|^2 - (\|\mathbf{v}\| \, \|\mathbf{u}\| \cos\theta)^2 \\
&= \|\mathbf{v}\|^2 \|\mathbf{u}\|^2 \sin^2\theta.
\end{aligned}$$

As a corollary, we have

7.11 $\|\mathbf{u} \times \mathbf{v}\|$ = area of parallelogram with two edges $\mathbf{u}$ and $\mathbf{v}$;

7.12 $(\mathbf{u} \times \mathbf{v}) \cdot \mathbf{w} = \pm$ volume of parallelepiped with three edges $\mathbf{u}, \mathbf{v}, \mathbf{w}$.

PROBLEMS

1. Compute

$$\mathbf{i} \times (\mathbf{i} + \mathbf{j} + \mathbf{k}); \qquad (2\mathbf{i} - \mathbf{j} + \mathbf{k}) \times (3\mathbf{i} + 3\mathbf{j} - \mathbf{k});$$

$$[2, 0, 1] \times [1, 1, 2]; \qquad \mathbf{k} \times (a_1\mathbf{i} + a_2\mathbf{j} + a_3\mathbf{k}).$$

2. Show $\mathbf{i} \times \mathbf{j} = \mathbf{k}$, $\mathbf{k} \times \mathbf{i} = \mathbf{j}$, $\mathbf{j} \times \mathbf{k} = \mathbf{i}$; $\mathbf{j} \times \mathbf{i} = -\mathbf{k}$, $\mathbf{i} \times \mathbf{k} = -\mathbf{j}$, $\mathbf{k} \times \mathbf{j} = -\mathbf{i}$.

3. From Problem 2 and the arithmetic of cross products deduce the original formula for $\mathbf{u} \times \mathbf{v}$, 7.1.

4. (a) Find a vector perpendicular to the plane containing $\mathbf{i} + \mathbf{j}$ and $2\mathbf{i} - \mathbf{j} - \mathbf{k}$.

(b) Find all vectors perpendicular to this plane.

5. Find a vector perpendicular to the plane containing the points $(0, 0, 0)$, $(1, 2, 3)$, and $(2, 1, 1)$.

6. Find a vector perpendicular to the plane containing the points $(2, 1, -1)$, $(1, 0, 1)$, and $(3, 3, 2)$.

7. Find a vector of length 1 perpendicular to the vectors $\mathbf{i} - \mathbf{j}$ and $\mathbf{i} + \mathbf{j} + \mathbf{k}$.

8. Show $(\mathbf{u} + a\mathbf{v}) \times \mathbf{v} = \mathbf{u} \times \mathbf{v}$.

9. Use Problem 8 and 7.11 to show that the parallelogram with edges $\mathbf{u}$ and $\mathbf{v}$ has the same area as the parallelogram with edges $\mathbf{u} + a\mathbf{v}$ and $\mathbf{v}$. Then give a geometric proof of this same fact, using area = base $\times$ altitude.

10. Use 7.8 to prove the Jacobi identity

$$\mathbf{u} \times (\mathbf{v} \times \mathbf{w}) + \mathbf{v} \times (\mathbf{w} \times \mathbf{u}) + \mathbf{w} \times (\mathbf{u} \times \mathbf{v}) = 0.$$

11. A partial proof of 7.8 can be constructed thus: $\mathbf{v} \times \mathbf{w}$ is perpendicular to the plane of $\mathbf{v}$ and $\mathbf{w}$, so any vector perpendicular to $\mathbf{v} \times \mathbf{w}$

will be in the plane of **v** and **w**, hence a linear combination of **v** and **w**. Therefore

$$\mathbf{u} \times (\mathbf{v} \times \mathbf{w}) = a\mathbf{v} + b\mathbf{w}.$$

Now use the condition that $\mathbf{u} \times (\mathbf{v} \times \mathbf{w})$ is perpendicular to **u** (dot product with **u** is 0) to show the pair $[a, b]$ is proportional to the pair $[\mathbf{u} \cdot \mathbf{w}, -\mathbf{u} \cdot \mathbf{v}]$.

12. Extend 7.9 thus:

$$\begin{aligned}(\mathbf{u} \times \mathbf{v}) \cdot \mathbf{w} &= (\mathbf{v} \times \mathbf{w}) \cdot \mathbf{u} = (\mathbf{w} \times \mathbf{u}) \cdot \mathbf{v} \\ &= -(\mathbf{v} \times \mathbf{u}) \cdot \mathbf{w} = -(\mathbf{w} \times \mathbf{v}) \cdot \mathbf{u} \\ &= -(\mathbf{u} \times \mathbf{w}) \cdot \mathbf{v}.\end{aligned}$$

8. VECTORS IN n-SPACE

The analytic aspect of vectors with tail at the origin—they are triples of numbers—is clearly meat for generalization. We say that a *vector in n-space* is an n-tuple of numbers. We denote by R^n the set of all such vectors in n-space (R stands for *real numbers*). If $n = 1, 2$, or 3 we can associate with a vector in R^n a geometric object: an arrow on the line, in the plane, or in 3-space, respectively (always with tail at the origin). For $n > 3$, we make no attempt to associate geometric aspects to our n-tuples, but when it is convenient we shall pretend that the same geometric considerations extend to them as apply to triples as arrows in 3-space.

This aspect of vectors allows us to represent as vectors many new quantities, physical or not. For example, if a class of students takes a battery of n tests, each student has n test scores; to each student corresponds an n-tuple; the student's performance (this n-tuple of scores) can appropriately be called a vector in n-space. "Phase space" is the scientist's phrase for n-space, used more or less in this sense.

Whether or not it is useful to think of a particular quantity as represented by a vector in space depends primarily on whether or not this quantity can meaningfully be subjected to certain standard vector operations, that is, whether, once the quantity has been identified with a vector in R^n, the operations of addition and multiplication by scalars (defined later) have some real significance. For example, this sum of two force vectors represents a force that will accelerate a particle exactly as much and in the same direction as would the two original forces if applied simultaneously. As another example, if

$$\mathbf{x} = [x_1, x_2, \ldots, x_n]$$

is the vector of test scores of an individual on n tests and $\mathbf{y}$ and $\mathbf{z}$ are the vectors of scores of two other individuals, then

$$\mathbf{a} = \tfrac{1}{3}(\mathbf{x} + \mathbf{y} + \mathbf{z})$$

is the n-tuple of average scores for these individuals.

In much of the rest of this course we shall work with R^n rather than R^3, but the reader is encouraged to concentrate always on the case $n = 3$. Vectors in R^n will be denoted with square brackets as before. For example, $[1, 2, 0, -1]$, $[2, \pi, 1, 0]$, and $[a_1, a_2, a_3, a_4]$ are vectors in R^4, at least if a_1, a_2, a_3, and a_4 are real numbers. A typical vector in R^n will be denoted by $[a_1, a_2, \ldots, a_n]$ or by a single boldface letter $\mathbf{a}$; a_1 is called the first component of $\mathbf{a}$, and a_p is the pth component for each $p = 1, 2, \ldots, n$. It will often be convenient to adopt the convention that if a vector is denoted by a boldface letter, its components are denoted by the same letter in italic type, with subscripts:

$$\mathbf{x} = [x_1, x_2, \ldots, x_n].$$

In R^3, of course, our notation has often been $[x, y, z]$ for a vector, rather than $[x_1, x_2, x_3]$. Since x-, y-, and z-coordinates seem more appealing than x_1-, x_2-, and x_3-coordinates, we shall retain this $[x, y, z]$ notation, but it obviously must yield to $[x_1, x_2, \ldots, x_n]$ in n-space. We hope the reader will not find it difficult to put up with this double notation in R^3.

In typing or handwriting, where one cannot very well use boldface, it is customary to use $\underline{x}$ or $\vec{x}$ where in print we use $\mathbf{x}$.

If $\mathbf{u} = [u_1, u_2, \ldots, u_n]$ and $\mathbf{v} = [v_1, v_2, \ldots, v_n]$ are vectors in R^n, we define

$$\mathbf{u} + \mathbf{v} = [u_1 + v_1, u_2 + v_2, \ldots, u_n + v_n]$$

and

$$a\mathbf{u} = [au_1, au_2, \ldots, au_n]$$

for all scalars a. We define linear combinations as before. The zero vector is $[0, 0, \ldots, 0]$ and will still be denoted by $\mathbf{0}$. Then all the ordinary arithmetic holds, especially 4.3 and 5.2. (The proof of all these is easy, except for the triangle inequality 5.2(c), for which see later.) The analogs of $\mathbf{i}$, $\mathbf{j}$, and $\mathbf{k}$ are $[1, 0, \ldots, 0]$, $[0, 1, 0, \ldots, 0]$, ..., $[0, 0, \ldots, 0, 1]$; there are n of these now, and we call them $\mathbf{i}_1, \mathbf{i}_2, \ldots, \mathbf{i}_n$. Every vector in R^n is a linear combination of $\mathbf{i}_1, \ldots, \mathbf{i}_n$ and in fact the coefficients of the linear combination are just the components of the vector:

$$[x_1, x_2, \ldots, x_n] = x_1\mathbf{i}_1 + x_2\mathbf{i}_2 + \cdots + x_n\mathbf{i}_n$$

(check this).

We can also define dot products and lengths:

$$[u_1, u_2, \ldots, u_n] \cdot [v_1, v_2, \ldots, v_n] = u_1v_1 + u_2v_2 + \cdots + u_nv_n,$$

$$\| [u_1, u_2, \ldots, u_n] \| = (\mathbf{u} \cdot \mathbf{u})^{1/2}$$
$$= (u_1^2 + u_2^2 + \cdots + u_n^2)^{1/2}.$$

There are other names for this dot product: "inner product" and "scalar product." This last name has the disadvantage of being easy to confuse with multiplication by scalars.

As you might expect, we say two vectors $\mathbf{u}$ and $\mathbf{v}$ are *orthogonal* if $\mathbf{u} \cdot \mathbf{v} = 0$. It is then easy to check the validity of 6.4, 6.7, and 6.8. Two inequalities are a little more difficult:

$$|\mathbf{v} \cdot \mathbf{w}| \leq \|\mathbf{v}\| \, \|\mathbf{w}\| \qquad \text{(Schwarz's inequality)},$$

$$\|\mathbf{v} + \mathbf{w}\| \leq \|\mathbf{v}\| + \|\mathbf{w}\| \qquad \text{(triangle inequality)}.$$

To prove the first, notice that for all a, $\| a\mathbf{v} + \mathbf{w} \|^2 \geq 0$. But

$$\| a\mathbf{v} + \mathbf{w} \|^2 = (a\mathbf{v} + \mathbf{w}) \cdot (a\mathbf{v} + \mathbf{w})$$
$$= a\mathbf{v} \cdot a\mathbf{v} + 2a\mathbf{v} \cdot \mathbf{w} + \mathbf{w} \cdot \mathbf{w}$$
$$= a^2 \|\mathbf{v}\|^2 + 2a(\mathbf{v} \cdot \mathbf{w}) + \|\mathbf{w}\|^2 = f(a).$$

Fix $\mathbf{v}$ and $\mathbf{w}$ and we have a quadratic polynomial in a (we call it $f(a)$) that is supposed never to be negative. The minimum value of the polynomial occurs when $f'(a) = 0$, that is, when $2a \|\mathbf{v}\|^2 + 2\mathbf{v} \cdot \mathbf{w} = 0$; that is, $a = -\mathbf{v} \cdot \mathbf{w} / \|\mathbf{v}\|^2$; the minimum value is then

$$\frac{(\mathbf{v} \cdot \mathbf{w})^2}{\|\mathbf{v}\|^2} - \frac{2(\mathbf{v} \cdot \mathbf{w})^2}{\|\mathbf{v}\|^2} + \|\mathbf{w}\|^2 = \|\mathbf{v}\|^{-2} (\|\mathbf{v}\|^2 \|\mathbf{w}\|^2 - (\mathbf{v} \cdot \mathbf{w})^2).$$

We demand that this be nonnegative, and we get

$$(\mathbf{v} \cdot \mathbf{w})^2 \leq \|\mathbf{v}\|^2 \|\mathbf{w}\|^2.$$

Taking square roots, we get Schwarz's inequality (compare Problem 8, Section 6).

Then $\|\mathbf{v} + \mathbf{w}\|^2 = (\mathbf{v} + \mathbf{w}) \cdot (\mathbf{v} + \mathbf{w}) = \mathbf{v} \cdot \mathbf{v} + 2\mathbf{v} \cdot \mathbf{w} + \mathbf{w} \cdot \mathbf{w} \leq \|\mathbf{v}\|^2 + 2\|\mathbf{v}\| \, \|\mathbf{w}\| + \|\mathbf{w}\|^2$ by Schwarz, but this equals $(\|\mathbf{v}\| + \|\mathbf{w}\|)^2$. Taking square roots again, we get the triangle inequality 5.2(c).

Thus all the algebraic operations and the concomitant properties of arrows in 3-space carry over to R^n.

We should like to call special attention to one of these vector spaces: the space of n-tuples with $n = 1$. An object in this vector space is a single number, $[a]$. Addition is just addition of numbers and multiplication by a scalar is just ordinary number multiplication. Thus the set of all scalars is being considered as a vector space. You might say scalars are vectors! Hopefully this is not infinitely confusing, because we shall have much use for this vector space.

PROBLEMS

1. If $\mathbf{i}_1, \mathbf{i}_2, \mathbf{i}_3, \mathbf{i}_4$ (in R^4) are the edges of a hypercube (tesseract) the diagonal is $\mathbf{i}_1 + \mathbf{i}_2 + \mathbf{i}_3 + \mathbf{i}_4$. Find the length of this diagonal. In R^n, $\mathbf{i}_1, \ldots, \mathbf{i}_n$ are the edges of a hypercube whose diagonal is $\mathbf{i}_1 + \mathbf{i}_2 + \cdots + \mathbf{i}_n$. Find the length of this diagonal. Compare Problem 15, Section 5.

2. In R^4, what are all the linear combinations of $\mathbf{i}_1, \mathbf{i}_2, \mathbf{i}_3$? Is this the set of all vectors in R^4?

3. Find all vectors in R^4 that are orthogonal to all three of the vectors $\mathbf{i}_1$, $\mathbf{i}_2$, and $\mathbf{i}_1 + \mathbf{i}_2 + \mathbf{i}_3 + \mathbf{i}_4$.

4. Find all vectors in R^4 that are orthogonal to $\mathbf{i}_1 + \mathbf{i}_2$, $\mathbf{i}_1 - \mathbf{i}_2$, $\mathbf{i}_1 + 2\mathbf{i}_3 - \mathbf{i}_4$.

5. Show that if a vector in R^n is orthogonal to two vectors $\mathbf{u}$ and $\mathbf{v}$, then it is orthogonal to every vector "in the plane of $\mathbf{u}$ and $\mathbf{v}$," that is, to every vector that is a linear combination of $\mathbf{u}$ and $\mathbf{v}$.

6. Show that addition and subtraction of vectors in R^1 amounts to addition and subtraction of numbers; multiplication of a vector by a scalar is just multiplication of numbers.

7. Show that $\| \mathbf{u} - \mathbf{v} \| \le \| \mathbf{u} \| + \| \mathbf{v} \|$ for every $\mathbf{u}$ and $\mathbf{v}$ in R^n.

8. Find all x_1 and x_2 such that

$$x_1[2, 3, 1, 1] + x_2[3, -1, -1, 2] = [0, 11, 5, -1].$$

Find all x_1 and x_2 such that

$$x_1[2, 3, 1, 1] + x_2[3, -1, -1, 2] = [5, 2, 0, 1].$$

9. Show that every vector $\mathbf{u}$ in R^4 is a linear combination of $\mathbf{i}_1, \mathbf{i}_2, \mathbf{i}_3, \mathbf{i}_4$, and that the coefficients in this linear combination are equal to $\mathbf{u} \cdot \mathbf{i}_1$, $\mathbf{u} \cdot \mathbf{i}_2$, $\mathbf{u} \cdot \mathbf{i}_3$, and $\mathbf{u} \cdot \mathbf{i}_4$.

9. STILL MORE GENERAL VECTOR SPACES

This section and the next are not essential to the material of the rest of this text. The student can read it as part of his general culture, returning to it occasionally as the course progresses. For the purposes of most of the rest of this text, the phrase *vector space*, which we define in this section, can be interpreted as the set of all n-tuples of numbers, or a suitable subset thereof (see Chapter 5, Section 1), with the understanding that triples of numbers can be identified with arrows in 3-space, as in Chapter 1. However, the beauty of linear algebra is that it casts light on mathematical systems that are quite different from sets of n-tuples, at least on the surface. Already we have reaped some profit from identifying triples (that is, n-tuples with $n = 3$) with arrows in 3-space, even though by no stretch of the imagination can we claim that an arrow *is* a triple of numbers. Once we admit that we shall sometimes be dealing with arrows and sometimes with n-tuples of numbers, we may as well go the whole hog and admit that exactly *what* the vectors are is not as important to us as how they behave; specifically, how they behave when they are added, multiplied by scalars, or dot-multiplied. We declare our right to refer to anything as a vector, provided it is an element of a set of objects called a vector space, defined as follows.

9.1 Definition

A *vector space* is any set of objects, with two operations: addition, and multiplication by real numbers; satisfying the following conditions for all objects $\mathbf{u}$, $\mathbf{v}$, $\mathbf{w}$ in the set and for all real numbers a, b.

(+0) Closure: $\mathbf{u} + \mathbf{v}$ is another object in the set.

(+1) Associative law: $(\mathbf{u} + \mathbf{v}) + \mathbf{w} = \mathbf{u} + (\mathbf{v} + \mathbf{w})$.

(+2) Commutative law: $\mathbf{u} + \mathbf{v} = \mathbf{v} + \mathbf{u}$.

(+3) Zero: There is an object $\mathbf{0}$ in the set such that $\mathbf{u} + \mathbf{0} = \mathbf{u}$.

(+4) Inverse: For each $\mathbf{u}$ there is an object $-\mathbf{u}$ in the set such that $\mathbf{u} + (-\mathbf{u}) = \mathbf{0}$.

(×0) Closure: $a\mathbf{u}$ is another object in the set.

(×1) $a(\mathbf{u} + \mathbf{v}) = a\mathbf{u} + a\mathbf{v}$.

(×2) $(a + b)\mathbf{u} = a\mathbf{u} + b\mathbf{u}$.

(×3) $(ab)\mathbf{u} = a(b\mathbf{u})$.

(×4) $1\mathbf{u} = \mathbf{u}$.

These axioms amount to the statement that the ordinary arithmetic properties of addition and multiplication by scalars will work for the addition and multiplication of the objects in the vector space.

In the preceding sections, we verified these axioms (usually explicitly, but sometimes implicitly) in the following two examples:

(1) The vector space whose objects are arrows in 3-space with tail at the origin, with the addition and multiplication by scalars defined geometrically in Sections 4 and 5.

(2) The vector space R^n whose objects are all n-tuples of real numbers, with addition and multiplication by scalars defined thus:

$$[a_1, \ldots, a_n] + [b_1, \ldots, b_n] = [a_1 + b_1, \ldots, a_n + b_n]$$

$$a[a_1, \ldots, a_n] = [aa_1, \ldots, aa_n].$$

In Chapter 5, Section 1, we shall study still another kind of vector space:

(3) Any subset of R^n that is closed under addition and multiplication by scalars. In fact, examples of this kind of vector space occur much earlier in the text; we leave their detection to the alertness of the reader. For example, take the set of all linear combinations of two fixed, nonparallel vectors in R^3. This is a vector space and is obviously "two-dimensional." Visualized geometrically, it will consist of all the vectors in the plane of the first two. But it is not R^2.

It can be "identified" with R^2, exactly as we identified the vector space of arrows with R^3 in Section 2, or the vector space of arrows in a plane with R^2 (Problem 13, Section 3): Choose two coordinates axes in the plane, then use this coordinate system to define a pair of components for each vector $\mathbf{v}$ in this plane. This pair of numbers is the vector in R^2 that is to be identified with $\mathbf{v}$.

The student should be aware that other kinds of vector spaces are also vitally important: for example,

(4) The vector space whose objects are forces where addition of two forces is performed by simultaneous application of the two forces, and so on. By choosing a coordinate system and defining components of forces, we can identify this vector space with R^3 just as we identified the space of arrows with R^3.

There is also another type of vector space whose objects are mathematical objects, and whose importance will become apparent to every student of differential equations and to everyone who goes more deeply into mathematics:

(5) Vector spaces whose objects are functions. For example, one vector space is composed of all functions **f** defined on the interval $0 \leq x \leq 1$ with addition defined as usual: $(\mathbf{f} + \mathbf{g})(x) = \mathbf{f}(x) + \mathbf{g}(x)$ for every x between 0 and 1. Similarly, $a\mathbf{f}$ is the function that at x has the functional value $(a)(\mathbf{f}(x))$. It is easy to check that all the properties $(+0), \ldots, (\times 4)$ are true.

This example is not so far divorced from R^n as one might think. After all, a function **f** is exactly determined by the functional values $\mathbf{f}(x)$, one for each real number x between 0 and 1, just as an n-tuple **v** is determined by its components $v_1, v_2, \ldots, v_n$, one for each integer between 1 and n. Thus these functions can be looked at as kinds of infinite-tuples. The addition and multiplication by scalars agrees well with this interpretation. In fact if we restrict ourselves to continuous functions there is even a dot product that treats the functional values of **f** as components of **f**:

$$\mathbf{f} \cdot \mathbf{g} = \int_0^1 \mathbf{f}(x)\mathbf{g}(x)\,dx.$$

Compare $\mathbf{v} \cdot \mathbf{w} = \sum_{p=1}^{n} v_p w_p$.

Various subsets of this vector space will also be vector spaces using the same addition and multiplication by scalars, for example:

(5a) The set of all continuous functions on the same interval $0 \leq x \leq 1$;

(5b) The set of all functions differentiable on the interval $0 \leq x \leq 1$;

(5c) The set of all quadratic polynomial functions. Note that this vector space is identifiable with the space of triples of numbers by making the polynomial $ax^2 + bx + c$ correspond to the triple $[a, b, c]$; this is quite a different correspondence from that suggested in the preceding paragraph.

(5d) The set of all functions **f** differentiable on the interval $0 \leq x \leq 1$ and satisfying the two conditions $\mathbf{f}''' - \mathbf{f}'' + \mathbf{f}' - \mathbf{f} = 0$, and $\mathbf{f}(0) = 0$. One of the accomplishments of the theory of differential equations is to show that all the objects in this vector space are linear combinations of $\mathbf{f}_1$ and $\mathbf{f}_2$ where $\mathbf{f}_1(x) = \sin x$ and $\mathbf{f}_2(x) = e^x - \cos x$.

Every n-dimensional vector space (see Chapter 5 for a careful definition of dimension) can be identified with R^n, as are Examples 1, 3, 4, 5(c) and 5(d). So in one sense, we lose little by restricting ourselves mainly to R^n, as we do in this text. However, there is one important sense in which R^n is not typical. In all other vector spaces, to get components of vectors, we have to pick a coordinate system, and any choice is as good as any other *a priori*. In R^n, by contrast, each vector already has a standard set of components, the entries in the n-tuple; R^n has a coordinate system already picked out: $\mathbf{i}_1$, $\mathbf{i}_2, \ldots, \mathbf{i}_n$. In this sense, R^n is different from all other vector spaces. We concentrate on R^n for simplicity, but we pay for this concentration later in the text, when we begin to study other coordinate systems in R^n in some detail. By that time it may be difficult to wrench ourselves away from this standard coordinate system with its standard components of vectors, and to develop the more flexible, more geometric approach that is often the most helpful aspect of vector algebra.

This is one reason—the existence of general vector spaces in the shadow of R^n—that we shall try as often as possible to treat R^n geometrically, as if it were a space of arrows or other nonnumerical objects.

PROBLEMS

1. Verify the vector space axioms for the space of arrows and the space of triples, either by direct proof or by citing theorems from the preceding sections.

2. Verify that the set of all vectors $x\mathbf{i} + y\mathbf{j} + z\mathbf{k}$ with x, y, and z satisfying $x + 2z = 0$ forms a vector space. Produce two vectors $\mathbf{v}_1$ and $\mathbf{v}_2$ whose linear combinations comprise this vector space.

3. Verify that the set of all vectors $x\mathbf{i} + y\mathbf{j} + z\mathbf{k}$ with both $x + 2z = 0$ and $2y + z = 0$ forms a vector space. Produce one vector $\mathbf{v}_1$ whose linear combinations comprise this vector space.

4. Verify the vector space axioms for the set of all continuous functions on the interval $0 \leq x \leq 1$, as in Example 5a.

5. Verify the vector space axioms for the set of differentiable functions as in Example 5b.

6. Check that every linear combination $\mathbf{f}$ of $\sin x$ and $e^x - \cos x$ does satisfy the conditions $\mathbf{f}(0) = 0$ and

$$\mathbf{f}''' - \mathbf{f}'' + \mathbf{f}' - \mathbf{f} = 0.$$

7. Show that all solutions $\mathbf{f}$ of the differential equation $\mathbf{f}' - \mathbf{f} = 0$ form a vector space. You can do this without finding the solutions. Can you then find the solutions and check once more that they form a vector space?

CHAPTER

2

Planes and Lines

1. PLANES

We have already had some experience finding equations of simple planes (Problems 2–4, Chapter 1, Section 1). The reader will recall that *an equation of a plane* is an equation involving x, y, z that is satisfied whenever (x, y, z) are the coordinates of a point on the plane and that is satisfied by the coordinates of no other points. In other words, the point (x, y, z) is on the plane if and only if the numbers x, y, z satisfy the equation. In still other words, the plane is the *graph of the equation*.

In accordance with our plan to replace points by their position vectors, we shall also refer to certain vector equations as equations of planes. For example, consider the equation

$$\mathbf{i} \cdot \mathbf{r} = 0,$$

where $\mathbf{r}$ is the position vector of a point. A point has a position vector $\mathbf{r}$ satisfying this equation if and only if $\mathbf{r}$ is perpendicular to $\mathbf{i}$, that is, in the yz-plane, which means the point is in the yz-plane. We say that $\mathbf{i} \cdot \mathbf{r} = 0$ is a vector equation of the yz-plane.

Write the vector equation out in component form to compare with the ordinary definition: If the point has coordinates x, y, z, then $\mathbf{r} = x\mathbf{i} + y\mathbf{j} + z\mathbf{k}$ and the equation $\mathbf{i} \cdot \mathbf{r} = 0$ becomes $x = 0$, which, as we know, is an equation of the yz-plane in the earlier sense.

Now we find the equation of an arbitrary plane. The easiest way to specify a plane geometrically is to specify a vector perpendicular to the plane (this specifies the direction or tilt of the plane but not its location) and one point on the plane (which pins down the location). Suppose we are given a vector $\mathbf{v}$ perpendicular to a certain plane and a point $P_0 = (x_0, y_0, z_0)$ on the plane. What is the condition on a point $P = (x, y, z)$ that will force P to lie on the plane? If P is on the plane, then the vector P_0P is in the plane, and hence is perpendicular to $\mathbf{v}$. Conversely, if P_0P is perpendicular to $\mathbf{v}$, then P_0P lies in the plane and so does P. (We have used the geometric principle that all the lines through P_0 and perpendicular to $\mathbf{v}$ comprise the plane that goes through P_0 and is perpendicular to $\mathbf{v}$.)

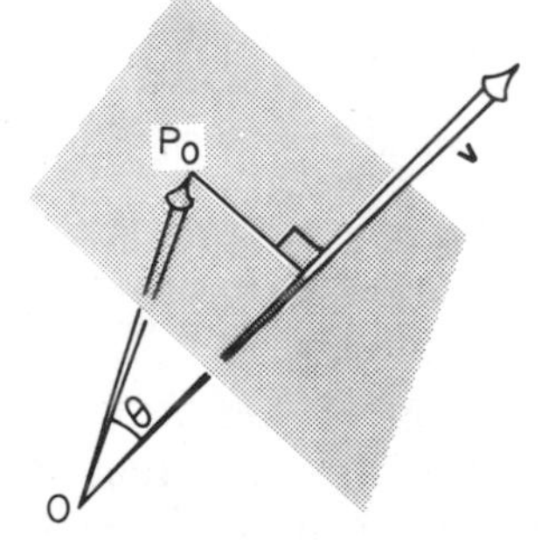

Figure 2.1

We now translate into an equation the geometric restriction on P, "the vector P_0P is perpendicular to $\mathbf{v}$." Let $\mathbf{r}$ denote the position vector of P (if the coordinates of P are (x, y, z), then $\mathbf{r} = x\mathbf{i} + y\mathbf{j} + z\mathbf{k}$), let $\mathbf{r}_0$ denote the position vector of the given point P_0 (write $\mathbf{r}_0 = x_0\mathbf{i} + y_0\mathbf{j} + z_0\mathbf{k}$), and let $\mathbf{v}$ be a vector perpendicular to the plane (say, $\mathbf{v} = a\mathbf{i} + b\mathbf{j} + c\mathbf{k}$). Then $P_0P = \mathbf{r} - \mathbf{r}_0$ (more accurately, P_0P is equivalent to $\mathbf{r} - \mathbf{r}_0$, according to Definition 2.4, Chapter 1) and by 6.2, Chapter 1, the required equation is $\mathbf{v} \cdot (\mathbf{r} - \mathbf{r}_0) = 0$. Summarizing and translating into components, we get

1.1 The plane which is perpendicular to the vector $\mathbf{v}$ and which goes through the point whose position vector is $\mathbf{r}_0$ is the set of all points with position vector $\mathbf{r}$ satisfying

$$\mathbf{v} \cdot (\mathbf{r} - \mathbf{r}_0) = 0.$$

In other words, this is a vector equation of the plane. In component form, the plane perpendicular to $[a, b, c]$ and through the point (x_0, y_0, z_0) is the set of points (x, y, z) satisfying

$$a(x - x_0) + b(y - y_0) + c(z - z_0) = 0.$$

This is an equation of the plane.

1.2 Every plane has an equation of the form $ax + by + cz = d$ for some numbers a, b, c, d, with a, b, c not all zero. Equivalently, every plane has an equation of the form $\mathbf{v} \cdot \mathbf{r} = d$ for some nonzero vector $\mathbf{v}$ and some scalar d. Conversely, every such equation is the equation of a plane; $\mathbf{v}$ will be perpendicular to the plane; a, b, c will be the components of a vector perpendicular to the plane.

Proof. The equation $\mathbf{v} \cdot (\mathbf{r} - \mathbf{r}_0) = 0$ in 1.1 is the same as (more accurately has the same graph as) $\mathbf{v} \cdot \mathbf{r} = \mathbf{v} \cdot \mathbf{r}_0$, which is of the form $\mathbf{v} \cdot \mathbf{r} = d$ if we take d to be the number $\mathbf{v} \cdot \mathbf{r}_0$. In component form, $a(x - x_0) + b(y - y_0) + c(z - z_0) = 0$ is the same as $ax + by + cz = d$ if we take $d = ax_0 + by_0 + cz_0$. As for the converse, $\mathbf{v} \cdot \mathbf{r} = d$ is the same as $\mathbf{v} \cdot (\mathbf{r} - \mathbf{r}_0) = 0$ if we choose $\mathbf{r}_0$ so that $\mathbf{v} \cdot \mathbf{r}_0 = d$, and that will be the equation of a plane by 1.1. It is possible to choose such an $\mathbf{r}_0$, for example $\mathbf{r}_0$ can be $s\mathbf{v}$ for a suitable scalar s, in fact for $s = d/(\mathbf{v} \cdot \mathbf{v})$. In components, $ax + by + cz = d$ has the same graph as $a(x - (d/a)) + by + cz = 0$, which graph is the plane through the point $(d/a, 0, 0)$ perpendicular to $a\mathbf{i} + b\mathbf{j} + c\mathbf{k}$. Of course, if $a = 0$, this argument will not work (though the preceding vector argument will), but since either b or c must then be not zero, the reader can easily repair the proof.

Example 1. The plane through $(1, -1, -1)$ perpendicular to $2\mathbf{i} + 3\mathbf{k}$ is $2(x - 1) + 0(y + 1) + 3(z + 1) = 0$, that is, $2x + 3z + 1 = 0$. We could also get this from 1.2, thus: We know the equation will be $2x + 0y + 3z = d$ and we only need to find d so that $(1, -1, -1)$ lies on the plane, that is, so that $x = 1, y = -1, z = -1$ satisfy the equation. This forces $d = -1$.

Example 2. The plane through the origin perpendicular to the z-axis (that is, the xy-plane; here P_0 is $(0, 0, 0)$ and $\mathbf{v} = \mathbf{k}$) has the vector equation $\mathbf{k} \cdot \mathbf{r} = 0$, or the scalar equation $0(x - 0) + 0(y - 0) + 1(z - 0) = 0$, which means $z = 0$. Clearly we could have used for $\mathbf{v}$ any vector parallel to $\mathbf{k}$ except $\mathbf{0}$. Such a vector is $s\mathbf{k}$ for some nonzero scalar s, and the resultant equation would have come out $s\mathbf{k} \cdot \mathbf{r} = 0$ or $sz = 0$, which does indeed have the same graph.

Example 3. The plane that is the perpendicular bisector of the line segment joining $(1, 2, 3)$ to $(-1, 2, 5)$ (here P_0 is the midpoint $(0, 2, 4)$ and $\mathbf{v} = (1 - (-1))\mathbf{i} + (2 - 2)\mathbf{j} + (3 - 5)\mathbf{k}$ is the vector along the line segment) is $2(x - 0) + 0(y - 2) - 2(z - 4) = 0$, or, equivalently, $x - z + 4 = 0$.

Example 4. Find the equation of the plane through the points $P_1 = (1, 0, 1)$, $P_2 = (2, 1, 3)$, $P_3 = (-4, 1, -1)$.

First solution. Use 1.1. We need a point on the plane (but we already have three given) and a vector perpendicular to the plane. We know two vectors in the plane: for example, P_1P_2 and P_1P_3, and we need a vector perpendicular to both. Since P_1P_2 is equivalent to $\mathbf{i} + \mathbf{j} + 2\mathbf{k}$ and P_1P_3 is equivalent to $-5\mathbf{i} + \mathbf{j} - 2\mathbf{k}$, a vector perpendicular to both is

$$P_1P_2 \times P_1P_3 = \begin{vmatrix} 1 & 2 \\ 1 & -2 \end{vmatrix} \mathbf{i} + \begin{vmatrix} 2 & 1 \\ -2 & -5 \end{vmatrix} \mathbf{j} + \begin{vmatrix} 1 & 1 \\ -5 & 1 \end{vmatrix} \mathbf{k}$$

$$= -4\mathbf{i} - 8\mathbf{j} + 6\mathbf{k}.$$

Final answer:

$$-4(x - 1) - 8(y - 0) + 6(z - 1) = 0$$

or

$$-4x - 8y + 6z = 2.$$

Equivalent answer:

$$2x + 4y - 3z + 1 = 0.$$

Second solution. Use 1.2. We know that the answer will be $ax + by + cz - d = 0$, but we do not know a, b, c, d. That $(1, 0, 1)$ must lie on the plane imposes a restriction on a, b, c, d; namely, $a(1) + b(0) + c(1) - d = 0$. Similarly, we obtain two more equations, getting a total of three equations in the four unknowns a, b, c, d:

$$\begin{aligned} a \phantom{{}+b} + c - d &= 0, \\ 2a + b + 3c - d &= 0, \\ -4a + b - c - d &= 0. \end{aligned}$$

Solve these for a, b, c, d. Of course there is no unique solution (Why?). The best we can hope for is to solve for three unknowns in terms of the fourth. By subtracting, we eliminate d from the first and second equations, then from the first and third. The original equations are then equivalent to (that is, have the same solutions as)

$$\begin{aligned} a \quad\; + c - d &= 0, \\ a + b + 2c \quad\; &= 0, \\ 5a - b + 2c \quad\; &= 0. \end{aligned}$$

Then eliminate c from the last two, and get another equivalent system

$$\begin{aligned} a \quad\; + c - d &= 0, \\ a + b + 2c \quad\; &= 0, \\ 4a - 2b \quad\; &= 0, \end{aligned}$$

which says $b = 2a$, $c = -(a + b)/2 = (-3/2)a$, $d = a + c = -\frac{1}{2}a$. We may choose any a except 0; if $a = 2$, we get our previous answer.

Example 5. Find the perpendicular distance from the origin to the plane $2x + 3y + 4z = 1$. Strategy: If we can find a point P_0 on the plane and a vector $\mathbf{v}$ perpendicular to the plane, then the distance from the origin O to the plane is the absolute value of $\| OP_0 \| \cos\theta$ where θ is the angle between OP_0 and $\mathbf{v}$ (see Fig. 2.1). Since $OP_0 \cdot \mathbf{v} = \| OP_0 \| \| \mathbf{v} \| \cos\theta$, the answer to our question is

$$\frac{| OP_0 \cdot \mathbf{v} |}{\| \mathbf{v} \|}.$$

Subproblem 1. Find $\mathbf{v}$ perpendicular to $2x + 3y + 4z = 1$. Answer:

$$2\mathbf{i} + 3\mathbf{j} + 4\mathbf{k}. \qquad \text{(Why?)}$$

Subproblem 2. Find a point P_0 on the plane. This is only difficult because there are so many correct answers. For example, choose any y and z and find x to match. One answer is $(\frac{1}{2}, 0, 0)$.

Final answer:

$$\frac{| \frac{1}{2}\mathbf{i} \cdot (2\mathbf{i} + 3\mathbf{j} + 4\mathbf{k}) |}{\| 2\mathbf{i} + 3\mathbf{j} + 4\mathbf{k} \|} = \frac{1}{(29)^{1/2}}.$$

PROBLEMS

1. Find an equation of the plane through the origin perpendicular to the vector $2\mathbf{i} - \mathbf{j} + \mathbf{k}$.

2. Check which of the following points are on the plane that goes through the point $(0, 1, -1)$ (that is, $\mathbf{r}_0 = \mathbf{j} - \mathbf{k}$) and are perpendicular to $2\mathbf{i} + \mathbf{k}$. Do it by checking the vector equation $\mathbf{v} \cdot (\mathbf{r} - \mathbf{r}_0) = 0$ as well as by using the appropriate scalar equation $ax + by + cz = d$:

$$(2, 0, 1); \qquad (-1, 0, 1); \qquad (2, 1, 1); \qquad (-1, 1, 1).$$

3. Show that the plane $2x + y - z = 1$ is perpendicular to the plane $x - y + z = 2$.

4. Find the angle between the planes $2x + y - 2z = 1$ and $x - 2y - 2z = 2$.

5. What is the relationship among the planes $2x + 3y + z = 0$, $2x + 3y + z = 1$, $2x + 3y + z = 2$?

6. Find the distance from the origin to the plane $x + 2y + 2z = 1$.

7. Find the distance from the origin to the plane $ax + by + cz = d$.

8. Find the distance from an arbitrary point (x_1, y_1, z_1) to the plane $ax + by + cz = d$.

9. Find an equation of the plane through the origin and the points $(2, 4, -5)$, $(6, -1, 1)$.

10. Find an equation of the plane through the points $(1, 1, 1)$, $(2, 0, 2)$, $(-1, 1, 2)$.

11. Find an equation of the plane parallel to the z-axis and passing through the points $(2, 1, 2)$ and $(1, 2, 1)$.

12. Tell how to see directly from an equation of the form $ax + by + cz = d$, whether the plane (a) passes through the origin; (b) is perpendicular to the x-axis; (c) is perpendicular to the y-axis; (d) is parallel to the x-axis; (e) is parallel to the y-axis; (f) is parallel to the xy-plane; (g) is parallel to the plane $x + 2y + 3z = 0$.

2. LINES

Just as the most convenient geometric data for specifying a plane are a point on the plane and a vector perpendicular to the plane, so the most convenient way of specifying a line is to give one point P_0 on the line and a nonzero vector $\mathbf{v}$ parallel to the line. Given these data, what condition on another point P will force P to lie on this line? The answer is that the vector P_0P must be parallel to $\mathbf{v}$ (why?). To reduce this to analytic form, let $\mathbf{r}$ be the position vector of an arbitrary point P on the line, $\mathbf{r}_0$ the position vector of the given P_0, and, as before, $\mathbf{v}$ the given vector parallel to the line. Our answer reduces to P_0P $(= \mathbf{r} - \mathbf{r}_0) = t\mathbf{v}$ for some scalar t or, equivalently,

$$\mathbf{r} = \mathbf{r}_0 + t\mathbf{v}. \tag{2.1}$$

If we write

$$\begin{aligned} \mathbf{r} &= x\mathbf{i} + y\mathbf{j} + z\mathbf{k}, \\ \mathbf{r}_0 &= x_0\mathbf{i} + y_0\mathbf{j} + z_0\mathbf{k}, \\ \mathbf{v} &= a\mathbf{i} + b\mathbf{j} + c\mathbf{k}, \end{aligned}$$

and use Chapter 1, 2.3, we can get an equivalent version without vectors:

$$\begin{aligned} x &= x_0 + at, \\ y &= y_0 + bt, \\ z &= z_0 + ct, \end{aligned} \tag{2.2}$$

where (x_0, y_0, z_0) are the coordinates of the given point P_0 on the line; a, b, c are the components of the vector $\mathbf{v}$ parallel to the line; and (x, y, z) are the coordinates of any other point on the line. These are *parametric equations* of the line. As t ranges over all real numbers, the (x, y, z) given by Eqs. (2.2) ranges over all the points on the line.

Example 1. Find equations of the line through the points (x_0, y_0, z_0) and (x_1, y_1, z_1). Here P_0 is (x_0, y_0, z_0) (or you could use (x_1, y_1, z_1)) and $\mathbf{v}$ is $(x_1 - x_0)\mathbf{i} + (y_1 - y_0)\mathbf{j} + (z_1 - z_0)\mathbf{k}$. The vector equation is

$$\mathbf{r} = \mathbf{r}_0 + (\mathbf{r}_1 - \mathbf{r}_0)t.$$

The scalar equations are

$$\begin{aligned} x &= x_0 + (x_1 - x_0)t, \\ y &= y_0 + (y_1 - y_0)t, \\ z &= z_0 + (z_1 - z_0)t. \end{aligned}$$

If we multiply out all parentheses, we can write the result more symmetrically:

$$\mathbf{r} = \lambda\mathbf{r}_0 + \mu\mathbf{r}_1 \qquad \text{with} \qquad \lambda + \mu = 1$$

(we have written λ for $1 - t$ and μ for t) or

$$x = \lambda x_0 + \mu x_1$$
$$y = \lambda y_0 + \mu y_1 \qquad \text{with} \qquad \lambda + \mu = 1.$$
$$z = \lambda z_0 + \mu z_1$$

Notice that the midpoint ($\lambda = \mu = \frac{1}{2}$), trisection point ($\lambda = \frac{2}{3}$, $\mu = \frac{1}{3}$), and so on, all appear as special points on this line, as they should. (Compare Problem 10, Chapter 1, Section 5.)

Example 2. Find parametric equations of the line of intersection of the two planes $x + y + z = 1$, $2x - y + z = 1$.

We need a vector $\mathbf{v}$ parallel to the line. Any vector perpendicular to the first plane, like $\mathbf{i} + \mathbf{j} + \mathbf{k}$ (why?) will be perpendicular to $\mathbf{v}$ since $\mathbf{v}$ is in this plane. For the same reason, $\mathbf{v}$ is perpendicular to $2\mathbf{i} - \mathbf{j} + \mathbf{k}$. Using cross products, we get one such $\mathbf{v}$,

$$\mathbf{v} = \begin{vmatrix} 1 & 1 \\ -1 & 1 \end{vmatrix} \mathbf{i} + \begin{vmatrix} 1 & 1 \\ 1 & 2 \end{vmatrix} \mathbf{j} + \begin{vmatrix} 1 & 1 \\ 2 & -1 \end{vmatrix} \mathbf{k}$$
$$= 2\mathbf{i} + \mathbf{j} - 3\mathbf{k}.$$

Now we need a point lying on both planes. There is an infinite number of them; we can get one by setting $z = 0$ and solving the equations of the planes for x and y. This gives $x_0 = \frac{2}{3}$, $y_0 = \frac{1}{3}$, so one point on the line is $(\frac{2}{3}, \frac{1}{3}, 0)$. The equations are then

$$\mathbf{r} = (\tfrac{2}{3}\mathbf{i} + \tfrac{1}{3}\mathbf{j}) + (2\mathbf{i} + \mathbf{j} - 3\mathbf{k})t$$

or

$$x = \tfrac{2}{3} + 2t,$$
$$y = \tfrac{1}{3} + t,$$
$$z = \quad - 3t.$$

Note that Example 2 starts by specifying a line as in Problems 9 and 10, Chapter 1, Section 1, as the set of points whose coordinates satisfy a pair of linear equations in x, y, z—no t. Inversely, given parametric equations of a line (Eq. (2.2)), it is clear (eliminating t) that

$$\frac{x - x_0}{a} = \frac{y - y_0}{b} = \frac{z - z_0}{c}. \tag{2.3}$$

These are called *symmetric equations* of the line.

Conversely, if three numbers x, y, z satisfy these two equations (we could consider (2.3) as three equations, but one of them would be superfluous, being a consequence of the other two), then there is a t for which Eqs. (2.2) hold, namely, $t =$ the common value of $(x - x_0)/a$, $(y - y_0)/b$, and $(z - z_0)/c$.

Equations (2.3) are of the type used in the problems in Chapter 1, Section 1. They represent the line as the intersection of two planes—in fact, as the intersection of a plane $(x - x_0)/a = (y - y_0)/b$ parallel to the z-axis (check this fact) with a plane $(y - y_0)/b = (z - z_0)/c$ parallel to the x-axis. Equations (2.3) have the advantage over other pairs of linear equations that the direction of the line can be read off instantly: the line is parallel to the vector $a\mathbf{i} + b\mathbf{j} + c\mathbf{k}$.

Example 3. We shall rework Example 2. Given the line with equations $x + y + z = 1$, $2x - y + z = 1$, eliminate x (add -2 times the first equation to the second) and eliminate y (add the equations). This gives

$$-3y - z = -1, \qquad 3x + 2z = 2.$$

Note that three numbers x, y, z satisfy the original pair of equations if and only if they satisfy this new pair; thus these are still equations of the same line. But this new pair can be written as symmetric equations by solving for z.

$$z = -3y + 1 = \frac{-3x + 2}{2}$$

or

$$\frac{x - \frac{2}{3}}{-\frac{2}{3}} = \frac{y - \frac{1}{3}}{-\frac{1}{3}} = \frac{z - 0}{1},$$

so that a vector parallel to the line is $(-\frac{2}{3})\mathbf{i} - (\frac{1}{3})\mathbf{j} + \mathbf{k}$ and a point on the line is $(\frac{2}{3}, \frac{1}{3}, 0)$. Parametric equations are

$$x = \tfrac{2}{3} - \tfrac{2}{3}t,$$
$$y = \tfrac{1}{3} - \tfrac{1}{3}t,$$
$$z = t.$$

Can you reconcile these parametric equations with those found in Example 2?

One defect of Eqs. (2.3) is that they are not available if one or more of a, b, c is zero. Geometrically, it is clear that if a line is parallel to the xy-plane (so $c = 0$) and we write it as the intersection of two planes parallel to axes, then the plane parallel to the x-axis or to the y-axis will in fact be parallel to the xy-plane and so cannot have an equation $(y - y_0)/b = \cdots$; its equation will be $z = z_0$. Similarly, if two of a, b, c are zero, say $b = c = 0$, then our line is parallel to the x-axis; a plane containing the line and parallel to the z-axis will be parallel to the xz-plane and will have an equation $y = y_0$. The equations replacing (2.3) will no longer by symmetric.

Example 4. Express the following line as an intersection of planes parallel to coordinate axes: the line is parallel to the vector $4\mathbf{i} - 5\mathbf{k}$ and goes through the point $(1, -2, 3)$. The answer is $(x - 1)/4 = (z - 3)/-5$ (as in the symmetric equations) and $y = -2$, replacing any equation that might involve the nonsensical $(y + 2)/0$.

3. VECTOR FUNCTIONS OF SCALARS

Equation (2.1), $\mathbf{r} = \mathbf{r}_0 + \mathbf{v}t$, is typical of many equations we shall have to deal with in the calculus of several variables. It expresses a vector $\mathbf{r}$ as a function of a scalar t ($\mathbf{r}_0$ and $\mathbf{v}$ are fixed). It is a function from R^1 to R^3. An important and helpful way of visualizing such a function is to think of the scalar as time and the vector as the position vector of a moving point. The function specifies the position of the point at each time. Of course Eq. (2.1) gives a very special kind of motion: the point moves in a straight line (that is what we mean by calling (2.1) an equation of a line) at constant speed. To check this last, consider the position of the point at one time, t:

$$\mathbf{r}_1 = \mathbf{r}_0 + t\mathbf{v}$$

and then the position at a later time $t + \Delta t$:

$$\mathbf{r}_2 = \mathbf{r}_0 + (t + \Delta t)\mathbf{v} = \mathbf{r}_0 + t\mathbf{v} + (\Delta t)\mathbf{v}.$$

The vector from the first point to the second is

$$\mathbf{r}_2 - \mathbf{r}_1 = (\Delta t)\mathbf{v}.$$

The distance covered is $\| \mathbf{r}_2 - \mathbf{r}_1 \| = | \Delta t | \, \| \mathbf{v} \|$. The speed is $\| \mathbf{r}_2 - \mathbf{r}_1 \| / | \Delta t | = \| \mathbf{v} \|$, which is the same at all times.

A "nonlinear" vector function of a scalar is given by this function, for example,

$$\mathbf{r} = \mathbf{i} \cos t.$$

The reader may verify that this represents an oscillating motion along the x-axis. Similarly, $\mathbf{r} = \mathbf{j} \sin t$ is an oscillation on the y-axis, but not in phase with the first oscillation. When the first oscillating point is at the center of its oscillation ($\mathbf{r} = 0, t = \pi/2, 3\pi/2, \ldots$), the second oscillating point is at an extreme point of its oscillation. The reader may also verify that the "sum" of these two motions

$$\mathbf{r} = \mathbf{i} \cos t + \mathbf{j} \sin t$$

is a circular motion at constant speed. For example, $\| \mathbf{r} \| = (\cos^2 t + \sin^2 t)^{1/2} = 1$ so the head of $\mathbf{r}$ is always 1 unit from the origin.

PROBLEMS

Find parametric and symmetric equations of the lines in Problems 1–7:

1. Through the origin, parallel to the vector $a\mathbf{i} + b\mathbf{j} + c\mathbf{k}$.

2. Through the point $(2, 1, 2)$, parallel to the x-axis.

3. Through the point $(1, -1, 2)$, parallel to the vector $3\mathbf{i} + 4\mathbf{j} - \mathbf{k}$.

4. Through the origin, perpendicular to the plane $2x - y + z = 3$.

5. Through the point $(2, 6, -1)$, perpendicular to the vectors $2\mathbf{i} + \mathbf{j} + \mathbf{k}$ and $\mathbf{i} - \mathbf{j} - \mathbf{k}$.

6. Through the point $(2, 1, 3)$, perpendicular to the plane $2x - y + 2z = 1$.

7. A horizontal line through the origin lying in the plane $2x + y + 3z = 0$.

8. In Problem 6, find the point of intersection of the line and the plane. Find the distance from the point $(2, 1, 3)$ to the plane by finding the distance from $(2, 1, 3)$ to this point of intersection. Compare with Problem 8, Section 1.

9. Find the intersection of the plane $2x + 2y - z = 5$ with the line $\mathbf{r} = \mathbf{i} + (2\mathbf{j} + \mathbf{k})t$.

10. Find the perpendicular distance from the origin to the line $x = 1 + 2t$, $y = 2 - 3t$, $z = 3t$. There are many ways to do this, but it should be done geometrically (in addition to any other way you may use) thus: If $\mathbf{r}_0$ is the position vector of a point on the line and $\mathbf{v}$ is a vector parallel to the line, then the required distance is $\| \mathbf{r}_0 \| \sin \theta$ where θ is the angle between $\mathbf{r}_0$ and $\mathbf{v}$. But $\| \mathbf{r}_0 \| \sin \theta = [\| \mathbf{r}_0 \|^2 - (\| \mathbf{r}_0 \| \cos \theta)^2]^{1/2}$ and we know how to find $\| \mathbf{r}_0 \| \cos \theta$, as $\mathbf{r}_0 \cdot \mathbf{v} / \| \mathbf{v} \|$. Another formula is $\| \mathbf{r}_0 \times \mathbf{v} \| / \| \mathbf{v} \|$. (Why?)

11. Find the perpendicular distance between the lines $x = 2 + 2t$, $y = 2 - t$, $z = 3t$ and $x = t$, $y = 2t$, $z = 1 - t$. Hint: Find a vector $\mathbf{w}$ perpendicular to both lines and a vector $\mathbf{u}$ joining a point on one line to a point on the other, then the answer will be numerically equal to the component of $\mathbf{u}$ along $\mathbf{w}$ (a geometric argument is needed here), that is, the answer is $|\mathbf{u} \cdot \mathbf{w}/\|\mathbf{w}\||$.

12. Find c so that the lines

$$\frac{x-1}{2} = \frac{y-3}{4} = \frac{z}{c} \quad \text{and} \quad \frac{x-2}{1} = \frac{y-4}{3} = \frac{z-2}{1}$$

intersect.

13. Let $\mathbf{u}$ and $\mathbf{v}$ be two fixed vectors, perpendicular to each other, but of equal length. Let $\mathbf{r}$ be the following function of t: $\mathbf{r} = \mathbf{u}\cos t + \mathbf{v}\sin t$. Show that the head of $\mathbf{r}$ moves in a circle. (Hint: $\|\mathbf{r}\| = (\mathbf{r} \cdot \mathbf{r})^{1/2}$ and use vector arithmetic.) What is the path of motion if $\|\mathbf{u}\| \neq \|\mathbf{v}\|$?

14. Show that the path of the motion $\mathbf{r} = \mathbf{i}t + \mathbf{k}t^2$ is a parabola.

15. What kind of motion is $\mathbf{r} = \mathbf{i}\cos t + \mathbf{j}\cos t$?

16. Find a so that the plane $ax + 2y + 3z = 1$ is parallel to the line $\mathbf{r} = \mathbf{r}_0 + (2\mathbf{i} + \mathbf{j} + \mathbf{k})t$.

17. Find a so that the plane $ax + 2y + 3z = 1$ is perpendicular to the line $x = 1 + 2t$, $y = 1 + 4t$, $z = 1 + 6t$.

18. Find a so that the plane $ax + 2y + 3z = 1$ is perpendicular to the line $(x - 1)/2 = (y - 2)/3 = (z - 3)/4$.

19. Find a so that the line $\mathbf{r} = \mathbf{i} + t(a\mathbf{j} + \mathbf{k})$ is perpendicular to the line $\mathbf{r} = \mathbf{i} + t(a\mathbf{i} + a\mathbf{j} - \mathbf{k})$.

20. Find a and b so that the line $(x - 1)/a = (y - 1)/b = (z - 1)/2$ is parallel to the line $2x + y + z = 1$, $x - y - 2z = 2$.

4. HALF-SPACES; INEQUALITIES

In Section 1 we saw that the set of all points (x, y, z) whose coordinates satisfy an equation $ax + by + cz = d$ is a plane provided a, b, and c are not all zero, which we shall assume throughout this section. What is the set of points whose coordinates satisfy the inequality $ax + by + cz \leq d$? Answer: The plane $ax + by + cz = d$ cuts all of space into two halves, and one of these halves is the set $ax + by + cz \leq d$. You can see this as follows: First, suppose $c > 0$. If a point (x, y, z) satisfies $ax + by + cz \leq d$, then by keeping x and y fixed and increasing z we can get a new point (x, y, z') with $ax + by + cz' = d$ (just take $z' = (d - ax - by)/c$; since $d \geq ax + by + cz$, we get $z' \geq z$). Geometrically, this means that all points satisfying the inequality are below the plane $ax + by + cz = d$. The argument is reversible: if a point (x, y, z) is below the plane, then some point (x, y, z') directly above $(z' \geq z)$ it is in the plane, $ax + by + cz' = d$ and $z' \geq z$ imply $ax + by + cz \leq d$. Thus the points whose coordinates satisfy the inequality are the points in the half-space below the plane $ax + by + cz = d$ (when $c > 0$).

If $c < 0$, the same argument shows that $ax + by + cz \leq d$ if and only if (x, y, z) is above the plane $ax + by + cz = d$ (in manipulating the inequalities we had to multiply or divide by c, which will reverse the inequalities if c is negative). Thus the points satisfying $ax + by + cz = d$ are the points of the half-space above the plane $ax + by + cz = d$ (when $c < 0$). If $c = 0$, the whole argument fails, and in fact the picture falls apart because $ax + by + cz = d$ is then a plane parallel to the z-axis, and if (x, y, z) is not on the plane, no point directly above or directly below it will be on the plane, either. However, in this case either a or b will be nonzero according to our assumption at the beginning of this section; if $a \neq 0$, for example, the same argument will work with x playing the role of z; $ax + by + cz \leq d$ describes either the left or right half-space depending on whether a is positive or negative.

Similarly, the set of all points (x, y) in R^2 that satisfy $ax + by \leq c$ will be a half-plane—the half-plane below the line $ax + by = c$ if $b > 0$, above if $b < 0$, and we leave the case $b = 0$ to the reader.

Example 1. $x + y + z \leq 1$ describes the half-space below the plane $x + y + z = 1$ (this is the plane through the three unit points on the coordinate axes); below, because if you start from a point on the plane and go straight down, you decrease z, and $x + y + z$ becomes less, so that $x + y + z \leq 1$. This is another version of the previous argument.

Example 2. $2x - 3z \leq 1$ describes the set of points above the plane $2x - 3z = 1$; above, because if you start on the plane and go up, z increases and $2x - 3z$ decreases, so that $2x - 3z \leq 1$; if you were to go down, $2x - 3z$ would become greater than 1.

In many applications of linear algebra, inequalities arise, but seldom one at a time. The problem is to find the points (x, y, z) satisfying several of these linear inequalities at the same time. Geometrically, this simply means finding the points that are simultaneously in several half-spaces, that is, the points that are in the intersection of several half-spaces. This intersection is some kind of a solid bounded by planes—a polyhedron.

Example 3. The set of points satisfying $x \geq 0$, $y \geq 0$, $z \geq 0$, and $x + y + z \leq 1$ form a tetrahedron or triangular pyramid. It is the intersection of four half-spaces: the half-space in front of the yz-plane, $x = 0$; the half-space to the right of the xz-plane; the half-space above the xy-plane; and the half-space under the tilted plane $x + y + z = 1$. The intersection of the first three half-spaces is just the first octant, and the intersection of this octant with the space under the tilted plane is the tetrahedron or triangular pyramid whose sides are in the three coordinate planes and whose base is the equilateral triangle cut out of the plane $x + y + z = 1$ by the first octant.

Example 4. The set of points satisfying $x \geq 0$, $y \geq 0$, and $x + 2z \leq 1$ is an infinitely long triangular prism extending to the right from the xz-plane; its bottom is the xy-plane; its back is the yz-plane; and its slanted top is the plane $x + 2z = 1$ (a plane parallel to the y-axis).

Example 5. The set of points (x, y) satisfying $x + y \leq 1$, $x - y \leq 0$, and $x \geq 0$ is the triangle (with its interior) bounded by the y-axis on the left, the line $x + y = 1$ on top and the line $y = x$ on the bottom. It is the intersection of the three half-planes: (i) below the line $x + y = 1$; (ii) above the line $y = x$; (iii) to the right of the line $x = 0$.

The subject of *linear programming* is concerned with finding the point whose coordinates satisfy some given linear inequalities as well as some linear equations (these restrictions so far demand that the point be in the intersection of some half-spaces and some planes; if there are two or fewer planes, there are probably still infinitely many points that satisfy the conditions) and maximize another linear function. An example will make this clearer.

Example 6. Maximize $2x + 3y + 4z$ subject to the conditions $x + y + z \leq 1$, $x > 0$, $y \geq 0$, $z \geq 0$; that is, among all the points in the pyramid described by the four inequalities (see Example 3), which one gives the maximum value to the function $2x + 3y + 4z$? There is a theorem at the base of all linear programming theory that says that this maximizing point must be one of the corners of the pyramid.

There are interesting, smooth, elementary routines for finding the right corner without actually looking at any pictures, and the interested reader should look into a book on linear programming for the details.† For a simple example like the present one, we can just find all the corners, evaluate $2x + 3y + 4z$ at each one of them, and choose the corner with the largest functional value. A corner of the pyramid is a point where three of the boundary planes intersect. The four boundary planes are $x = 0$, $y = 0$, $z = 0$, and $x + y + z = 1$. Solving three of them simultaneously will give the coordinates of one corner. Each set of three equations will then give one corner—there will be four corners in all. The three equations we get by omitting the last of the four have a common solution $x = 0$, $y = 0$, $z = 0$ (the origin). Omitting the third equation we get $x = 0$, $y = 0$, and $x + y + z = 1$, so $z = 1$; this corner is $(0, 0, 1)$. Omitting the second equation we get the point $(0, 1, 0)$. Omitting the first, we get $(1, 0, 0)$. These four points give the following functional values to $2x + 3y + 4z$:

point:	$(0, 0, 0)$	$(0, 0, 1)$	$(0, 1, 0)$	$(1, 0, 0)$
$2x + 3y + 4z$:	0	4	3	2

Hence the desired point is $(0, 0, 1)$, giving the maximum value of 4.

Example 7. Maximize $2x + 3y + 4z$ subject to the conditions

$$x + 2z - 2 = 0, \quad x + y - 1 = 0, \quad x \geq 0, \quad y \geq 0, \quad z \geq 0.$$

The figure consisting of allowable points is the intersection of two planes and three half-spaces. The intersection of the three half-spaces is the first octant, the two planes intersect in a line, which intersects the first octant in a line segment. The "corners" of this line segment are the endpoints, the points where the line meets the boundary of the first octant. Thus our algebraic problem is, once more, to solve simultaneously the two equations of the line with the equation of one of the boundary planes of the first octant. If we solve $x + 2z = 2$, $x + y = 1$, and $x = 0$, we get the point $(0, 1, 1)$; if instead of $x = 0$ we use $y = 0$, we get $(1, 0, \frac{1}{2})$; using $z = 0$, we get $(2, -1, 0)$. But a line segment cannot have three endpoints—not every intersection of the line with one of the boundary planes is an endpoint of the segment in question. In fact, the third point is not on the line segment at all; it does not satisfy all five conditions (the two equations and three inequalities); it violates $y \geq 0$. Thus the two candidates are $(0, 1, 1)$

† For example, A. S. Barsov, "What Is Linear Programming?" Heath, Boston, Massachusetts, 1964.

and $(1, 0, \frac{1}{2})$. Calculating $2x + 3y + 4z$ at these two points, we get the functional values 7 and 4, respectively. The first point $(0, 1, 1)$ is the answer.

Example 8. A manufacturer produces two products, X and Y, but he cannot produce arbitrary amounts of both products.

(1) His daily storage facilities can handle 1000 units of X or 500 units of Y, or any appropriate mixture of the two. Translation: if x is the number of units of X produced per day and y the number of units of Y, then $x + 2y \leq 1000$.

(2) Y is a by-product of X. One unit of X produces ingredients for two units of Y. Translation: $y \leq 2x$.

(3) It is uneconomical to produce less than 100 units of Y per day. Translation: $y \geq 100$.

If the profit is one dollar per unit of X and three dollars per unit of Y, how much of each should be produced to maximize the profit? Translation: maximize $x + 3y$.

The mathematical problem is now to maximize $x + 3y$ subject to three linear inequalities. The inequalities describe a triangle and its interior. The corners are found by solving two equalities simultaneously. $x + 2y = 1000$ and $y = 2x$ give $x = 200$, $y = 400$ as one corner; $x + 2y = 1000$ and $y = 100$ give $(800, 100)$; $y = 2x$ and $y = 100$ give $(50, 100)$. The profit $x + 3y$ at these three corners is 1400, 1100, and 350, respectively. The first corner is the correct one: 200 units of X and 400 units of Y per day.

This problem is so transparent that the answer may have been obvious to you from the beginning; but it is probably also clear that problems of this kind can become very complex, so that a smooth routine or algorithm would be most welcome.

PROBLEMS

1. Describe these half-spaces. Sketch a picture if possible.

(a) $x \leq 0$. (b) $x \leq 1$.

(c) $z \geq 2$. (d) $x + 2y \leq 1$.

(e) $x - y + z \leq 1$. (f) $x - y - z \leq 1$.

2. (a) Describe the figure consisting of all points satisfying all these conditions: $z \geq 0$, $2y - z \geq 0$, $2y + z \leq 4$, $x \geq 0$, and $x \leq 2$. Sketch the figure. Find all its corners.

(b) Do the same for the figure given by the conditions $x + y + z = 1$, $x \geq 0$, $y \geq 0$, $z \geq 0$.

(c) Do the same for $x + y + z = 1$, $x \geq 0$, $y \geq 0$, $z \geq 0$, and $y \leq \frac{1}{2}$.

(d) Do the same for $x + y + z \leq 1$, $x \geq 0$, $y \geq 0$, $z \geq 0$, $x = y$, and $6x + z = 2$.

(e) Do the same for $x + y \geq 0$, $x + y + z \leq 1$, $x + y - z \leq 0$, and $x = y$.

3. Maximize the function $2x - 3y + z$ on each of the five sets of points in problem 2.

4. A factory will buy and burn 100 tons of coal. Low-sulfur coal ($\frac{1}{2}$ percent sulfur) costs c_1 dollars per ton and high-sulfur coal (4 percent sulfur) costs c_2 dollars per ton and $c_1 > c_2$. The law requires that the factory burn a mixture that contains at most 2 percent sulfur.

(a) How much of each kind of coal should the factory buy to minimize the total cost and still be within the law? Note: as long as $c_1 > c_2$, your answer is independent of these prices. This is plausible; the cheapest thing is to buy as little of the high-priced coal as the law will allow.

(b) What if the low-sulfur coal were cheaper?

5. The following table gives the number of units of vitamin A and vitamin C in one gram of each of three foods, and the price of these foods in cents per gram.

	Vitamin A	Vitamin C	Price
Carrots	120	0.06	0.05
Oranges	2	0.50	0.03
Grapefruit	0	0.40	0.02

(a) What amounts of these foods will provide the minimum daily requirement of these vitamins—5000 units of vitamin A and 75 units (milligrams) of vitamin C—at minimum cost?

(b) What would be the answer to part (a) if the price of grapefruit rose to equal the price of oranges?

CHAPTER

3

Linear Functions

1. DEFINITION

Just as our principal study in calculus has been of functions of real numbers, so, now that we have learned something of vectors, we turn to the study of functions of vectors.

As we know, a function is a rule associating to each object (in our case, to each vector) from a certain set of objects called the *domain* of the function (in our case the domain is R^n) another object, called the functional value, usually in a different set. When we use several functions in one discussion, we need distinguishing names for them, and we shall usually use letters like f, g, h as such names. If f is the name of a function, it is standard notation to use $f(\mathbf{v})$ for the functional value that f associates to the vector $\mathbf{v}$. One intuitive version of this function concept visualizes each function as a machine that, when presented with an object $\mathbf{v}$ from the domain, converts it into the associated object $f(\mathbf{v})$. This intuition is probably most closely associated with the synonym *operator* for *function*. The correct idea of a function presents it as a formula in the sense of a recipe rather than in the sense of a collection of mathematical symbols. Thus, to specify a function of vectors, we must tell what it does to every vector in the domain; for each $\mathbf{v}$ we must give an explicit formula for $f(\mathbf{v})$, and this formula may be verbal as well as mathematical, provided only it is specific and single valued.

Functions from R^1 to R^1 are just functions that associate to each real number another real number; these are the major concern of first courses in calculus. Functions from R^2 to R^1 are just the functions of two variables: such a function associates to every pair of real numbers $[x, y]$ a real number $f([x, y])$, and there is no real difference in ideas if we omit the square brackets and write $f(x, y)$ for the functional value, as is usually done. For our purposes we shall usually write $\mathbf{v}$ for $[x, y]$ and write $f(\mathbf{v})$ for the functional value.

Functions from R^n to R^1 are just functions of n real variables with real functional values. A function from R^n to R^m is a function whose domain is R^n and whose functional values are also vectors, but in R^m. The reader should study all examples carefully for the language and usage that are going to be employed here and elsewhere.

To compensate for the increase in complexity of the objects on which the functions work (vectors are somewhat more complex than numbers) we impose a simplifying assumption on the functions:

1.1 Definition

A linear function or a *linear mapping* or a *linear transformation* or a *linear operator* from n-space to m-space is a function f that associates to each vector in R^n a vector in R^m in such a way that, for every two vectors $\mathbf{v}$ and $\mathbf{w}$ in R^n and every number a,

$$f(\mathbf{v} + \mathbf{w}) = f(\mathbf{v}) + f(\mathbf{w})$$

and

$$f(a\mathbf{v}) = af(\mathbf{v}).$$

Equivalently,

$$f(a\mathbf{v} + b\mathbf{w}) = af(\mathbf{v}) + bf(\mathbf{w})$$

for all vectors $\mathbf{v}$ and $\mathbf{w}$ in R^n and all scalars a and b. (See Problem 15 for the proof of equivalence.)

We shall not study nonlinear functions in this course (except for some quadratic functions in Chapter 8). In later courses, when we do, the prime device will consist in approximating each nonlinear function by a constant plus a linear function. For real-valued functions of one variable (functions from R^1 to R^1) this is exactly what differential calculus does: It approximates a curve (graph of a nonlinear function) by its tangent line (graph of a constant plus a linear function; see Example 4 for the present meaning of "linear function" from R^1 to R^1).

Example 1. For each vector $\mathbf{v}$ in R^3 define $f(\mathbf{v})$ to be the vector obtained by rotating $\mathbf{v}$ around the z-axis through a 120-degree angle, in the positive sense (the sense that carries the positive x-axis past the positive y-axis). This f is a function from R^3 to R^3. It is linear because if $\mathbf{v}$ and $\mathbf{w}$ are any two vectors and $\mathbf{v} + \mathbf{w}$ is their sum, we have a picture like Fig. 3.1.

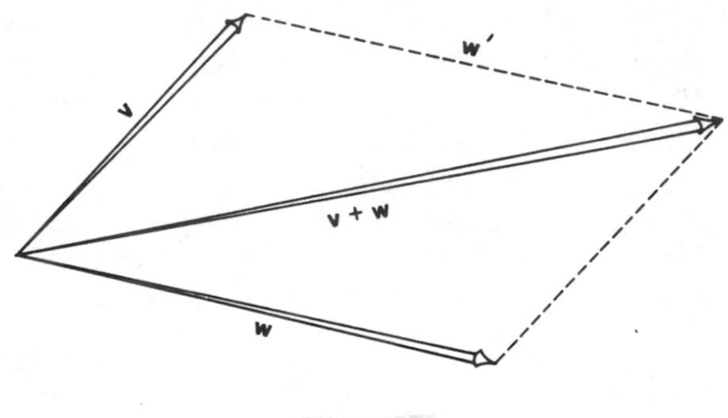

Figure 3.1

Since rotation is a rigid motion, if we rotate this whole picture, the parallelogram remains a parallelogram, the two sides become $f(\mathbf{v})$ and $f(\mathbf{w})$, and the diagonal becomes $f(\mathbf{v} + \mathbf{w})$. But the rotated diagonal is the sum of the two rotated sides, by the definition of vector addition; so

$$f(\mathbf{v} + \mathbf{w}) = f(\mathbf{v}) + f(\mathbf{w}).$$

A similar argument shows that

$$f(a\mathbf{v}) = af(\mathbf{v}).$$

Example 2. Define a function g from R^3 to R^3 thus: For each $\mathbf{v}$ in R^3, $g(\mathbf{v})$ is the vector obtained by projecting $\mathbf{v}$ on the xy-plane (project the head and tail of $\mathbf{v}$ to get the head and tail of $g(\mathbf{v})$).

Example 3. Define a function h from R^3 to R^1 by the formula

$$h(\mathbf{v}) = v_1 + 2v_2 - 4v_3,$$

where we are using the convention that v_1, v_2, and v_3 denote the components of $\mathbf{v}$. Then $h(\mathbf{i}) = 1$ because the components of $\mathbf{i}$ are 1, 0, and 0; and $h(\mathbf{i} + \mathbf{j} - \mathbf{k}) = 1 + 2 \cdot 1 - 4(-1) = 7$; etc. This function is linear, because if $\mathbf{w} = [w_1, w_2, w_3]$ is any other vector in R^3, then

$$\mathbf{v} + \mathbf{w} = [v_1 + w_1, v_2 + w_2, v_3 + w_3],$$

$$\begin{aligned} h(\mathbf{v} + \mathbf{w}) &= (v_1 + w_1) + 2(v_2 + w_2) - 4(v_3 + w_3) \\ &= (v_1 + 2v_2 - 4v_3) + (w_1 + 2w_2 - 4w_3) \\ &= h(\mathbf{v}) + h(\mathbf{w}); \end{aligned}$$

and since $a\mathbf{v}$ has components av_1, av_2, av_3, we have

$$\begin{aligned} h(a\mathbf{v}) &= av_1 + 2(av_2) - 4av_3 \\ &= a(v_1 + 2v_2 - 4v_3) \\ &= ah(\mathbf{v}). \end{aligned}$$

Another version of this function is $h(\mathbf{v}) = [1, 2, -4] \cdot \mathbf{v}$.

Example 4. $f(x) = 3x$ defines a linear function from R^1 to R^1 because

$$f(x + y) = 3(x + y) = 3x + 3y = f(x) + f(y)$$

and

$$f(ax) = 3(ax) = a(3x) = af(x)$$

for all x, y, and a. In general, for any fixed scalar s (instead of 3), $f(x) = sx$ defines a linear function from R^1 to R^1. *Note that* $F(x) = 3x + 1$ *is not linear* in the present definition since

$$F(x + y) = 3(x + y) + 1,$$

but

$$F(x) + F(y) = (3x + 1) + (3y + 1)$$

and these are never equal. (Actually, to spoil linearity it would be enough for them to differ for even one x and one y.) Similarly,

$$F(ax) = 3(ax) + 1 \neq a(3x + 1) = aF(x),$$

at least when $a \neq 1$. However, the fact that $F(x + y) \neq F(x) + F(y)$ is enough to spoil linearity and we do not need to check this second half of the definition.

Example 5. Define functions from R^3 to R^2 by

$$f(\mathbf{v}) = [v_1, v_2]$$

$$g(\mathbf{v}) = [2v_1 - 3v_2, 3v_1 - 2v_3]$$

$$h(\mathbf{v}) = [a_{11}v_1 + a_{12}v_2 + a_{13}v_3, a_{21}v_1 + a_{22}v_2 + a_{23}v_3]$$

for fixed numbers $a_{11}, a_{12}, a_{13}, a_{21}, a_{22}, a_{23}$. (Here we are using the convention of Chapter 1, Section 8 that $\mathbf{v}$ denotes the vector $[v_1, v_2, v_3]$.) Check that h is linear: Since

$$\mathbf{v} + \mathbf{w} = [v_1 + w_1, v_2 + w_2, v_3 + w_3]$$

$$\begin{aligned} h(\mathbf{v} + \mathbf{w}) &= [\sum_{p=1}^{3} a_{1p}(v_p + w_p), \sum_{p=1}^{3} a_{2p}(v_p + w_p)] \\ &= [\sum a_{1p}v_p + \sum a_{1p}w_p, \sum a_{2p}v_p + \sum a_{2p}w_p] \\ &= [\sum a_{1p}v_p, \sum a_{2p}v_p] + [\sum a_{1p}w_p, \sum a_{2p}w_p] \\ &= h(\mathbf{v}) + h(\mathbf{w}) \end{aligned}$$

and

$$\begin{aligned} h(a\mathbf{v}) &= [\sum a_{1p}(av_p), \sum a_{2p}(av_p)] \\ &= a[\sum a_{1p}(v_p), \sum a_{2p}(v_p)] \\ &= ah(\mathbf{v}). \end{aligned}$$

The reader should check that for special choices of $a_{11}, \ldots, a_{23}$, this h reduces to f and to g, so that f and g are also linear.

Example 6. There are two standard linear functions from R^n to R^n that are simple but very important. The *identity* function I is defined by $I(\mathbf{v}) = \mathbf{v}$ for every vector $\mathbf{v}$ in R^n; this function leaves every vector unchanged. The linearity is trivial to check. The *zero* function 0 is defined by $0(\mathbf{v}) = \mathbf{0}$ for every $\mathbf{v}$; this is a constant function since it associates the same functional value to all vectors. Its linearity is also easy to check.

PROBLEMS

1. Show that the function g in Example 2 is linear. Compute $g(\mathbf{i})$, $g(\mathbf{j})$, $g(\mathbf{k})$, $g(\mathbf{i} + \mathbf{j})$ (and compare it with $g(\mathbf{i}) + g(\mathbf{j})$), and $g(x_1\mathbf{i} + x_2\mathbf{j} + x_3\mathbf{k})$.

2. Verify the linearity of I and 0 in Example 6.

3. (a) Let $f(\mathbf{v})$ be defined to be $\mathbf{k} \cdot \mathbf{v}$ for every $\mathbf{v}$ in R^3. Show that f is a linear mapping from 3-space to 1-space.

(b) Let **u** be a fixed but unspecified vector in R^3 (we say, "Let **u** be an arbitrary vector in R^3") and define a function g from R^3 to R^1 by $g(\mathbf{v}) = \mathbf{u} \cdot \mathbf{v}$. Show that g is linear.

(c) Same as part (b) but change R^3 to R^n everywhere.

4. (a) For every vector $\mathbf{v} = x\mathbf{i} + y\mathbf{j} + z\mathbf{k}$ in R^3, define $f(\mathbf{v})$ to be $2x - y - z$, an element of R^1. Show that f is a linear function from R^3 to R^1.

(b) With **v** as in part (a), define a new function g by $g(\mathbf{v}) = 2x - y - z + 1$. Show that g is not linear in our sense.

(c) If a, b, and c are fixed (but arbitrary) numbers, and if h is the function from R^3 to R^1 defined by $h(\mathbf{v}) = ax + by + cz$ for every vector $v = x\mathbf{i} + y\mathbf{j} + z\mathbf{k}$, show that h is linear.

(d) Let φ be the function from R^3 to R^1 defined by $\varphi(\mathbf{v}) = ax + by + cz + d$; show that φ is linear if and only if $d = 0$.

(e) State and prove the analog of part (c) for functions from R^n to R^1.

(f) Translate part (c) into part (b) of Problem 3.

(g) All linear functions from R^3 to R^1 are given by the formula in part (c) for suitable choice of the numbers a, b, c. That is, for every linear function F from R^3 to R^1, there are numbers a, b, and c such that $F(\mathbf{v}) = ax + by + cz$ for every $\mathbf{v} = x\mathbf{i} + y\mathbf{j} + z\mathbf{k}$. Hint: a must be $F(\mathbf{i})$, etc.

(h) Prove a converse of Problem 3(b): Every linear function from R^3 to R^1 is of the form $F(\mathbf{v}) = \mathbf{u} \cdot \mathbf{v}$ for some suitable vector **u**, not depending on **v**. Hint: this is the same as part (g).

5. Let c be a fixed scalar, and define the mapping h from 3-space to 3-space as multiplication by c, that is, $h(\mathbf{v}) = c\mathbf{v}$ for each vector **v**. Is h a linear mapping? Why? Define a similar function from R^n to R^n. Check for linearity.

6. Consider the mapping f from R^3 to R^3 defined by $f(\mathbf{v}) = \mathbf{v} + \mathbf{i}$. Is this a linear mapping from 3-space to 3-space? Why?

7. For every linear function f from R^n to R^m, prove that $f(\mathbf{0}) = \mathbf{0}$.

8. Prove that the only constant linear function from R^n to R^m is the zero function.

9. Consider the mapping $\| \ \|$ from 3-space to 1-space, that is, $f(\mathbf{v}) = \| \mathbf{v} \|$. Is this a linear mapping? Why?

10. Let g be the function from R^2 to R^1 defined by $g([x, y]) = x^2 + y^2$ (or $g(\mathbf{v}) = x^2 + y^2$ for every $v = xi + yj$). Show that g is not linear. In fact, it satisfies neither of the two axioms in 1.1.

11. Verify $f(a\mathbf{v}) = af(\mathbf{v})$ for all a and $\mathbf{v}$ where f is the function in Example 1.

12. If $\mathbf{u}$ is a fixed vector in R^3 and f is the function $f(\mathbf{v}) = \mathbf{u} \times \mathbf{v}$ from R^3 to R^3, show f is linear.

13. If $\mathbf{u}$ is any fixed vector in R^3 and if, for every $\mathbf{v}$ in R^3 $g(\mathbf{v})$ denotes the projection (vector) of $\mathbf{v}$ on $\mathbf{u}$, show that g is a linear function. This is a function from R^3 to R^3, although of course the functional values (the set of outputs of the function) do not comprise all of R^3. There is also a related function from R^3 to R^1: $h(\mathbf{v}) =$ the component (scalar) of $\mathbf{v}$ along $\mathbf{u}$, that is, $\| \mathbf{u} \|^{-1} \mathbf{u} \cdot \mathbf{v}$. Show that h is also linear.

14. Let $\mathbf{w}_1$, $\mathbf{w}_2$, $\mathbf{w}_3$ be three fixed vectors in R^3, and define $f([x_1, x_2, x_3])$ to be $x_1\mathbf{w}_1 + x_2\mathbf{w}_2 + x_3\mathbf{w}_3$. Show that this is a linear function from 3-space to 3-space, and that, for suitable choices of $\mathbf{w}_1$, $\mathbf{w}_2$, $\mathbf{w}_3$, we get the functions in Examples 1 and 2 as special cases.

15. Show that a function from R^n to R^m is linear if and only if, for all scalars a and b and for all vectors $\mathbf{v}$ and $\mathbf{w}$ in R^n, $f(a\mathbf{v} + b\mathbf{w}) = af(\mathbf{v}) + bf(\mathbf{w})$.

16. If f is a function from R^n to R^m, and if g is a function from R^m to R^p, then the *composite* function (call it h) is a function from R^n to R^p: $h(\mathbf{v}) = g(f(\mathbf{v}))$ for every $\mathbf{v}$ in R^n. Show that h is linear if f and g are.

17. If g is a linear function from 3-space to 1-space and if $g(\mathbf{i}) = 1$, $g(\mathbf{j}) = 2$, and $g(\mathbf{k}) = 3$, find $g(2\mathbf{i} - \mathbf{j} - \mathbf{k})$, $g(\mathbf{i} + 2\mathbf{j} - 3\mathbf{k})$, $g(x\mathbf{i} + y\mathbf{j} + z\mathbf{k})$. Find all vectors $\mathbf{v} = x\mathbf{i} + y\mathbf{j} + z\mathbf{k}$ such that $g(\mathbf{v}) = 0$.

18. If h is a linear function from 3-space to 1-space and if $h(\mathbf{i} + \mathbf{j}) = 1$, $h(\mathbf{i} - \mathbf{j}) = 0$, and $h(\mathbf{i} + 2\mathbf{j} + \mathbf{k}) = 2$, find $h(\mathbf{i})$, $h(\mathbf{j})$, and $h(\mathbf{k})$.

19. If f is a linear function from 3-space to 3-space and if $f(\mathbf{i}) = \mathbf{i} + \mathbf{j}$, $f(\mathbf{j}) = \mathbf{i} - \mathbf{j}$, and $f(\mathbf{k}) = \mathbf{i} + \mathbf{j} + \mathbf{k}$, find $f(2\mathbf{i} - \mathbf{j} - \mathbf{k})$, $f(\mathbf{i} + 2\mathbf{j} - 3\mathbf{k})$, and $f(x\mathbf{i} + y\mathbf{j} + z\mathbf{k})$. Find all vectors $\mathbf{v} = x\mathbf{i} + y\mathbf{j} + z\mathbf{k}$ such that $f(\mathbf{v}) = \mathbf{0}$.

20. For each $\mathbf{v}$ in R^3 let $F(\mathbf{v})$ be defined as the projection of $\mathbf{v}$ on the plane $2x + 3y + z = 0$. What are the components of the vector $F(\mathbf{v})$?

Show that F is a linear function. A geometrical proof is probably easier than a proof using components, but both are instructive.

21. Let $h(\mathbf{v}) = g(f(\mathbf{v}))$ where f rotates each vector 90 degrees around the z-axis in the positive direction and g rotates each vector 90 degrees around the y-axis (carrying $\mathbf{i}$ into $\mathbf{k}$—this information fixes the direction of rotation). Compute $h(\mathbf{i})$, $h(\mathbf{j})$, and $h(\mathbf{k})$. This h is in fact a rotation. Find the axis of rotation by finding a nonzero vector $\mathbf{v}$ such that $h(\mathbf{v}) = \mathbf{v}$.

22. Let ξ be the function from 3-space to 3-space that rotates every vector 90 degrees around the x-axis and carries $\mathbf{j}$ into $\mathbf{k}$ (this determines the direction of rotation). Let η be a similar rotation about the y-axis carrying $\mathbf{i}$ into $\mathbf{k}$. Let ζ be a rotation about the z-axis carrying $\mathbf{i}$ into $\mathbf{j}$. Finally let f be the composite of ζ, η, and ξ in that order. That is,

$$f(\mathbf{v}) - \xi(\eta(\zeta(\mathbf{v}))).$$

(a) Compute $f(\mathbf{i})$, $f(\mathbf{j})$, and $f(\mathbf{k})$.

(b) Show $f(\mathbf{i}) = \eta(\mathbf{i})$, $f(\mathbf{j}) = \eta(\mathbf{j})$, $f(\mathbf{k}) = \eta(\mathbf{k})$.

(c) Using the fact that f and η are linear (Example 1 and Problem 10), show that part (b) implies $f = \eta$, that is, $f(\mathbf{v}) = \eta(\mathbf{v})$ for every $\mathbf{v}$ in 3-space. (In the notation of Section 4, $\xi \circ \eta \circ \zeta = \eta$.)

2. MATRICES

A matrix (pl., matrices) is just a rectangular array of numbers. For example, these are matrices:

$$\begin{pmatrix} 1 & 0 \\ 0 & 1 \end{pmatrix}; \quad \begin{pmatrix} 1 & 2 & \pi \\ \sqrt{3} & 0 & -4 \end{pmatrix}; \quad \begin{pmatrix} 1 \\ 4 \\ 6 \end{pmatrix}; \quad \begin{pmatrix} 1 & 4 & 6 \end{pmatrix}.$$

The most general matrix may be denoted thus:

$$\begin{pmatrix} a_{11} & a_{12} & \cdots & a_{1n} \\ a_{21} & a_{22} & \cdots & a_{2n} \\ \vdots & \vdots & \vdots & \vdots \\ a_{m1} & a_{m2} & \cdots & a_{mn} \end{pmatrix} \tag{2.1}$$

where each a_{pq} denotes a number. (This means a_{pq} is a number when p is any of the integers $1, 2, \ldots, m$ and q is any of the integers $1, 2, \ldots, n$.) This matrix (2.1) is sometimes denoted by just (a_{pq}).

The numbers in one horizontal row comprise a *row* of the matrix; each row of the matrix can be thought of as an n-tuple, as a vector in R^n. The numbers in one vertical column of the matrix comprise a *column* of the matrix; each column is also a vector. If a matrix has m rows and n columns, it is called an m by n matrix. The numbers in the matrix are called the *entries* of the matrix, and the number in the pth row and qth column is called the *p, q-entry* of the matrix. The 1, 1-entry, the 2, 2-entry, and so on are called the *diagonal entries* of the matrix and they comprise *the diagonal* of the matrix. The p, q-entry in Eq. (2.1) is a_{pq}. The diagonal entries are a_{11}, a_{22}, a_{33}, ... Note that these are pairs of subscripts; a_{11} is read "a sub one one," not "a sub eleven."

Example 5 in the preceding section used 2 by 3 matrices implicitly to give examples of linear functions from R^3 to R^2. This is quite general:

2.1 Definition

If $\mathbf{A}$ is any m by n matrix, define a function from R^n to R^m thus: to each n-tuple $\mathbf{v} = [v_1, \ldots, v_n]$ associate the m-tuple $\mathbf{Av}$ defined by row-by-column operation of $\mathbf{A}$ on $\mathbf{v}$: the pth component of $\mathbf{Av}$ is the dot product of the pth row of $\mathbf{A}$ with the vector $\mathbf{v}$. In more detail, if $\mathbf{A}$ is the matrix in Eq. (2.1) and $\mathbf{v} = [v_1, \ldots, v_n]$ is a vector in R^n, then $\mathbf{Av}$ is the m-tuple $[w_1, \ldots, w_m]$ where

$$w_p = \sum_{q=1}^{n} a_{pq}v_q \qquad \text{for each} \quad p = 1, \ldots, m;$$

that is,

$$\begin{aligned} w_1 &= a_{11}v_1 + a_{12}v_2 + \cdots + a_{1n}v_n \\ w_2 &= a_{21}v_1 + a_{22}v_2 + \cdots + a_{2n}v_n \\ &\vdots \\ w_m &= a_{m1}v_1 + a_{m2}v_2 + \cdots + a_{mn}v_n. \end{aligned}$$

If $\mathbf{v}$ is carried into $\mathbf{w}$ by the matrix $\mathbf{A}$ as in definition 2.1, then, as indicated in that definition, we write $\mathbf{w} = \mathbf{Av}$. This same statement is also written

$$\begin{pmatrix} a_{11} & a_{12} & \cdots & a_{1n} \\ a_{21} & a_{22} & \cdots & a_{2n} \\ \vdots & \vdots & \vdots & \vdots \\ a_{m1} & a_{m2} & \cdots & a_{mn} \end{pmatrix} \begin{pmatrix} v_1 \\ v_2 \\ \vdots \\ v_n \end{pmatrix} = \begin{pmatrix} w_1 \\ w_2 \\ \vdots \\ w_m \end{pmatrix}.$$

Example 1. The matrices used in Section 1, Example 5 are:

$$\text{matrix of } f = \begin{pmatrix} 1 & 0 & 0 \\ 0 & 1 & 0 \end{pmatrix}$$

because

$$\begin{pmatrix} 1 & 0 & 0 \\ 0 & 1 & 0 \end{pmatrix}\begin{pmatrix} v_1 \\ v_2 \\ v_3 \end{pmatrix} = \begin{pmatrix} v_1 \\ v_2 \end{pmatrix};$$

$$\text{matrix of } g = \begin{pmatrix} 2 & -3 & 0 \\ 3 & 0 & -2 \end{pmatrix}$$

because

$$\begin{pmatrix} 2 & -3 & 0 \\ 3 & 0 & -2 \end{pmatrix}\begin{pmatrix} v_1 \\ v_2 \\ v_3 \end{pmatrix} = \begin{pmatrix} 2v_1 - 3v_2 \\ 3v_1 - 2v_3 \end{pmatrix};$$

matrix of h is (a_{pq}), that is,

$$\begin{pmatrix} a_{11} & a_{12} & a_{13} \\ a_{21} & a_{22} & a_{23} \end{pmatrix}.$$

Similarly, Example 3, Section 1 is the linear function defined by the matrix $(1 \quad 2 \quad -4)$ because

$$(1 \quad 2 \quad -4)\begin{pmatrix} v_1 \\ v_2 \\ v_3 \end{pmatrix} = (v_1 + 2v_2 - 4v_3).$$

Example 4 in Section 1 used the 1 by 1 matrix (3). Example 2 in Section 1 can be defined by the matrix

$$\begin{pmatrix} 1 & 0 & 0 \\ 0 & 1 & 0 \\ 0 & 0 & 0 \end{pmatrix}$$

because, analytically, this projection function sends $[v_1, v_2, v_3]$ into $[v_1, v_2, 0]$, and this matrix, used according to the procedure in 2.1, gives the same result.

It turns out, as we might expect from these examples, that every linear function is described by a matrix.

First, consider a linear mapping f from 3-space to 3-space. How will f act on a typical vector $\mathbf{v}$? We can write

$$\mathbf{v} = v_1\mathbf{i} + v_2\mathbf{j} + v_3\mathbf{k},$$

and then

$$\begin{aligned} f(\mathbf{v}) &= f(v_1\mathbf{i} + v_2\mathbf{j} + v_3\mathbf{k}) \\ &= v_1 f(\mathbf{i}) + v_2 f(\mathbf{j}) + v_3 f(\mathbf{k}), \end{aligned}$$

because f is linear. Thus, to know how f acts on every vector $\mathbf{v}$, it suffices to know how f acts on the three vectors $\mathbf{i}$, $\mathbf{j}$, $\mathbf{k}$. Suppose

$$f(\mathbf{i}) = a_{11}\mathbf{i} + a_{21}\mathbf{j} + a_{31}\mathbf{k},$$

$$f(\mathbf{j}) = a_{12}\mathbf{i} + a_{22}\mathbf{j} + a_{32}\mathbf{k},$$

and

$$f(\mathbf{k}) = a_{13}\mathbf{i} + a_{23}\mathbf{j} + a_{33}\mathbf{k}.$$

Then

$$\begin{aligned} f(\mathbf{v}) = {} & (a_{11}v_1 + a_{12}v_2 + a_{13}v_3)\mathbf{i} + (a_{21}v_1 + a_{22}v_2 + a_{23}v_3)\mathbf{j} \\ & + (a_{31}v_1 + a_{32}v_2 + a_{33}v_3)\mathbf{k}. \end{aligned}$$

Instead of memorizing this formula, we convert it to an algorithm: (a) Write the components of $f(\mathbf{i})$ in a vertical column; do the same with the components of $f(\mathbf{j})$ and $f(\mathbf{k})$. Write these columns next to each other, to get a 3 by 3 matrix

$$\begin{pmatrix} a_{11} & a_{12} & a_{13} \\ a_{21} & a_{22} & a_{23} \\ a_{31} & a_{32} & a_{33} \end{pmatrix}.$$

(b) Compute each component of $f(\mathbf{v})$ by taking the dot product of the corresponding *row* of the matrix with $\mathbf{v}$. We usually write the components of $\mathbf{v}$ in a column just to the right of the matrix and consider the matrix as operating on this column, just as f operates on $\mathbf{v}$. This operation is called *row-by-column multiplication* as in 2.1.

Example 2. Let f be the rotation of 30 degrees about the z-axis similar to Example 1, Section 1. We compute its matrix: $f(\mathbf{i})$ = the vector of length 1, in the xy-plane, 30 degrees from the x-axis, which therefore has components (check this in a picture) $[1 \cos 30°, 1 \sin 30°, 0]$, or, to put it differently, $f(\mathbf{i}) = (\frac{1}{2}\sqrt{3})\mathbf{i} + (\frac{1}{2})\mathbf{j}$. Similarly, $f(\mathbf{j})$ = the vector 1 unit long in the xy-plane, 30 degrees away from the y-axis or 120 degrees away from the x-axis = $(\cos 120°)\mathbf{i} + (\sin 120°)\mathbf{j} + 0\mathbf{k} = (-\frac{1}{2})\mathbf{i} + (\frac{1}{2}\sqrt{3})\mathbf{j}$. Finally, $f(\mathbf{k}) = \mathbf{k}$. Thus we can see that the columns of the matrix are

$$\begin{pmatrix} \frac{1}{2}\sqrt{3} \\ \frac{1}{2} \\ 0 \end{pmatrix}, \quad \begin{pmatrix} -\frac{1}{2} \\ \frac{1}{2}\sqrt{3} \\ 0 \end{pmatrix}, \quad \text{and} \quad \begin{pmatrix} 0 \\ 0 \\ 1 \end{pmatrix}.$$

The matrix is

$$\begin{pmatrix} \frac{1}{2}\sqrt{3} & -\frac{1}{2} & 0 \\ \frac{1}{2} & \frac{1}{2}\sqrt{3} & 0 \\ 0 & 0 & 1 \end{pmatrix}.$$

Now to compute $f(\mathbf{v})$ for any $\mathbf{v} = x\mathbf{i} + y\mathbf{j} + z\mathbf{k}$, we can perform row-by-column multiplication of the matrix on the column of components of $\mathbf{v}$ thus:

$$\begin{pmatrix} \frac{1}{2}\sqrt{3} & -\frac{1}{2} & 0 \\ \frac{1}{2} & \frac{1}{2}\sqrt{3} & 0 \\ 0 & 0 & 1 \end{pmatrix} \begin{pmatrix} x \\ y \\ z \end{pmatrix} = \begin{pmatrix} \frac{1}{2}\sqrt{3}\,x - \frac{1}{2}y + 0z \\ \frac{1}{2}x + \frac{1}{2}\sqrt{3}\,y + 0z \\ 0x + 0y + 1z \end{pmatrix}$$

so

$$f(\mathbf{v}) = [\tfrac{1}{2}\sqrt{3}\,x - \tfrac{1}{2}y, \ \tfrac{1}{2}x + \tfrac{1}{2}\sqrt{3}\,y, \ z].$$

As a special case, let us compute $f(\mathbf{i})$:

$$\begin{pmatrix} \frac{1}{2}\sqrt{3} & -\frac{1}{2} & 0 \\ \frac{1}{2} & \frac{1}{2}\sqrt{3} & 0 \\ 0 & 0 & 1 \end{pmatrix} \begin{pmatrix} 1 \\ 0 \\ 0 \end{pmatrix} = \begin{pmatrix} \frac{1}{2}\sqrt{3} + 0 + 0 \\ \frac{1}{2} + 0 + 0 \\ 0 + 0 + 0 \end{pmatrix} = \begin{pmatrix} \frac{1}{2}\sqrt{3} \\ \frac{1}{2} \\ 0 \end{pmatrix},$$

which means

$$f(\mathbf{i}) = \tfrac{1}{2}\sqrt{3}\,\mathbf{i} + \tfrac{1}{2}\mathbf{j},$$

which we already knew.

The reader will notice that this gives the formula for the coordinates of the point (x', y', z') into which the point (x, y, z) is rotated by this 30-degree rotation since x', y', z' are the components of $f(\mathbf{v})$ if we think of $f(\mathbf{v})$ as a position vector:

$$\begin{aligned} x' &= \tfrac{1}{2}\sqrt{3}\,x - \tfrac{1}{2}y, \\ y' &= \tfrac{1}{2}x + \tfrac{1}{2}\sqrt{3}\,y, \\ z' &= z. \end{aligned}$$

This analysis works just as well for all linear functions from R^n to R^m, as we shall see in 2.3.

2.2 Definition

If f is a linear function from R^n to R^m, then the *matrix of* f is the m by n matrix whose columns are the m-tuples $f(\mathbf{i}_1), f(\mathbf{i}_2), \ldots, f(\mathbf{i}_n)$.

2.3 Theorem

The two definitions 2.1 and 2.2 establish a one-to-one correspondence between the m by n matrices and the linear functions from R^n to R^m.

Proof. There are only two things at stake: the fact that the function defined in 2.1 is linear (we leave this to Problem 2) and the meaning of a one-to-one correspondence. If we write $f \to \mathbf{A}$ to mean that $\mathbf{A}$ is the matrix of f as in definition 2.2 and if we write $\mathbf{A} \to g$ to mean that g is the function determined by $\mathbf{A}$ as in definition 2.1, then "these two arrows define a one-to-one correspondence" means two things: (1) If $f \to \mathbf{A}$ and $\mathbf{A} \to g$, then $f = g$; and (2) if $\mathbf{A} \to g$ and $g \to \mathbf{B}$, then $\mathbf{A} = \mathbf{B}$. If one-to-one correspondence means something different to the reader, he should be able to square it with this definition. For example, if we can show (1) and (2), we will know that every matrix $\mathbf{A}$ is the matrix of one linear mapping (namely, that f such that $\mathbf{A} \to f$, because we will have proved that $f \to \mathbf{A}$) and only one linear mapping (because if also $g \to \mathbf{A}$, then we will have proved that $\mathbf{A} \to g$ whereas $\mathbf{A} \to f$, so $f = g$).

For the proof of (2) we must show that if $\mathbf{A} \to g$, then $g(\mathbf{i}_p)$ (which is then the pth column of $\mathbf{B}$) equals the pth column of $\mathbf{A}$, for each p. Now $g(\mathbf{i}_p)$ is computed by row-by-column multiplication of the matrix $\mathbf{A}$ by the column whose entries are all zero except the pth, which is 1. We leave it to the reader to carry out this computation; the result is indeed the pth column of $\mathbf{A}$.

For the proof of (1) we assume $f \to \mathbf{A}$ and $\mathbf{A} \to g$. We have just shown that if $\mathbf{A} \to g$, then $g(\mathbf{i}_p)$ is the pth column of $\mathbf{A}$; but if $f \to \mathbf{A}$, then the pth column of $\mathbf{A}$ is $f(\mathbf{i}_p)$. Hence $f(\mathbf{i}_p) = g(\mathbf{i}_p)$ for every $p = 1, 2, \ldots, n$. Since f and g are both linear,

$$\begin{aligned} f(v_1\mathbf{i}_1 + \cdots + v_n\mathbf{i}_n) &= v_1 f(\mathbf{i}_1) + \cdots + v_n f(\mathbf{i}_n) \\ &= v_1 g(\mathbf{i}_1) + \cdots + v_n g(\mathbf{i}_n) \\ &= g(v_1\mathbf{i}_1 + \cdots + v_n\mathbf{i}_n). \end{aligned}$$

But all vectors in R^n are of the form $v_1\mathbf{i}_1 + \cdots + v_n\mathbf{i}_n$, so $f(\mathbf{v}) = g(\mathbf{v})$ for all $\mathbf{v}$ in R^n. This is what is meant by $f = g$.

We take special note of the linear functions from R^n to R^m when m or n, or both, are 1:

2.4 Every linear function f from R^1 to R^1 is defined by $f(x) = ax$ for some number a (compare Example 4, Section 1). This is true because such a linear function must be described by a 1 by 1 matrix (a_{11}), and we let $a = a_{11}$; row-by-column multiplication of 1 by 1 matrices is just ordinary multiplication of scalars. We note that the function $g(x) = ax + b$ is not linear in our sense unless $b = 0$ (see Example 4, Section 1 again).

2.5 Every linear function f from R^1 to R^n is defined by $f(x) = x\mathbf{a}$ for some vector $\mathbf{a}$ (where $\mathbf{a}$ does not depend on x; of course, the various vectors we might use for $\mathbf{a}$ give rise to the various f's). This is true because f is described by an n by 1 matrix

$$\begin{pmatrix} a_{11} \\ \vdots \\ a_{n1} \end{pmatrix}.$$

We take $\mathbf{a}$ to be the vector $[a_{11}, \ldots, a_{n1}]$. When this matrix multiplies a 1-tuple x, the row-by-column product is

$$\begin{pmatrix} a_{11}x \\ \vdots \\ a_{n1}x \end{pmatrix}.$$

Thus

$$f(x) = [a_{11}x, \ldots, a_{n1}x] = x\mathbf{a}.$$

Notice that 2.4 is the special case of 2.5 where $n = 1$.

2.6 Every linear function f from R^n to R^1 is defined by $f(\mathbf{v}) = \mathbf{a} \cdot \mathbf{v}$ for some vector $\mathbf{a}$ in R^n (compare Example 3, Section 1 and Problems 3 and 4 in that section), because such an f has a 1 by n matrix $(a_{11}, \ldots, a_{1n})$ and we may take $\mathbf{a} = [a_{11}, \ldots, a_{1n}]$. The row-by-column product that gives $f(\mathbf{v})$ is now just the dot product of $\mathbf{a}$ and $\mathbf{v}$.

Example 3. Row-by-column multiplication can also have concrete significance in some practical problems.

For example, suppose each of three countries, X, Y, and Z annually produces amounts of four different products (in tons weight) as in the following table:

Country	Product 1	Product 2	Product 3	Product 4
X	a_{11}	a_{12}	a_{13}	a_{14}
Y	a_{21}	a_{22}	a_{23}	a_{24}
Z	a_{31}	a_{32}	a_{33}	a_{34}

Let $\mathbf{A}$ denote the 3 by 4 matrix with pq-entry a_{pq}.

Suppose the value of product 1 is v_1 dollars per ton, and the value of product 2 is v_2 dollars per ton, etc., and let $\mathbf{v}$ be the vector $[v_1, v_2, v_3, v_4]$. Then if we write $\mathbf{v}$ as a column vector and compute $\mathbf{Av}$ by row-by-column multiplication, we get a triple of numbers. These three numbers are the total value of products 1–4 produced by each of the three countries. For example, $a_{11}v_1$ is the value of product 1 produced in country X, $a_{12}v_2$ is the value of product 2 produced in the same country, etc., so $a_{11}v_1 + a_{12}v_2 + a_{13}v_3 + a_{14}v_4$ is the total value of these products produced in country X.

PROBLEMS

1. If f is a linear function from R^a to R^b, how many rows has its matrix? How many columns? Is the matrix b by a or a by b? What is the significance for a function that its matrix is square?

2. For every 3 by 3 matrix $\mathbf{A}$, show that the function $g(\mathbf{v}) = \mathbf{Av}$ defined in 2.1 is linear. Do the same for every m by n matrix.

3. (a) Given the matrix

$$\begin{pmatrix} 2 & 1 & 3 \\ 0 & 1 & 4 \\ -1 & 0 & 1 \end{pmatrix},$$

let f denote the corresponding linear function on 3-space. Compute $f(\mathbf{i}), f(\mathbf{j}), f(\mathbf{k}), f(2\mathbf{i} + 3\mathbf{j} + \mathbf{k}), f(2\mathbf{i} - 3\mathbf{j})$.

(b) Let F be the function whose matrix is

$$\begin{pmatrix} 1 & 1 & -1 \\ 2 & 0 & 2 \\ -1 & 1 & 1 \end{pmatrix}.$$

Compute $F(\mathbf{v})$ if $\mathbf{v} = [1, 1, 1]$; if $\mathbf{v} = [2, 1, 3]$; if $\mathbf{v} = [1, -1, 1]$.

4. (a) Compute $g(\mathbf{v})$ if the matrix of g is

$$\begin{pmatrix} 1 & 3 & 0 & -1 \\ 2 & 0 & 4 & 3 \end{pmatrix}$$

and if $\mathbf{v} = [1, 1, 1, 1]$; if $\mathbf{v} = [2, -1, 0, 1]$; if $\mathbf{v} = [1, 0, 0, 0]$; if $\mathbf{v} = [0, 0, 1, 0]$.

(b) If the matrix of a linear function G is

$$\begin{pmatrix} 1 & 0 & 1 \\ 0 & 2 & 0 \end{pmatrix},$$

compute $G(\mathbf{i})$, $G(\mathbf{j})$, $G(\mathbf{k})$, and $G(2\mathbf{i} + 2\mathbf{j} + 2\mathbf{k})$.

5. (a) If the matrix of a linear function h is

$$\begin{pmatrix} 2 & 1 \\ 3 & 2 \\ 4 & 3 \\ 5 & 4 \end{pmatrix}$$

then h is a function from R^n to R^m; what are m and n? What is $h(\mathbf{i}_1)$? $h(\mathbf{i}_2)$? $h(2\mathbf{i}_1 + 3\mathbf{i}_2)$?

(b) H is the function whose matrix is

$$\begin{pmatrix} 1 & 0 \\ 0 & 1 \\ 1 & 1 \\ -1 & 1 \end{pmatrix}.$$

Find $h(\mathbf{v})$ if $\mathbf{v} = [1, 0], [0, 1], [1, -1]$, and $[2, 7]$.

6. Consider the mapping g from R^3 to R^3 that sends each arrow $\mathbf{v}$ into its projection on the xy-plane (Example 2, Section 1). Write the matrix of g and use it to compute $g(\mathbf{i} + 2\mathbf{j} + 3\mathbf{k})$, $g(\mathbf{i} - 2\mathbf{j} + 3\mathbf{k})$, $g(2\mathbf{i} - \mathbf{k})$. Compare with the same functional values computed geometrically from the definition of g.

7. Consider the mapping h from R^3 to R^3 that rotates each arrow 90 degrees around the x-axis (carrying the positive y-axis into the positive z-axis) and then stretches the result by a factor of 2. Write the matrix of h; then use the matrix to compute $h(\mathbf{i} + 2\mathbf{j} + 3\mathbf{k})$, $h(\mathbf{i} - 2\mathbf{j} + 3\mathbf{k})$, $h(2\mathbf{i} - \mathbf{k})$, and compare with the same functional values computed geometrically from the definition of h.

8. Find the matrix of the identity function I from R^n to R^n (see Example 6, Section 1). This is called the *n-rowed identity matrix* and is usually denoted by $\mathbf{I}_n$ or just $\mathbf{I}$. Check by row-by-column multiplication that $\mathbf{Iv} = \mathbf{v}$ for all $\mathbf{v}$.

9. Find the matrix of the zero function from R^n to R^n (see Example 6, Section 1). This is called the *n-rowed zero matrix* and is denoted by $\mathbf{O}_n$ or just by $\mathbf{O}$. Check by row-by-column multiplication that $\mathbf{Ov} = \mathbf{0}$ for every $\mathbf{v}$ in R^n.

10. Let f be the function from R^n to R^n that consists of multiplication by a scalar c; that is, $f(\mathbf{v}) = c\mathbf{v}$ for all $\mathbf{v}$. Find the matrix of f. This is called a *scalar* matrix. (Note that when $c = 1$ you get the identity matrix.) Check that row-by-column multiplication by this matrix multiplies every n-tuple by the scalar c.

11. A matrix with entries a_{pq} $(p = 1, \ldots, n, q = 1, \ldots, n)$ is called a *diagonal* matrix if $a_{pq} = 0$ whenever $p \neq q$; that is, only the diagonal entries are nonzero. (Notice that the identity matrix, the zero matrix, and all scalar matrices are diagonal.) Consider the linear mapping f from R^3 to R^3 that has the diagonal matrix

$$\begin{pmatrix} 1 & 0 & 0 \\ 0 & 1 & 0 \\ 0 & 0 & 2 \end{pmatrix}.$$

Draw $a\mathbf{i} + b\mathbf{j}$ and $f(a\mathbf{i} + b\mathbf{j})$ for typical a and b; draw $a\mathbf{i} + b\mathbf{j} + c\mathbf{k}$ and $f(a\mathbf{i} + b\mathbf{j} + c\mathbf{k})$ for typical a, b, and c.

12. Show that a linear mapping on 3-space has a diagonal matrix if and only if the mapping sends every vector parallel to an axis into a vector parallel to that same axis.

13. An n by n matrix (a_{pq}) is called *triangular* if $a_{pq} = 0$ whenever $p < q$. Write the most general 3 by 3 triangular matrix and consider the corresponding function f on R^3. Show that (1) if $\mathbf{v}$ is parallel to the z-axis, then so is $f(\mathbf{v})$, and (2) if $\mathbf{v}$ is parallel to the yz-plane, then so is $f(\mathbf{v})$. Conversely, show that if a linear mapping from R^3 to R^3 has properties (1) and (2), then its matrix is triangular.

14. Find the matrices of the linear mappings in Example 5, Section 1.

15. Find the matrices of the linear functions ξ, η, ζ, and f in Problem 22, Section 1.

16. Find the matrix of h in Problem 18, Section 1 and use it to verify that $h(\mathbf{i} + \mathbf{j}) = 1$, $h(\mathbf{i} - \mathbf{j}) = 0$, and $h(\mathbf{i} + 2\mathbf{j} + \mathbf{k}) = 2$. Find $h(x\mathbf{i} + y\mathbf{j} + z\mathbf{k})$.

17. Let f be a linear function from R^2 to R^1 such that $f([1, 1]) = 1$ and $f([2, -1]) = -1$. Find all vectors $\mathbf{v}$ such that $f(\mathbf{v}) = 0$.

18. Find the matrix of the linear function from R^1 to R^3 that sends $[1]$ into $2\mathbf{i} + \mathbf{j} + \mathbf{k}$. What is the set of all $f(x)$ as x runs through R^1? What is the set of all x in R^1 such that $f(x) = \mathbf{0}$?

19. Answer the same questions as in Problem 18 for the function g such that $g(\pi) = 2\mathbf{i} + \mathbf{j} + \mathbf{k}$.

The following two problems involve vector spaces of functions as in Chapter 1, Section 9.

20. Let V be the vector space of all quadratic polynomials. If $\mathbf{v}$ is the polynomial $ax^2 + bx + c$, let the components of $\mathbf{v}$ be $[a, b, c]$. With this notion of components find the 3 by 3 matrix of the differentiation operator $f(\mathbf{v}) = d\mathbf{v}/dx$.

21. Let V be as in Problem 20 and let W be the space of 1-tuples. Find the 1 by 3 matrix of the linear mapping

$$f(\mathbf{v}) = \int_0^1 \mathbf{v}(x)\,dx.$$

3. SUMS AND SCALAR MULTIPLES OF LINEAR MAPPINGS

3.1 Definition

Given two linear mappings f and g, both mapping R^n to R^m, we define $f + g$ just as we usually define addition of functions: $f + g$ sends each vector $\mathbf{v}$ into $f(\mathbf{v}) + g(\mathbf{v})$. (Of course, this last plus sign indicates addition of vectors in R^m.)

3.2 If f and g are linear, so is $f + g$.

Proof

$$\begin{aligned}(f+g)(\mathbf{v}+\mathbf{w}) &= f(\mathbf{v}+\mathbf{w}) + g(\mathbf{v}+\mathbf{w}) && \text{(definition of } f+g)\\ &= f(\mathbf{v}) + f(\mathbf{w}) + g(\mathbf{v}) + g(\mathbf{w}) && (f \text{ and } g \text{ linear})\\ &= f(\mathbf{v}) + g(\mathbf{v}) + f(\mathbf{w}) + g(\mathbf{w}) && (+ \text{ commutative})\\ &= (f+g)(\mathbf{v}) + (f+g)(\mathbf{w}) && \text{(definition of } f+g),\end{aligned}$$

$$\begin{aligned}(f+g)(a\mathbf{v}) &= f(a\mathbf{v}) + g(a\mathbf{v})\\ &= af(\mathbf{v}) + ag(\mathbf{v})\\ &= a(f(\mathbf{v}) + g(\mathbf{v}))\\ &= a(f+g)(\mathbf{v}).\end{aligned}$$

3.3 Definition

Given a linear mapping f from R^n to R^m and a scalar a, we define af to be the mapping sending $\mathbf{v}$ into $af(\mathbf{v})$ (the operation f followed by multiplication in R^m by the scalar a).

3.4 If f is linear and a is a scalar, then af is also linear.

Of course, now that we have defined sums and scalar multiples of linear mappings from R^n to R^m, we define a linear combination of two such mappings f and g as any mapping of the form $af + bg$ where a and b are scalars; and linear combinations of several functions, all of which must be functions from the same R^n to the same R^m, are the sums of scalar multiples of the given functions.

Everything we do for linear mappings we shall imitate with matrices. Hence we make the following definition:

3.5 Definition

The *sum* of two matrices is the matrix of the sum of the two corresponding linear mappings.

3.6 The sum of two matrices is defined if and only if both matrices have the same size (both are m by n). In this case, the sum is obtained simply by adding corresponding entries.

Proof. Let f and g denote the linear mapping whose matrices are $\mathbf{A}$ and $\mathbf{B}$, respectively. Then the qth column of $\mathbf{A} + \mathbf{B}$ thought of as a vector is the vector $(f + g)\ (\mathbf{i}_q)$ where $\mathbf{i}_q$ is the qth basis vector of R^n. But by definition of $f + g$,

$$(f + g)\ (\mathbf{i}_q) = f(\mathbf{i}_q) + g(\mathbf{i}_q),$$

so the qth column of $\mathbf{A} + \mathbf{B}$ is the vector sum of the qth column of $\mathbf{A}$ and the qth column of $\mathbf{B}$. Since the pq-entry of a matrix is just the pth component of the qth column, the pq-entry in $\mathbf{A} + \mathbf{B}$ is the sum of the pq-entries in $\mathbf{A}$ and in $\mathbf{B}$.

Example 1.

$$\begin{pmatrix} 2 & 1 & 3 \\ 0 & 1 & 4 \\ 5 & -1 & 1 \end{pmatrix} + \begin{pmatrix} 0 & 0 & 1 \\ -1 & 0 & 0 \\ 2 & 2 & 2 \end{pmatrix} = \begin{pmatrix} 2 & 1 & 4 \\ -1 & 1 & 4 \\ 7 & 1 & 3 \end{pmatrix}.$$

3.7 Definition

The product of a scalar a and a matrix $\mathbf{A}$ is the matrix of the product of a by the linear mapping corresponding to $\mathbf{A}$. A linear combination of matrices is a sum of scalar multiples of the matrices.

3.8 $a\mathbf{A}$ is the matrix obtained by multiplying all the entries in $\mathbf{A}$ by a.

Example 2.

$$3\begin{pmatrix} 2 & -1 \\ -1 & 1 \\ 2 & 2 \end{pmatrix} = \begin{pmatrix} 6 & -3 \\ -3 & 3 \\ 6 & 6 \end{pmatrix}.$$

PROBLEMS

1. Let f be the mapping from R^3 to R^3 that is the projection on the xy-plane (Example 2, Section 1) and g be projection on the z-axis. Describe the following mappings: $f + g$, $-g$ (that is, $(-1)g$), $f - g$ (that is, $f + (-g)$), $2f + 2g$, $2f + g$. Find their matrices.

2. Let f be a rotation of 90 degrees around the z-axis and g be the identity. What is $f - g$? Compute $(f - g)$ $(\mathbf{i})$. Compute the matrix of $f - g$.

3. Prove 3.4.

4. Prove 3.8.

5. Let $\mathbf{E}_{pq}$ be the m by n matrix which has 1 as its p, q-entry and 0 everywhere else. Show that every m by n matrix is a linear combination of $\mathbf{E}_{11}, \mathbf{E}_{12}, \ldots, \mathbf{E}_{1n}, \mathbf{E}_{21}, \mathbf{E}_{22}, \ldots, \mathbf{E}_{2n}, \ldots, \mathbf{E}_{m1}, \mathbf{E}_{m2}, \ldots, \mathbf{E}_{mn}$.

The following two problems refer to the concept of a general vector space in Chapter 1, Section 9.

6. Show that the m by n matrices form a vector space.

7. Show that the set of all linear mappings from R^n to R^m is a vector space using the addition and multiplication by scalars just defined. (The special case where $m = 1$, the set of all linear functions from R^n to the scalars, is called the *dual space* of R^n. Its elements are called dual vectors. They are represented by 1 by n matrices and so are n-tuples in a sense; they comprise the rows of an m by n matrix in contrast to the ordinary n-tuples on which the matrix acts, which are usually written as columns.)

4. COMPOSITES OF LINEAR MAPPINGS AND PRODUCTS OF MATRICES

If f and g are linear functions from n-space to 1-space, we can multiply them to get a product function $h(\mathbf{v}) = f(\mathbf{v})g(\mathbf{v})$, since $f(\mathbf{v})$ and $g(\mathbf{v})$ are numbers. But this product function is no longer linear (it is "quadratic"), so we postpone any such multiplication to a later treatment of nonlinear functions. Similarly, if f and g are functions from n-space to m-space, we reject the function $f(\mathbf{v}) \cdot g(\mathbf{v})$ for now. There is, however, one very good combination of linear functions that always gives a linear function as a result—composition.

4.1 Definition

If f is a linear mapping from R^n to R^m and g is a linear mapping from R^m to R^l, the *composite* $g \circ f$ is the function from R^n to R^l sending each $\mathbf{v}$ in R^n into $g(f(\mathbf{v}))$ in R^l.

Note that $g \circ f$ is the operation that consists of applying first f, then g; the notation is rather backward: the *first* operator is the *right*-hand one and the second is the left-hand one.

4.2 If f and g are as in the foregoing definition, then $g \circ f$ is linear (Problem 16, Section 1).

4.3 Definition

If $\mathbf{A}$ is an m by n matrix and $\mathbf{B}$ is an l by m matrix, then the *product* $\mathbf{BA}$ is the matrix of the composite of the linear mappings corresponding to $\mathbf{B}$ and $\mathbf{A}$ (it is an l by n matrix).

4.4 The entry in the pth row, qth column of $\mathbf{BA}$ is the dot product of the pth row of $\mathbf{B}$ by the qth column of $\mathbf{A}$.

Proof. If f corresponds to $\mathbf{A}$ and g to $\mathbf{B}$ then the qth column of $\mathbf{BA}$ contains the components of $g(f(\mathbf{i}_q))$ where $\mathbf{i}_q = [0, \ldots, 0, 1, 0, \ldots, 0]$ as usual. But we recall that to compute g of the vector $f(\mathbf{i}_q)$ we take the matrix of g, namely, $\mathbf{B}$, and operate on the column vector of components of $f(\mathbf{i}_q)$ as in 2.1. The pth entry produced by this operation is, by definition, the dot product of the pth row of $\mathbf{B}$ with the components of $f(\mathbf{i}_q)$, that is, with the qth column of $\mathbf{A}$.

Another way we sometimes see 4.4 phrased is: if $\mathbf{A} = (a_{pq})$ (that is, the entry in the pth row, qth column is a_{pq} for each p between 1 and m and for each q between 1 and n) and if $\mathbf{B} = (b_{rs})$ $(1 \leq r \leq l,$ $1 \leq s \leq m)$, then

$$\mathbf{BA} = \left(\sum_{s=1}^{m} b_{rs}a_{sq}\right) \qquad (1 \leq r \leq l, 1 \leq q \leq n).$$

Example 1.

$$\begin{pmatrix} 2 & 1 & 1 \\ 3 & 2 & 1 \\ 4 & 0 & 2 \end{pmatrix}\begin{pmatrix} 1 & 1 & 0 \\ 2 & 2 & 4 \\ 1 & 6 & 7 \end{pmatrix} = \begin{pmatrix} 5 & 10 & 11 \\ 8 & 13 & 15 \\ 6 & 16 & 14 \end{pmatrix}$$

because $[2, 1, 1] \cdot [1, 2, 1] = 5$, $[2, 1, 1] \cdot [1, 2, 6] = 10$, and so on.

Example 2.

$$\begin{pmatrix} 3 & 1 & 0 \\ 0 & 0 & 1 \end{pmatrix}\begin{pmatrix} -1 & 2 \\ 1 & 0 \\ 0 & 0 \end{pmatrix} = \begin{pmatrix} -2 & 6 \\ 0 & 0 \end{pmatrix}$$

because

$$\begin{aligned} 3 \times -1 + 1 \times 1 + 0 \times 0 &= -2, \\ 3 \times 2 + 1 \times 0 + 0 \times 0 &= 6, \\ 0 \times -1 + 0 \times 1 + 0 \times 0 &= 0, \\ 0 \times 2 + 0 \times 0 + 1 \times 0 &= 0. \end{aligned}$$

4.5 Let $\mathbf{A}$ be an n by n diagonal matrix with all its diagonal entries distinct;

$$\mathbf{A} = \begin{pmatrix} a_1 & 0 & \cdots & 0 \\ 0 & a_2 & \cdots & 0 \\ \cdot & \cdot & \cdots & \cdot \\ 0 & 0 & \cdots & a_n \end{pmatrix},$$

with $a_p \neq a_q$ whenever $p \neq q$. Then another matrix $\mathbf{B}$ will commute with $\mathbf{A}$ (that is, $\mathbf{AB} = \mathbf{BA}$) if and only if $\mathbf{B}$ is an n by n diagonal matrix.

Proof. In order for $\mathbf{BA}$ and $\mathbf{AB}$ to be defined, $\mathbf{B}$ must be n by n. (Why?) so let

$$\mathbf{B} = (b_{pq}) = \begin{pmatrix} b_{11} & b_{12} & \cdots & b_{1n} \\ \cdot & \cdot & \cdots & \cdot \\ b_{n1} & b_{n2} & \cdots & b_{nn} \end{pmatrix}.$$

The (r, s)-entry of $\mathbf{AB}$ is

$$[0, \ldots, a_r, \ldots, 0] \cdot [b_{1s}, b_{2s}, \ldots, b_{ns}] = a_r b_{rs}.$$

The (r, s)-entry of **BA** is

$$[b_{r1}, b_{r2}, \ldots, b_{rn}] \cdot [0, \ldots, a_s, \ldots, 0] = b_{rs}a_s.$$

Then **BA** = **AB** if and only if $a_r b_{rs} = b_{rs} a_s$ for all r and s. If $r = s$, this is a vacuous condition. If $r \neq s$, this gives $(a_r - a_s)b_{rs} = 0$, which implies $b_{rs} = 0$, because $a_r \neq a_s$. Thus the matrices that commute with **A** are the diagonal matrices.

Note that this shows **AB** $\neq$ **BA** for many matrices (for example, for **A** as in 4.5 and **B** not diagonal); the commutative law does not hold for multiplication of matrices. Several of the exercises will also demonstrate this fact. This is not surprising, since operators do not usually commute. As one famous algebraist put it, "Putting on your socks then putting on your shoes is not the same as putting on your shoes then putting on your socks."

4.6 The distributive laws hold for linear mappings and for matrices; composition distributes over addition of linear mappings and matrix multiplication distributes over addition of matrices.

Proof. $f \circ (g + h) = f \circ g + f \circ h$ because any vector $\mathbf{v}$ on which g and h operate is sent by $f \circ g + f \circ h$ into

$$f(g(\mathbf{v})) + f(h(\mathbf{v})) = f(g(\mathbf{v}) + h(\mathbf{v})),$$

which is where $f \circ (g + h)$ sends $\mathbf{v}$. Similarly, $(f + g) \circ h$ sends $\mathbf{v}$ into

$$\begin{aligned}(f + g)(h\,\mathbf{v})) &= f(h(\mathbf{v})) + g(h(\mathbf{v})) \\ &= (f \circ h)(\mathbf{v}) + (g \circ h)(\mathbf{v}),\end{aligned}$$

which is where $\mathbf{v}$ is sent by $f \circ h + g \circ h$. We leave the matrix proof to the reader (the easiest way is to deduce it from the linear mapping result).

4.7 The associative law holds for composites of linear mappings and for products of matrices.

Proof. Let f be a linear mapping of R^n into R^m, g a linear mapping of R^m into R^l, and h a linear mapping of R^l into R^k. We are to show $(h \circ g) \circ f = h \circ (g \circ f)$; that is, that every vector $\mathbf{v}$ in R^n is sent to the same vector in R^k by $(h \circ g) \circ f$ as by $h \circ (g \circ f)$. Where does $h \circ (g \circ f)$ send $\mathbf{v}$? $g \circ f$ sends it to $g(f(\mathbf{v}))$ so the composite $h \circ (g \circ f)$ sends $\mathbf{v}$ into $h(g(f(\mathbf{v})))$. Similarly $(h \circ g) \circ f$ sends $\mathbf{v}$ to $(h \circ g)(f(\mathbf{v})) = h(g(f(\mathbf{v})))$ which is indeed the same.

If **A**, **B**, **C** are matrices such that (**CB**)**A** is defined (the number of columns in **C** equals the number of rows in **B**, and the number of columns in **B** equals the number of rows in **A**), let f be the function whose matrix is **A**, g be the function whose matrix is **B**, and h be the function whose matrix is **C**. Then the matrix of $(h \circ g) \circ f$ is (**CB**)**A** and the matrix of $h \circ (g \circ f)$ is **C**(**BA**). Since these two composite functions are equal, as we have just proved, so are the two products of matrices, by 2.3.

4.8 Let f be a linear mapping from R^n to R^m, g a linear mapping from R^m to R^l, and a any scalar. Then

$$g \circ (af) = a(g \circ f) = (ag) \circ f.$$

If **A** is an m by n matrix, **B** is an l by m matrix, and a is a scalar, then

$$\mathbf{B}(a\mathbf{A}) = a(\mathbf{BA}) = (a\mathbf{B})\mathbf{A}.$$

We leave the proof to Problems 2 and 6.

If f is a linear mapping of n-space to itself, then we may define f^2 to be $f \circ f$, $f^3 = f^2 \circ f$, and so on. The associative law implies that $f^2 \circ f = f \circ f^2$, so that both possible definitions of f^3 agree; also $f^3 \circ f = f^2 \circ f^2 = f \circ f^3$ so that all possible definitions of f^4 are the same, and so on. Note that the associative law implies that all powers of f commute with one another. All these remarks are also true for square matrices. (Why only for *square* matrices?) See Problem 25 for the laws of exponents.

Finally, we remark that much of what we have said in Sections 3 and 4 applies equally well to nonlinear functions from R^n to R^m, in particular the definitions 3.1, 3.3, and 4.1, the associative law 4.7, and half of statements 4.6 and 4.8:

$$(f + g) \circ h = f \circ h + g \circ h$$

and

$$a(g \circ f) = (ag) \circ f.$$

PROBLEMS

1. Prove 4.2; that is, repeat Problem 16, Section 1.

2. Prove 4.8.

3. Show that if f is a linear mapping from R^n to R^m and a is a scalar, then af may also be thought of as a composite of two linear mappings (compare Problem 5, Section 1).

4. Compute the following matrix products:

$$\begin{pmatrix} 1 & 1 \\ -1 & -1 \end{pmatrix}\begin{pmatrix} 2 & 2 \\ 1 & 3 \end{pmatrix}; \quad \begin{pmatrix} 1 & 2 & 3 \\ 4 & 5 & 6 \\ 7 & 8 & 9 \end{pmatrix}\begin{pmatrix} 1 & 2 & 3 \\ 4 & 5 & 6 \\ 7 & 8 & 9 \end{pmatrix};$$

$$\begin{pmatrix} 1 & 0 & 0 & 1 \\ 2 & -1 & 1 & 0 \\ 3 & 0 & 1 & 0 \end{pmatrix}\begin{pmatrix} 2 & 0 \\ 1 & -1 \\ 1 & -1 \\ 1 & -1 \end{pmatrix};$$

$$\begin{pmatrix} 2 & 3 \\ 4 & 1 \end{pmatrix}\begin{pmatrix} 1 & -3 \\ -4 & 2 \end{pmatrix}; \quad \begin{pmatrix} a & b \\ c & d \end{pmatrix}\begin{pmatrix} d & -b \\ -c & a \end{pmatrix}.$$

5. The n-rowed identity matrix **I** is the matrix of the identity function on R^n (compare Problem 8, Section 2). Directly from the definition show that, for any other n by n matrix **A**, **IA** = **AI** = **A**. Then show the same thing using matrix multiplication. The identity mapping and the identity matrix thus act as identity elements for composition and matrix multiplication, respectively.

6. Let **A** be the n by n scalar matrix c**I** (compare Problem 10, Section 2). Show, using matrix multiplication, that for every other n by n matrix **B**, **AB** = **BA** = c**B**. Then show the same thing by translating everything into linear mappings. Use this and Problems 3 and 7 to give another proof of 4.8.

7. Prove the converse of Problem 6: If **B** is a 2 by 2 matrix such that **BA** = **AB** for every 2 by 2 matrix **A**, then **B** = c**I** for some scalar c. Hint: Let

$$\mathbf{B} = \begin{pmatrix} a & b \\ c & d \end{pmatrix}$$

and see what **BA** = **AB** means when

$$\mathbf{A} = \begin{pmatrix} 0 & 1 \\ 0 & 0 \end{pmatrix}.$$

Then try other **A**'s with three entries equal to 0. Do the same for n by n matrices.

8. Given a 2 by 2 matrix

$$\mathbf{A} = \begin{pmatrix} a & b \\ c & d \end{pmatrix}$$

with $ad - bc \neq 0$, what are the entries in the matrix **B** that satisfy **BA** = **I**? **B** is called the *inverse* of **A**. Show that this same **B** satisfies **AB** = **I**. What if $ad - bc = 0$?

9. The matrix

$$\begin{pmatrix} 2 & 0 & 0 \\ 0 & 2 & 0 \\ 0 & 0 & 0 \end{pmatrix}$$

represents the mapping that takes a vector, projects it on the xy-plane, then doubles the projection. Since the mapping is described as a composite, this matrix should be the product of two matrices. Write the two matrices and check that their product is the given matrix.

10. Check that

$$\mathbf{AB} \neq \mathbf{BA} \quad \text{if} \quad \mathbf{A} = \begin{pmatrix} 0 & 1 \\ 0 & 0 \end{pmatrix} \quad \text{and} \quad \mathbf{B} = \begin{pmatrix} 0 & 0 \\ 1 & 0 \end{pmatrix}.$$

11. Check that the composite of linear mappings is not commutative by showing geometrically that $f \circ g \neq g \circ f$ when f is rotation 90 degrees around the x-axis and g is rotation by 90 degrees around the z-axis. Verify this same fact using matrices.

12. Show that every two n by n diagonal matrices commute.

13. Show that multiplication on the left by a diagonal matrix merely multiplies each row of the second matrix by the corresponding diagonal entry of the diagonal matrix. What is the result of multiplying on the right by a diagonal matrix? Use this to find more examples of noncommutative multiplication.

14. Compute the second, third, and fourth powers of the matrix

$$\begin{pmatrix} 1/\sqrt{2} & -1/\sqrt{2} & 0 \\ 1/\sqrt{2} & 1/\sqrt{2} & 0 \\ 0 & 0 & 1 \end{pmatrix}.$$

Interpret these matrices as the matrices of geometric mappings (they are rotations) and verify geometrically that they are the powers of the original one.

15. Prove the associative law for 2 by 2 matrices directly by matrix multiplication.

16. Find a 2 by 2 matrix A that has the property $\mathbf{A}^2 + \mathbf{I} = 0$. Hint:

Geometrically it is not difficult to find a mapping f such that $(f \circ f)(\mathbf{v}) = -\mathbf{v}$ for all $\mathbf{v}$.

17. What geometric properties of a mapping f will cause its matrix to have the following forms? (The dots indicate unspecified entries.)

$$\begin{pmatrix} \cdot & \cdot & 0 \\ \cdot & \cdot & 0 \\ \cdot & \cdot & 1 \end{pmatrix}; \quad \begin{pmatrix} \cdot & \cdot & 0 \\ \cdot & \cdot & 0 \\ \cdot & \cdot & 0 \end{pmatrix}; \quad \begin{pmatrix} \cdot & \cdot & 0 \\ \cdot & \cdot & 0 \\ 0 & 0 & \cdot \end{pmatrix}; \quad \begin{pmatrix} \cdot & \cdot & 0 & 0 \\ \cdot & \cdot & 0 & 0 \\ 0 & 0 & \cdot & \cdot \\ 0 & 0 & \cdot & \cdot \end{pmatrix}.$$

18. Refer to Problem 5, Section 3 for the definition of $\mathbf{E}_{pq}$. Take $m = n$ so the $\mathbf{E}_{pq}$'s will be square n by n matrices. Show $\mathbf{E}_{pq}\mathbf{E}_{rs} = 0$ if $q \neq r$ and $= \mathbf{E}_{ps}$ if $q = r$.

19. With the same conventions as in the preceding problem, show that $\mathbf{E}_{pq}\mathbf{A}$ is a matrix with only one nonzero row, the pth, and that that row is the same as the qth row of $\mathbf{A}$.

20. With the same conventions as in Problem 18, show that $\mathbf{A}\mathbf{E}_{pq}$ is a matrix with only one nonzero column, the qth, and that that column is the same as the pth column of $\mathbf{A}$.

21. Show that a diagonal n by n matrix is a linear combination of $\mathbf{E}_{11}, \ldots, \mathbf{E}_{nn}$, and hence give another proof of Problem 13, using Problems 19 and 20.

22. Prove the addition formulas for the sine and cosine thus: Show that the matrix of rotation in R^2 through an angle α is

$$\mathbf{A}_\alpha = \begin{pmatrix} \cos\alpha & -\sin\alpha \\ \sin\alpha & \cos\alpha \end{pmatrix}.$$

It follows geometrically that $\mathbf{A}_\alpha\mathbf{A}_\beta = \mathbf{A}_{\alpha+\beta}$; use matrix multiplication to compute $\mathbf{A}_\alpha\mathbf{A}_\beta$, and compare with $\mathbf{A}_{\alpha+\beta}$.

23. Show that, if the pth row of matrix $\mathbf{A}$ is $\mathbf{0}$, then the pth row of $\mathbf{AB}$ is $\mathbf{0}$ for every matrix $\mathbf{B}$. Similarly, if the qth column of $\mathbf{B}$ is $\mathbf{0}$, then the qth column of $\mathbf{AB}$ is $\mathbf{0}$ for every $\mathbf{A}$. Give examples to show that the first statement becomes false if you replace the word "row" by "column."

24. Show that every product of diagonal matrices is diagonal (Problem 11, Section 2) and every product of triangular matrices is triangular (Problem 13, Section 2).

25. Prove the laws of exponents for linear mappings from R^n to R^n, and for square matrices:

$$f^p \circ f^q = f^{p+q} \qquad \text{for all positive integers } p \text{ and } q;$$

$$(f^p)^q = f^{pq} \qquad \text{for all positive integers } p \text{ and } q;$$

$$(f \circ g)^p = f^p \circ g^p \qquad \text{for all } p \textit{ if } f \textit{ commutes with } g.$$

See also Problem 3, Section 5 following.

26. (a) Take a fixed (but arbitrary) plane through the origin in R^3 and define f from R^3 to R^3 to be perpendicular projection on this plane; for every $\mathbf{v}$ in R^3, $f(\mathbf{v})$ is the vector projection of $\mathbf{v}$ on the fixed plane. Show $f^2 = f$ geometrically.

(b) Suppose f is projection on the xz-plane. Find the matrix, $\mathbf{A}$ of f and show by computation that $\mathbf{A}^2 = \mathbf{A}$.

(c) Suppose f is projection on the plane $x + y + 2z = 0$. Find the matrix of f and show that it equals its own square.

27. Let g be reflection in a fixed plane through the origin in R^3. That is, for every vector $\mathbf{v}$ in the plane, $g(\mathbf{v}) = \mathbf{v}$; for every vector $\mathbf{w}$ perpendicular to the plane, $g(\mathbf{w}) = -\mathbf{w}$; every vector in R^3 is a sum $\mathbf{v} + \mathbf{w}$ of a vector in the plane and a vector perpendicular to the plane, and $g(\mathbf{v} + \mathbf{w})$ is defined to be $\mathbf{v} - \mathbf{w}$.

(a) Show $g^2 = I$.

(b) Suppose g is reflection in the xz-plane. Find the matrix of g and compute the square of this matrix.

(c) Suppose g is reflection in the plane $x + y + 2z = 0$. Find the matrix of g and compute the square of the matrix.

(d) Give another proof of (a) by using the f in Problem 26: Every vector $\mathbf{x}$ is $f(\mathbf{x}) + (\mathbf{x} - f(\mathbf{x}))$ and $f(\mathbf{x})$ is in the plane while $\mathbf{x} - f(\mathbf{x})$ is perpendicular to the plane. Hence $g(\mathbf{x}) = f(\mathbf{x}) - (\mathbf{x} - f(\mathbf{x})) = (2f - I)(\mathbf{x})$. So $g = 2f - I$ and you can multiply out $g \circ g$ and use $f \circ f = f$.

28. Let $\mathbf{r}$ be a fixed vector in R^3 and define $h(\mathbf{v}) = \mathbf{r} \times \mathbf{v}$ for every $\mathbf{v}$ in R^3. This h is a linear mapping from R^3 to R^3.

(a) Use Chapter 1, 7.8 to show $h^3 =$ scalar times h (the scalar is $-\|\mathbf{r}\|^2$).

(b) Take $\mathbf{r} = \mathbf{i}$ and compute the matrix $\mathbf{A}$ of h. Compute $\mathbf{A}^3$ and show it is $-\mathbf{A}$.

29. Three countries, which we number 1, 2, and 3, produce four kinds of products which we number products 1, 2, 3, and 4. The number of tons of product number q produced by country number p is denoted by t_{pq} for $p = 1, 2, 3$ and $q = 1, 2, 3, 4$. Let $\mathbf{T}$ be the 3 by 4 matrix (t_{pq}).

Each product requires imports of two raw materials which we shall call materials 1 and 2. One ton of product q requires the import of m_{qr} tons of material number r ($q = 1, 2, 3, 4$ and $r = 1, 2$). Let $\mathbf{M}$ be the 4 by 2 matrix whose qr-entry is m_{qr}.

Show that the pr-entry in the matrix $\mathbf{TM}$ is the amount of raw material number r imported for these purposes in country p. Compare Example 3, Section 3.

30. If a_{pq} denotes the number of nonstop airplane flights from city number p to city number q and $\mathbf{A}$ is the matrix (a_{pq}), show that the pq-entry in $\mathbf{A}^2$ is the number of one-stop combination flights (that is two-flight combination trips) from city p to city q. Show also that the pq-entry in $\mathbf{A}^n$ is the number of flights from city p to city q involving $n - 1$ stops.

31. (a) Let $\mathbf{A}$ be a square matrix such that the sum of the entries in every row is 1. Show that all powers of $\mathbf{A}$ have the same property. Hint: Watch $\mathbf{A}$ operate on $[1, 1, \ldots, 1]$. If the entries of such a matrix $\mathbf{A}$ are nonnegative besides, they may faithfully be interpreted as probabilities: Picture a box with n slots and a jumping bean that every minute jumps from one slot to another (possibly to the same one). Let a_{pq} be the probability that he jumps from slot number p to slot number q. Then these a_{pq}'s are the entries of a matrix $\mathbf{A}$. The entry in the pth row, qth column of $\mathbf{A}^n$ will then be the probability of finding the bean in slot q after n minutes, if he started in slot p.

(b) Let $\mathbf{A}$ be the matrix

$$\begin{pmatrix} \frac{2}{3} & \frac{1}{3} \\ \frac{1}{3} & \frac{2}{3} \end{pmatrix}.$$

Find $\mathbf{A}^n$ for every n. Find

$$\lim_{n\to\infty} \mathbf{A}^n.$$

What does this mean in terms of the probability interpretation of $\mathbf{A}^n$ in part (a)?

5. INVERSES

5.1 Definition

If f is a function from R^n to R^m, and g is a function from R^m to R^n, we say g is an *inverse* of f (and we write $g = f^{-1}$) if $f \circ g$ and $g \circ f$ are both identity mappings (one is the identity from R^n to R^n, the other from R^m to R^m). If f has an inverse, we say it is *invertible* (some authors use the synonym *nonsingular*).

This concept of inverse function is the same as that met in other contexts. For example, the exponential function (a nonlinear function from R^1 to the set of positive real numbers) has an inverse, namely, the logarithm function: If f denotes the exponential function, $f(x) = e^x$, and if g denotes the logarithm (to the base e) $g(y) = \log y$, then $f \circ g$ is the identity:

$$f(g(y)) = e^{\log y} = y \qquad \text{for all positive real numbers } y;$$

and $g \circ f$ is the identity:

$$g(f(x)) = \log e^x = x \qquad \text{for all } x.$$

Here is another nonlinear example: if f is the square function $f(x) = x^2$ and g is the square root function $g(y) = \sqrt{y}$, then g is only a one-sided inverse of f; $f \circ g$ is the identity, but $g \circ f$ is not quite the identity since $\sqrt{x^2} = | x |$; this can be repaired by restricting the domain of f to the set of positive real numbers. This restricted function does have a true inverse, namely, the square root function. Similar *ad hoc* devices make a restricted sine function invertible, and its inverse is $\sin^{-1}$ (not to be confused with $1/\sin x$).

One of the pleasant aspects of linear functions is that they never give rise to the trouble that we just met in handling squares and sines. We shall show in Chapter 4, 2.7 that if f is a linear function and $f \circ g = I$, then this same g has the property $g \circ f = I$. We shall take this for granted in the rest of this section.

We can rephrase definition 5.1 thus:

5.2 If f is invertible, $\mathbf{w} = f(\mathbf{v})$ means the same as $\mathbf{v} = f^{-1}(\mathbf{w})$; f^{-1} is an operator that undoes what f does.

Proof. If $\mathbf{w} = f(\mathbf{v})$ then

$$f^{-1}(\mathbf{w}) = f^{-1}(f(\mathbf{v})) = (f^{-1} \circ f)(\mathbf{v}) = I(\mathbf{v}) = \mathbf{v}.$$

Conversely, if $\mathbf{v} = f^{-1}(\mathbf{w})$, then

$$f(\mathbf{v}) = f(f^{-1}(\mathbf{w})) = (f \circ f^{-1})(\mathbf{w}) = I(\mathbf{w}) = \mathbf{w}.$$

We use 5.2 to find f^{-1}. For example, let f be the linear function whose matrix is

$$\begin{pmatrix} 2 & -1 & 3 \\ 1 & -2 & 0 \\ 4 & 0 & 0 \end{pmatrix}.$$

The equation $f(\mathbf{v}) = \mathbf{w}$ is the system

$$\begin{aligned} 2v_1 - v_2 + 3v_3 &= w_1, \\ v_1 - 2v_2 \qquad &= w_2, \\ 4v_1 \qquad\qquad &= w_3. \end{aligned}$$

We are asked to solve for $\mathbf{v}$ as a function of $\mathbf{w}$, since that function will be f^{-1}. This system of equations is easy to solve; solve the third equation for v_1, then using this formula for v_1, solve the second equation for v_2 in terms of the w's, and finally solve the first equation for v_3. We get

$$\begin{aligned} v_1 &= \tfrac{1}{4}w_3, \\ v_2 &= -\tfrac{1}{2}w_2 + \tfrac{1}{8}w_3, \\ v_3 &= \tfrac{1}{3}w_1 - \tfrac{1}{6}w_2 - \tfrac{1}{8}w_3. \end{aligned}$$

These are the formulas describing f^{-1}; this is the recipe that describes $\mathbf{v} = f^{-1}(\mathbf{w})$ as a function of $\mathbf{w}$. Written in matrix form,

$$\begin{pmatrix} v_1 \\ v_2 \\ v_3 \end{pmatrix} = \begin{pmatrix} 0 & 0 & \frac{1}{4} \\ 0 & -\frac{1}{2} & \frac{1}{8} \\ \frac{1}{3} & -\frac{1}{6} & -\frac{1}{8} \end{pmatrix} \begin{pmatrix} w_1 \\ w_2 \\ w_3 \end{pmatrix}$$

so this last 3 by 3 matrix is the matrix of f^{-1}.

One further rephrasing of definition 5.1, which makes it clear that a function is invertible if and only if it is a one-to-one correspondence between the vectors in R^n and the vectors in R^m, is this:

5.3 A function f from R^n to R^m is invertible if and only if for every vector $\mathbf{w}$ in R^m there is one and only one vector $\mathbf{v}$ in R^n such that $f(\mathbf{v}) = \mathbf{w}$. The function that associates to each $\mathbf{w}$ the one and only one $\mathbf{v}$ just described is f^{-1}.

Proof. Suppose for each $\mathbf{w}$ there is one and only one $\mathbf{v}$ with $f(\mathbf{v}) = \mathbf{w}$, and define $g(\mathbf{w})$ to be this $\mathbf{v}$. If we can show $g \circ f = I$ and $f \circ g = I$, then f will be invertible and g will be its inverse. But this is easy enough: $(g \circ f)(\mathbf{v}) = g(f(\mathbf{v})) =$ that vector $\mathbf{v}'$ such that $f(\mathbf{v}') = f(\mathbf{v})$; since there we are supposing that there is only one $\mathbf{v}'$ that will do, and since $\mathbf{v}$ will do, we know $\mathbf{v} = \mathbf{v}'$, that is, $(g \circ f)(\mathbf{v}) = \mathbf{v}$. Similarly, $(f \circ g)(\mathbf{w}) = f(g(\mathbf{w})) = f(\mathbf{v})$ where $\mathbf{v}$ is chosen so that $f(\mathbf{v}) = \mathbf{w}$; hence $(f \circ g)(\mathbf{w}) = \mathbf{w}$.

Conversely, suppose f is invertible and f^{-1} denotes its inverse. For every vector $\mathbf{w}$ in R^m, choose $\mathbf{v} = f^{-1}(\mathbf{w})$; then

$$f(\mathbf{v}) = f(f^{-1}(\mathbf{w})) = (f \circ f^{-1})(\mathbf{w}) = I(\mathbf{w}) = \mathbf{w},$$

and $\mathbf{v}$ is the only vector with this property: for if also $f(\mathbf{v}') = \mathbf{w}$, then $f^{-1}(f(\mathbf{v}')) = f^{-1}(\mathbf{w}) = \mathbf{v}$, but $f^{-1} \circ f = I$, so $f^{-1}(f(\mathbf{v}')) = \mathbf{v}'$, proving $\mathbf{v} = \mathbf{v}'$.

5.4 If a function f is invertible, it has only one inverse function. If f is invertible and linear, its inverse is also linear.

Proof. Suppose g and h are both inverses of f, that is,

$$g \circ f = I, \qquad f \circ g = I,$$

$$h \circ f = I, \qquad f \circ h = I.$$

We need only the first and last of these equations, together with the associative law of composition 4.6:

$$h = I \circ h = (g \circ f) \circ h = g \circ (f \circ h) = g \circ I = g.$$

To show f^{-1} is linear if f is, we need to prove $f^{-1}(\mathbf{v} + \mathbf{w}) = f^{-1}(\mathbf{v}) + f^{-1}(\mathbf{w})$ and $f^{-1}(a\mathbf{v}) = af^{-1}(\mathbf{v})$ for all $\mathbf{v}$, $\mathbf{w}$, and a. By 5.3, to show two vectors $\mathbf{x}$ and $\mathbf{y}$ are equal, it is enough to show $f(\mathbf{x}) = f(\mathbf{y})$. We do this with $\mathbf{x} = f^{-1}(\mathbf{v} + \mathbf{w})$ and $\mathbf{y} = f^{-1}(\mathbf{v}) + f^{-1}(\mathbf{w})$. Then $f(\mathbf{x}) = f(f^{-1}(\mathbf{v} + \mathbf{w})) = \mathbf{v} + \mathbf{w}$ and $f(\mathbf{y}) = f(f^{-1}(\mathbf{v}) + f^{-1}(\mathbf{w})) = f(f^{-1}(\mathbf{v})) + f(f^{-1}(\mathbf{w})) = \mathbf{v} + \mathbf{w}$, using 5.1 and the linearity of f. We leave it to the reader to produce a similar proof that $f^{-1}(a\mathbf{v}) = af^{-1}(\mathbf{v})$.

5.5 Definition

If **A** and **B** are matrices, we say **B** is an *inverse* of **A** (and write $\mathbf{B} = \mathbf{A}^{-1}$) if **AB** and **BA** are both identity matrices. If **A** has an inverse, we say **A** is *invertible* (synonym, nonsingular).

Clearly, a matrix is invertible if and only if the corresponding linear mapping is invertible. And if **A** is the matrix of f, then $\mathbf{A}^{-1}$ is the matrix of f^{-1}.

An invertible matrix must be square. We shall prove this statement in Chapter 5, 4.13; *a linear mapping from R^n to R^m cannot be invertible unless $m = n$.* Of course, the converse is not true; there are many square matrices and many linear functions from R^n to R^n that are not invertible, as we shall see in this and the next chapter.

5.6 Every composite of invertible linear mappings is invertible and its inverse is the product of the inverses *in reverse order:*

$$(g \circ f)^{-1} = f^{-1} \circ g^{-1}.$$

The same statement is true for matrices and matrix multiplication.

Proof. By the definition of invertible, we need only check that $f^{-1} \circ g^{-1}$ composed with $g \circ f$ on both left and right gives the identity:

$$\begin{aligned}(f^{-1} \circ g^{-1}) \circ (g \circ f) &= f^{-1} \circ (g^{-1} \circ g) \circ f = f^{-1} \circ I \circ f \\ &= f^{-1} \circ f = I, \\ (g \circ f) \circ (f^{-1} \circ g^{-1}) &= g \circ (f \circ f^{-1}) \circ g^{-1} = g \circ I \circ g^{-1} \\ &= g \circ g^{-1} = I.\end{aligned}$$

Note that 5.6 does not say $(g \circ f)^{-1} = g^{-1} \circ f^{-1}$. In view of the noncommutativity of composition, this is not surprising. The shoes and socks analogy quoted before in 4.6 is again relevant. If f is the operation of putting on socks and g is the operation of putting on shoes, then $g \circ f$ is the operation of putting on socks, then shoes. The inverse operation consists in taking off first shoes, then socks, that is, $f^{-1} \circ g^{-1}$, not $g^{-1} \circ f^{-1}$.

5.7 If f is invertible, so is f^{-1}, and $(f^{-1})^{-1} = f$.

Proof. The statements $f \circ f^{-1} = I$ and $f^{-1} \circ f = I$, which say that f^{-1} is the inverse of f, also say that f is the inverse of f^{-1}.

In the next two chapters we shall develop more techniques for determining whether a given linear function or matrix is invertible and, if it is, how to find the inverse.

PROBLEMS

1. Find the inverse of each of the following matrices:

$$\begin{pmatrix} 1 & 2 \\ 3 & 5 \end{pmatrix}; \quad \begin{pmatrix} 2 & 1 & 1 \\ 0 & 2 & 4 \\ -1 & -1 & -1 \end{pmatrix};$$

$$\begin{pmatrix} 1 & 0 & 2 & 3 \\ 4 & 1 & 0 & 1 \\ -1 & -1 & 1 & 1 \\ 0 & 0 & -3 & -2 \end{pmatrix}; \quad \begin{pmatrix} \sin\theta & \cos\theta & 0 \\ -\cos\theta & \sin\theta & 0 \\ 0 & 0 & 1 \end{pmatrix}.$$

2. Which of the following linear mappings are invertible? What are their inverses?

(a) Rotation about the z-axis through an angle α;

(b) Multiplication by a scalar a;

(c) Projection on the xy-plane;

(d) Reflection in the xy-plane.

3. Show that the laws of exponents hold for powers of invertible linear mappings. First define $f^{-p} = (f^{-1})^p$ and $f^0 = I$. Then

(a) show that $f^{-p} = (f^p)^{-1}$;

(b) show $f^p f^q = f^{p+q}$ and $(f^p)^q = f^{pq}$ for all integers p and q, positive, negative, or zero;

(c) show $(f \circ g)^p = f^p \circ g^p$ for all p if and only if f and g commute. See Problem 25, Section 4 for part of (b) and (c).

4. Consider the diagonal matrix

$$\begin{pmatrix} x_1 & 0 \\ 0 & x_2 \end{pmatrix}.$$

(a) What restrictions on x_1 and x_2 are necessary and sufficient for the matrix to be invertible? If the matrix is invertible, what is its inverse?

(b) Answer the same questions for the triangular matrix

$$\begin{pmatrix} x_1 & 0 \\ y & x_2 \end{pmatrix}.$$

(c) Answer the same questions for the most general n by n diagonal matrix. Show that the inverse of a diagonal matrix (if it has one) is also diagonal.

(d) Answer the same questions for the most general n by n triangular matrix, but do not compute all the entries in the inverse matrix; find the diagonal entries and show that the inverse of an invertible triangular matrix is triangular.

5. Solve the following equations for x_1 and x_2, assuming $a_{11}a_{22} - a_{21}a_{12} \neq 0$:

$$a_{11}x_1 + a_{12}x_2 = y_1,$$

$$a_{21}x_1 + a_{22}x_2 = y_2.$$

Hence, write a formula for

$$\begin{pmatrix} a_{11} & a_{12} \\ a_{21} & a_{22} \end{pmatrix}^{-1}.$$

6. Use the method of the preceding problem to find a formula for the inverse of the general 3 by 3 matrix.

7. If $\mathbf{A}$ is an invertible n by n matrix, define the following function F of n by n matrices: $F(\mathbf{B}) = \mathbf{A}^{-1}\mathbf{BA}$. Show that $F(\mathbf{BC}) = F(\mathbf{B})F(\mathbf{C})$, $F(\mathbf{B} + \mathbf{C}) = F(\mathbf{B}) + F(\mathbf{C})$, and $F(\mathbf{B}^t) = (F(\mathbf{B}))^t$ for all integers t, including negative integers if $\mathbf{B}$ is invertible.

8. Let f be rotation about the origin in R^2 through an angle α and let g be reflection in the y-axis, that is, $g(x\mathbf{i} + y\mathbf{j}) = -x\mathbf{i} + y\mathbf{j}$. Show geometrically that $g \circ f \circ g$ is rotation through an angle $-\alpha$, and so $(g \circ f \circ g)^{-1} = f$. Compute the matrices of g, f, and $g \circ f \circ g$. Compute the matrices of g^{-1}, f^{-1}, and $(g \circ f \circ g)^{-1}$. Compute this last again, using 5.6. Verify again that $(g \circ f \circ g)^{-1} = f$.

9. Using f and g as in Problem 8 and the result of Problem 8 that $(g \circ f \circ g)^{-1} = f$, show without further matrix computation that $(g \circ f)^2$ is the identity. Note that $g^2 \circ f^2$ is not the identity (except for special α's). This is more evidence that the commutative law fails, because of Problem 3(c).

10. If f is invertible, and if $f^{-1} \circ g \circ f = h$, show that $g = f \circ h \circ f^{-1}$.

11. If $f^2 = 0$, show that $I + f$ is invertible and its inverse is $I - f$. Hence, show

$$\begin{pmatrix} 1 & a \\ 0 & 1 \end{pmatrix}^{-1} = \begin{pmatrix} 1 & -a \\ 0 & 1 \end{pmatrix}.$$

6. KERNEL AND IMAGE

Given a linear mapping f from R^n to R^m we define the *kernel* of f as a certain subset of R^n: the set of all $\mathbf{v}$ such that $f(\mathbf{v}) = \mathbf{0}$. The *image* of f is a certain subset of R^m: the set of all $f(\mathbf{v})$ as $\mathbf{v}$ ranges over R^n. The image of f is also called the *range* of f, and the kernel of f is often called the *null space* of f. Notice that both the kernel and the image are sets of vectors, and that the kernel is a set of vectors from the domain of f (a set of input vectors into the f machine) and that the image is a set of functional values of f (the set of all functional values, in fact; the set of outputs of the f machine).

Before we translate the foregoing into matrices and other analytic personalities, let us look at some examples with an eye to verifying the statement, "The bigger the kernel, the smaller the image." This will be proved explicitly and quantitatively in Chapter 5.

Example 1. Let f be the rotation in Example 1, Section 1. The kernel of f is $\mathbf{0}$, the image of f is the set of all vectors in R^3.

Example 2. Let f be the projection on the xy-plane (Example 2, Section 1). The kernel of f is the set of all vectors on the z-axis (all linear combinations of the one vector $\mathbf{k}$) The image of f is the set of all vectors in the xy-plane (all linear combinations of $\mathbf{i}$ and $\mathbf{j}$).

Example 3. If h is the projection on the x-axis, that is, $h(a\mathbf{i} + b\mathbf{j} + c\mathbf{k}) = a\mathbf{i}$, then the kernel of h is the set of vectors in the yz-plane (all linear combinations of $\mathbf{j}$ and $\mathbf{k}$), the image of h is the set of vectors along the x-axis (all linear combinations of $\mathbf{i}$).

Example 4. If a linear mapping from R^n to R^m has kernel equal to all vectors in R^n, then its image consists of $\mathbf{0}$ alone; it is the zero function that is described in Example 6, Section 1 and Problem 9, Section 2.

Regarding the dual personality, let f be a linear mapping from n-tuples to m-tuples and let $\mathbf{A} = (a_{ij})$ be its matrix. The kernel of f is the set of n-tuples $\mathbf{x}$ such that $\mathbf{Ax} = \mathbf{0}$, that is, the set of all n-tuples $\mathbf{x} = [x_1, \ldots, x_n]$ such that

$$\begin{pmatrix} a_{11} & \cdots & a_{1n} \\ \vdots & \cdot & \vdots \\ a_{m1} & \cdots & a_{mn} \end{pmatrix} \begin{pmatrix} x_1 \\ \vdots \\ x_n \end{pmatrix} = \begin{pmatrix} 0 \\ \vdots \\ 0 \end{pmatrix};$$

that is, the set of all $[x_1, \ldots, x_n]$ such that all these equations are true:

$$\begin{aligned} a_{11}x_1 + \cdots + a_{1n}x_n &= 0, \\ &\vdots \\ a_{m1}x_1 + \cdots + a_{mn}x_n &= 0. \end{aligned} \tag{6.1}$$

6.1 The kernel is the set of solutions of the system of simultaneous linear equations (6.1).

It is not clear which occurs more frequently in practice, the kernel of a linear mapping or its alter ego, the set of all solutions of a system of linear equations (6.1).

We leave it to the reader to verify that the image of f is the set of m-tuples $[b_1, \ldots, b_m]$ for which the system of equations

$$\begin{aligned} a_{11}x_1 + \cdots + a_{1n}x_n &= b_1 \\ &\vdots \\ a_{m1}x_1 + \cdots + a_{mn}x_n &= b_m \end{aligned} \tag{6.2}$$

has some solution. Note that the kernel has to do with linear equations with 0 on the right-hand side; the image concerns linear equations with any numbers on the right-hand side.

Finally we notice that (6.2) can be written

$$[b_1, \ldots, b_m] = x_1[a_{11}, \ldots, a_{m1}] + x_2[a_{12}, \ldots, a_{m2}] + \cdots + x_n[a_{1n}, \ldots, a_{mn}];$$

that is:

6.2 The image of f is the set of all linear combinations of the columns of the matrix of f.

The kernel is not usually described as a set of linear combinations; the problem of solving the system (6.1) might be thought of exactly as the problem of so expressing the kernel. We shall make much of this in the next chapter.

PROBLEMS

1. (a) Let f be the mapping from R^3 to R^1 defined by

$$f(\mathbf{v}) = (2\mathbf{i} + \mathbf{j}) \cdot \mathbf{v}.$$

Find the kernel and the image of f.

(b) Let $\mathbf{u}$ be any fixed (but unspecified) nonzero vector in R^3 and let g be the mapping defined by $g(\mathbf{v}) = \mathbf{u} \cdot \mathbf{v}$. Find the kernel and the image of g.

2. (a) Let F be the mapping from R^1 to R^3 defined by

$$F([t]) = t(\mathbf{i} + 3\mathbf{j} - \mathbf{k}).$$

Find the kernel and the image of F.

(b) Let $\mathbf{u}$ be any fixed (but unspecified) nonzero vector in R^3 and let G be the mapping from R^1 to R^3 defined by $G([t]) = t\mathbf{u}$. Find the kernel and the image of G.

3. Find the images of the following linear mappings. Where possible, do this in two ways, geometrically and analytically.

(a) $f: R^3 \to R^3, f(\mathbf{v})$ = projection of $\mathbf{v}$ on the x-axis.

(b) $g: R^3 \to R^0, g(\mathbf{v})$ = rotation by $\pi/2$ around the z-axis.

(c) $h = g \circ f$ with g, f as in parts (a) and (b).

(d) $h' = f \circ g$ with g, f as in parts (a) and (b).

(e) The mapping whose matrix is

$$\begin{pmatrix} 2 & 0 \\ 1 & 3 \end{pmatrix}.$$

(f) The mapping whose matrix is

$$\begin{pmatrix} 2 & 1 \\ 4 & 2 \end{pmatrix}.$$

(g) The mapping whose matrix is

$$\begin{pmatrix} 2 & 1 & 0 \\ 3 & 2 & 0 \\ 4 & 1 & 0 \end{pmatrix}.$$

4. Find the kernels of the linear mappings in Problem 3. Where possible, do this in two ways, geometrically and analytically.

5. Prove: If f is a linear mapping from R^n to R^m, then two vectors $\mathbf{u}, \mathbf{u}'$ in R^n are carried into the same vector by f if and only if $\mathbf{u} = \mathbf{u}' +$ a vector in the kernel of f.

6. If f is a linear mapping from R^3 to R^1, show that the kernel of f is a plane through the origin, unless $f = 0$ (that is, $f(\mathbf{u}) = 0$ for every $\mathbf{u}$), in which case the kernel of f is all of R^3. (Use matrices.)

7. Show that every plane through the origin is the kernel of some linear mapping from R^3 to R^1.

8. If f is a linear mapping from R^3 to R^2, show that the kernel of f is a line through the origin—or, in exceptional circumstances, a plane through the origin or all of R^3. How can you recognize these exceptional circumstances by looking at the matrix of f?

9. Show that every line through the origin is the kernel of some linear mapping from R^3 to R^2. (Hint: Use nonparametric equations).

10. Show that the image of a linear mapping from R^1 to R^3 is a line through the origin, unless the mapping is 0, in which case the image is just the zero vector.

11. Show that every line through the origin in R^3 is the image of a linear mapping from R^1 to R^3. (Use parametric equations.)

12. Show that the image of every linear mapping from R^2 to R^3 is a plane through the origin—or, in exceptional circumstances, a line through the origin or the zero vector alone. How can you recognize these exceptional circumstances by looking at the matrix of the mapping?

13. Show that every plane through the origin is the image of a linear mapping from R^2 to R^3. This gives "parametric equations of the plane": all points on the plane are given by

$$x = a_{11}u + a_{12}v, \qquad y = a_{21}u + a_{22}v, \qquad z = a_{31}u + a_{32}v$$

as $[u, v]$ ranges over R^2.

14. Show that parametric equations of a plane not necessarily through the origin are

$$x = a_{11}u + a_{12}v + x_0, \quad y = a_{21}u + a_{22}v + y_0, \quad z = a_{31}u + a_{32}v + z_0,$$

where the a's and x_0, y_0, z_0 are numbers having the following geometric significance: (x_0, y_0, z_0) is one point on the plane; $[a_{11}, a_{21}, a_{31}]$ and $[a_{12}, a_{22}, a_{32}]$ are two vectors parallel to the plane.

CHAPTER

4

Solution of Equations

1. SOLUTION PROCESS; REDUCTION TO ECHELON FORM

We shall study the system of equations

$$\begin{aligned} a_{11}x_1 + a_{12}x_2 + \cdots + a_{1n}x_n &= b_1, \\ a_{21}x_1 + a_{22}x_2 + \cdots + a_{2n}x_n &= b_2, \\ &\vdots \\ a_{m1}x_1 + a_{m2}x_2 + \cdots + a_{mn}x_n &= b_m, \end{aligned} \tag{1.1}$$

which we shall abbreviate to $\sum_{q=1}^{n} a_{pq}x_q = b_p$ $(p = 1, \ldots, m)$. We shall think of this system in two other ways, one as a matrix equation

$$\begin{pmatrix} a_{11} & \cdots & a_{1n} \\ a_{21} & \cdots & a_{2n} \\ \cdot & \cdots & \cdot \\ a_{m1} & \cdots & a_{mn} \end{pmatrix} \begin{pmatrix} x_1 \\ x_2 \\ \vdots \\ x_n \end{pmatrix} = \begin{pmatrix} b_1 \\ b_2 \\ \cdot \\ b_m \end{pmatrix}; \tag{1.2}$$

another as an equation

$$f(\mathbf{x}) = \mathbf{b}, \tag{1.3}$$

where f is the linear function from R^n to R^m whose matrix is the m by n matrix (a_{pq}) displayed in (1.2), where $\mathbf{x}$ is the vector $[x_1, \ldots, x_n]$ in R^n, and where $\mathbf{b}$ is the vector $[b_1, \ldots, b_m]$ in R^m. The matrix (a_{pq}) is called the *matrix of coefficients* of the system (1.1). The m by $n + 1$ matrix

$$\begin{pmatrix} a_{11} & \cdots & a_{1n} & b_1 \\ a_{21} & \cdots & a_{2n} & b_2 \\ \cdot & \cdots & \cdot & \cdot \\ a_{m1} & \cdots & a_{mn} & b_m \end{pmatrix}$$

is called the *augmented matrix* of the system.

A *solution*, or *solution vector*, of the system is an n-tuple of numbers $[x_1, \ldots, x_n]$ that makes all the equations true statements. The usual problem is to find whether solutions exist, and if so, to find all of them. We already know how to solve this problem: If $a_{11} \neq 0$, we add a suitable multiple of the first equation to the second to get a new equation not involving x_1 (make the coefficient of x_1 zero); similarly, we add multiples of the first equation to the third, fourth, ..., mth equations, getting $m - 1$ equations, none of which involve x_1; this gives $m - 1$ equations in $n - 1$ unknowns. Repeating the technique, eliminating x_2, then x_3, and so on, we get down to just one unknown or one equation.

The following example shows the technique:

Example 1. To find all $[x_1, x_2, x_3]$ satisfying

$$2x_1 + 3x_2 + 4x_3 = 1, \tag{1.4}$$

$$x_1 + x_2 - x_3 = 0, \tag{1.5}$$

$$4x_1 - x_2 + 2x_3 = 0, \tag{1.6}$$

add $-\frac{1}{2}$ times Eq. (1.4) to Eq. (1.5) and -2 times Eq. (1.4) to Eq. (1.6) to get two new equations not involving x_1:

$$-\tfrac{1}{2}x_2 - 3x_3 = -\tfrac{1}{2}, \tag{1.7}$$

$$-7x_2 - 6x_3 = -2. \tag{1.8}$$

If $[x_1, x_2, x_3]$ satisfies Eqs. (1.4)–(1.6), it must also satisfy Eqs. (1.7) and (1.8). In fact, we can get a converse, too, in the following way: $[x_1, x_2, x_3]$ satisfies Eqs. (1.4)–(1.6) if and only if it satisfies

$$2x_1 + 3x_2 + 4x_3 = 1 \qquad \text{(same as Eq. (1.4))}, \tag{1.9}$$

$$-\tfrac{1}{2}x_2 - 3x_3 = -\tfrac{1}{2} \qquad \text{(same as Eq. (1.7))}, \tag{1.10}$$

$$-7x_2 - 6x_3 = -2 \qquad \text{(same as Eq. (1.8))}, \tag{1.11}$$

because if Eqs. (1.9) and (1.10) hold, so do Eqs. (1.4
because (1.5) = (1.10) + $\frac{1}{2}$(1.9) and, similarly, if E
(1.11) hold, so does Eq. (1.6).

The two equations (1.10) and (1.11) can be handled in this sa
way: We add −14 times (1.10) to (1.11) to eliminate x_2; we get a new system that still has the same solutions as Eqs. (1.9)–(1.11)

$$2x_1 + 3x_2 + 4x_3 = 1, \tag{1.12}$$

$$-\tfrac{1}{2}x_2 - 3x_3 = -\tfrac{1}{2}, \tag{1.13}$$

$$36x_3 = 5. \tag{1.14}$$

This system is easy to solve. Equation (1.14) says $x_3 = \frac{5}{36}$; Eq. (1.13) then says $x_2 = 1 - 6x_3 = \frac{1}{6}$; and, finally, Eq. (1.12) gives $x_1 = \frac{1}{2}(1 - 3x_2 - 4x_3) = -\frac{1}{36}$. Thus there is exactly one solution to Eqs. (1.4)–(1.6), namely, the vector $[-\frac{1}{36}, \frac{1}{6}, \frac{5}{36}]$.

We formalize this process so that it can be carried out with any size system of linear equations, without the excess writing involved in carrying along all the x's and the equal signs: Given a system (1.1), we allow ourselves three manipulations on the rows of the augmented matrix which convert it to the augmented matrix of a new (and, hopefully, easier) system of equations having exactly the same solutions:

Manipulation 1. Add a multiple of one row to another.

Manipulation 2. Interchange two rows.

Manipulation 3. Multiply one row by a nonzero scalar.

Manipulation 2 has the effect of interchanging two of the equations. Manipulation 3 multiplies one equation by a nonzero number. Clearly these do not change the solutions any. Manipulation 1, adding s times the pth row to the qth row, replaces the pth and qth equations (call them E_p and E_q) by the two equations E_p and $E_q + sE_p$. Any solution of this first pair of equations will clearly be a solution of the second pair, and conversely, since the first pair may be retrieved from the second by the same process: $E_q = (E_q + sE_p) + (-s)E_p$. This proves the following theorem.

1.1 Applying a sequence of Manipulations 1, 2, 3 to the augmented matrix of a system of equations gives an augmented matrix of a system of equations having the same solutions as the original system.

An effective program of manipulations is described as follows:

If the augmented matrix has a nonzero entry in the first column, use Manipulation 2 to put this nonzero entry into the upper left corner, and use Manipulation 3 to make this entry equal to 1. Then use Manipulation 1 repeatedly, adding suitable multiples of this new first row to each of the other rows to make the first column

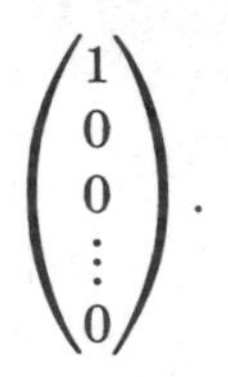

If the original matrix has no nonzero entry in the first column, we pass to the first column where there *is* a nonzero entry and perform the sequence of operations just presented. We have arrived at a matrix

$$\begin{pmatrix} 0 & 0 & \cdots & 0 & 1 & * & \cdots & * \\ 0 & 0 & \cdots & 0 & 0 & * & \cdots & * \\ \vdots & \vdots & & \vdots & \vdots & \vdots & & \vdots \\ 0 & 0 & \cdots & 0 & 0 & * & \cdots & * \end{pmatrix}$$

where the *'s indicate entries about which we say nothing at the moment, and where the columns of 0's may be absent.

We next look at the matrix of *'s and find the first column having a nonzero entry outside the first row. Use Manipulation 2 to put this entry in the second row, Manipulation 3 to make it 1, then Manipulation 1 to make all other entries in this column 0. We have arrived at

$$\begin{pmatrix} 0 & 0 & \cdots & 0 & 1 & * & \cdots & * & 0 & * & \cdots & * \\ 0 & 0 & \cdots & 0 & 0 & 0 & \cdots & 0 & 1 & * & \cdots & * \\ \vdots & \vdots & & \vdots & \vdots & \vdots & & \vdots & \vdots & \vdots & & \vdots \\ 0 & 0 & \cdots & 0 & 0 & 0 & \cdots & 0 & 0 & * & \cdots & * \end{pmatrix}.$$

Continuing in this way we arrive at an *echelon form*, that is, a matrix of the form

$$\begin{pmatrix} 0 & \cdots & 0 & 1 & * & \cdots & * & 0 & * & \cdots & * & 0 & \cdots & * & 0 & * & \cdots & * \\ 0 & \cdots & 0 & 0 & 0 & \cdots & 0 & 1 & * & \cdots & * & 0 & \cdots & * & 0 & * & \cdots & * \\ 0 & \cdots & 0 & 0 & 0 & \cdots & 0 & 0 & 0 & \cdots & 0 & 1 & \cdots & * & 0 & * & \cdots & * \\ \vdots & & \vdots & \vdots & \vdots & & \vdots & \vdots & \vdots & & \vdots & \vdots & & \vdots & \vdots & \vdots & & \vdots \\ 0 & \cdots & 0 & 0 & 0 & \cdots & 0 & 0 & 0 & \cdots & 0 & 0 & \cdots & 0 & 1 & * & \cdots & * \\ 0 & \cdots & 0 & 0 & 0 & \cdots & 0 & 0 & 0 & \cdots & 0 & 0 & \cdots & 0 & 0 & 0 & \cdots & 0 \\ \vdots & & \vdots & \vdots & \vdots & & \vdots & \vdots & \vdots & & \vdots & \vdots & & \vdots & \vdots & \vdots & & \vdots \\ 0 & \cdots & 0 & 0 & 0 & \cdots & 0 & 0 & 0 & \cdots & 0 & 0 & \cdots & 0 & 0 & 0 & \cdots & 0 \end{pmatrix}$$

where the *'s are numbers about which we say nothing.

Explicitly, an *echelon form* is a matrix whose columns, reading from left to right, are as follows: first, some columns of zeros (by "some" we mean to include the possibility that there are none); then a "distinguished" column vector equal to $\mathbf{i}_1 = [1, 0, \ldots, 0]$; then some columns with zeros everywhere except possibly in the first row; then a distinguished column vector equal to $\mathbf{i}_2 = [0, 1, 0, \ldots, 0]$; then some columns with zeros everywhere except possibly in the first two rows; then a distinguished column equal to $\mathbf{i}_3$; and so on.

An echelon form is also called a *Hermite normal form,* although the ideas were essentially known to Gauss about 1800; Hermite's dates are 1822–1901. The reduction process to echelon form is named after Gauss; it is called Gauss–Jordan reduction process (Jordan: 1838–1922).

We repeat that if a matrix **A** is reduced to echelon form **B** by Manipulations 1, 2, 3, then the system of equations with augmented matrix **A** has the same solutions as the system of equations with augmented matrix **B**.

The solutions of a system of equations corresponding to a matrix in echelon form are easy to find. If we keep on one side of the equations all the unknowns that correspond to distinguished columns (call this set of unknowns *Group 1*), and move all the other unknowns to the other side (call this set of unknowns *Group 2*), we have thereby *solved for the unknowns in Group 1 in terms of the unknowns in Group 2.* (See Example 2.) That is, for any choice of numbers substituted for the unknowns in Group 2, we have formulas determining the numbers to be used for the unknowns in Group 1. The resulting set of n numbers is a solution vector (of both the echelon form and the original system of equations), and all solutions are obtained in this way.

A sharper look at these equations with augmented matrix in echelon form shows that, although it is generally correct that we can solve for the unknowns in Group 1 in terms of those in Group 2, we have missed an important eventuality where this statement is wrong. Each of the nonzero equations expresses one of the x's in Group 1 in terms of the x's in Group 2 and constants, *except possibly the last* of the nonzero equations. This equation also expresses one of the x's in Group 1 (the last one) in terms of the other x's and a constant, unless this last nonzero row of the echelon form is $[0, 0, \ldots, 0, 1]$, that is, unless the last column is one of the distinguished columns. In this case, the last nonzero equation is

$$0x_1 + \cdots + 0x_n = 1,$$

which is false for every $[x_1, \ldots, x_n]$. In this case, no vector can be a solution vector; the original system of equations has no solutions at all. We would like to display this fact.

1.2 Consider a system of equations with augmented matrix **A** and suppose **A** can be reduced to an echelon form **B** by Manipulations 1, 2, 3. Then the original system of equations has no solutions if and only if the last column of **B** is one of the distinguished columns.

Another special circumstance that we should notice is the case when there are no x's in Group 2 at all; that is, when all the columns except the last are distinguished. In this case, every x is completely determined, and the system of equations has at most one solution vector. In any other case, some unknowns can be chosen arbitrarily, and we will have either no solutions, as in 1.2, or an infinite number of solutions. In Section 2 and again in Chapter 5, we shall come back to this analysis of how many solutions there are.

Example 2.

$$\begin{aligned}
x_1 + 2x_2 + 3x_3 - 2x_4 \phantom{{}- 2x_5} &= 0,\\
2x_1 + 4x_2 + 6x_3 - 5x_4 - 2x_5 &= -3,\\
2x_1 + 4x_2 + 6x_3 + 8x_5 + x_6 &= 12,\\
5x_4 + 10x_5 + x_6 &= 15.
\end{aligned}$$

The augmented matrix is

$$\begin{pmatrix} 1 & 2 & 3 & -2 & 0 & 0 & 0\\ 2 & 4 & 6 & -5 & -2 & 0 & -3\\ 2 & 4 & 6 & 0 & 8 & 1 & 12\\ 0 & 0 & 0 & 5 & 10 & 1 & 15 \end{pmatrix}.$$

Add -2 times the first row to the second and -2 times the first row to the third to get

$$\begin{pmatrix} 1 & 2 & 3 & -2 & 0 & 0 & 0\\ 0 & 0 & 0 & -1 & -2 & 0 & -3\\ 0 & 0 & 0 & 4 & 8 & 1 & 12\\ 0 & 0 & 0 & 5 & 10 & 1 & 15 \end{pmatrix}.$$

The first three columns are now in final form. Multiply the second row by -1. Add -4 times the second row to the third, and -5 times the second row to the fourth, and 2 times the second row to the first. Finally, add -1 times the resulting third row to the fourth, to get the echelon form

$$\begin{pmatrix} 1 & 2 & 3 & 0 & 4 & 0 & 6 \\ 0 & 0 & 0 & 1 & 2 & 0 & 3 \\ 0 & 0 & 0 & 0 & 0 & 1 & 0 \\ 0 & 0 & 0 & 0 & 0 & 0 & 0 \end{pmatrix}.$$

The corresponding system of equations is

$$\begin{aligned} x_1 + 2x_2 + 3x_3 \quad + 4x_5 \quad &= 6, \\ x_4 + 2x_5 \quad &= 3, \\ x_6 &= 0, \\ 0 &= 0. \end{aligned}$$

Group 1 consists of x_1, x_4, and x_6; Group 2 consists of x_2, x_3, and x_5. The full set of solutions (of both this echelon system and the original system) is described by

$$\begin{aligned} x_1 &= 6 - 2x_2 - 3x_3 - 4x_5, \\ x_4 &= 3 \qquad\qquad\qquad - 2x_5, \\ x_6 &= 0. \end{aligned}$$

Another description of the solutions is: all vectors of the form $[6 - 2x_2 - 3x_3 - 4x_5, x_2, x_3, 3 - 2x_5, x_5, 0]$ where x_2, x_3, x_5 are arbitrary numbers.

Here is a third description of this same set of vectors. Write the vector $[6 - 2x_2 - 3x_3 - 4x_5, x_2, x_3, 3 - 2x_5, x_5, 0]$ as a sum of four vectors, the first including all terms with no x's in them, the second including all terms with just x_2's, the third having only the x_3's, and the fourth all the x_5's:

$$[6, 0, 0, 3, 0, 0] + [-2x_2, x_2, 0, 0, 0, 0] + [-3x_3, 0, x_3, 0, 0, 0] + [-4x_5, 0, 0, -2x_5, x_5, 0].$$

Factoring out scalars x_2, x_3, x_5 from the last three vectors, we have expressed the solutions as

$$[6, 0, 0, 3, 0, 0] + \text{all linear combinations of } [-2, 1, 0, 0, 0, 0], [-3, 0, 1, 0, 0, 0], \text{ and } [-4, 0, 0, -2, 1, 0],$$

which is a more geometric version of the set of solutions—some three-dimensional "plane" in R^6.

1.3 If an m by n matrix **A** is changed to a matrix **B** by Manipulations 1, 2, and 3 (in particular, if **B** is an echelon form associated to **A**), if **A** is the matrix of a linear mapping f and **B** is the matrix of a linear mapping g, then f and g have the same kernel.

Proof. The kernel of f is the set of all vectors $\mathbf{x}$ satisfying $f(\mathbf{x}) = \mathbf{0}$. If we write this vector equation out as in Eq. (1.1) or as in Chapter 3, Eq. (6.1) we get a system of linear equations $\sum_{q=1}^{n} a_{pq}x_q = 0$ for the components $x_1, \ldots, x_n$ of $\mathbf{x}$. The matrix of coefficients is **A**; the augmented matrix is **A** with an extra column of 0's attached on the right edge; denote this augmented matrix by (**A** **O**). Similarly, the kernel of g is the set of all vectors $\mathbf{x}$ that are solutions of the system with augmented matrix (**B** **O**). Manipulations of rows that convert **A** to **B** will convert (**A** **O**) to (**B** **O**) (check that Manipulations 1, 2, and 3 leave that last column of zeros unchanged). By 1.1, the solutions of the system with augmented matrix (**A** **O**) are the same as the solutions of the system with augmented matrix (**B** **O**). Hence the kernel of f is the same as the kernel of g.

Example 3. Find the kernel of the linear mapping from R^4 to R^3 whose matrix is

$$\begin{pmatrix} 2 & 2 & 0 & -1 \\ -2 & 6 & 8 & -11 \\ 2 & 6 & 4 & -7 \end{pmatrix}.$$

This amounts to finding all solution vectors of

$$\begin{aligned} 2x_1 + 2x_2 \qquad\quad - \quad x_4 &= 0, \\ -2x_1 + 6x_2 + 8x_3 - 11x_4 &= 0, \\ 2x_1 + 6x_2 + 4x_3 - \ \ 7x_4 &= 0. \end{aligned}$$

We can appeal to 1.1, however, which tells us first to reduce the augmented matrix and then to solve the equations with that coefficient matrix. Note that 1.1 does not require us to adhere rigidly to the routine spelled out earlier, nor even to reduce all the way to echelon form. Also in a system like this, where the constant terms are all zero, we can just as well manipulate the coefficient matrix instead of the augmented matrix and save writing the extra column of constants; they stay zero under all the manipulations (compare the proof of 1.3).

So we add the first row to the second and add minus the first row to the third to get

$$\begin{pmatrix} 2 & 2 & 0 & -1 \\ 0 & 8 & 8 & -12 \\ 0 & 4 & 4 & -6 \end{pmatrix}.$$

Then add $(-\frac{1}{2})$ times the second row to the third to get a bottom row consisting of zeros. The kernel of this final linear mapping (which is the same as the required kernel) is the set of solutions of

$$\begin{aligned} 2x_1 + 2x_2 \qquad\quad - \quad x_4 &= 0, \\ 8x_2 + 8x_3 - 12x_4 &= 0, \\ 0 &= 0. \end{aligned}$$

We can take x_3 and x_4 arbitrarily and solve for x_1 and x_2:

$$\begin{aligned} x_2 &= -x_3 + \tfrac{3}{2}x_4, \\ x_1 &= \frac{(-2x_2 + x_4)}{2} \\ &= x_3 - x_4. \end{aligned}$$

The required kernel is the set of all

$$[x_3 - x_4, -x_3 + \tfrac{3}{2}x_4, x_3, x_4]$$

or all

$$x_3[1, -1, 1, 0] + x_4[-1, \tfrac{3}{2}, 0, 1]$$

or all linear combinations of

$$[1, -1, 1, 0] \qquad \text{and} \qquad [-1, \tfrac{3}{2}, 0, 1].$$

Example 4. Find the inverse of the matrix

$$\begin{pmatrix} 2 & 1 & 0 \\ 4 & -1 & 5 \\ 1 & -1 & 2 \end{pmatrix}.$$

The given matrix is the matrix of a linear mapping that we shall call f. We are to find the matrix of a linear mapping g such that the statement $\mathbf{y} = f(\mathbf{x})$ is equivalent to $\mathbf{x} = g(\mathbf{y})$. In other words, if

$$\begin{aligned} 2x_1 + x_2 \qquad\quad &= y_1, \\ 4x_1 - x_2 + 5x_3 &= y_2, \\ x_1 - x_2 + 2x_3 &= y_3, \end{aligned}$$

we are asked to express $[x_1, x_2, x_3]$ as a function of $[y_1, y_2, y_3]$; that is, to solve these equations for $\mathbf{x}$. The augmented matrix is

$$\begin{pmatrix} 2 & 1 & 0 & y_1 \\ 4 & -1 & 5 & y_2 \\ 1 & -1 & 2 & y_3 \end{pmatrix}.$$

Interchange the first and last rows, add -4 times the new first row to the second, -2 times this same first row to the third. Then add -1 times the resulting second row to the third, multiply the second row by $\frac{1}{3}$, the third row by -1. Finally, add the second row to the first, -1 times the third row to the first, and the third row to the second, getting

$$\begin{pmatrix} 1 & 0 & 0 & y_1 - \frac{2}{3}y_2 + \frac{5}{3}y_3 \\ 0 & 1 & 0 & -y_1 + \frac{4}{3}y_2 - \frac{10}{3}y_3 \\ 0 & 0 & 1 & -y_1 + y_2 - 2y_3 \end{pmatrix},$$

whose solutions are already displayed by the equations themselves:

$$x_1 = y_1 - \tfrac{2}{3}y_2 + \tfrac{5}{3}y_3,$$

$$x_2 = -y_1 + \tfrac{4}{3}y_2 - \tfrac{10}{3}y_3,$$

$$x_3 = -y_1 + y_2 - 2y_3$$

or

$$\begin{pmatrix} x_1 \\ x_2 \\ x_3 \end{pmatrix} = \begin{pmatrix} 1 & -\frac{2}{3} & \frac{5}{3} \\ -1 & \frac{4}{3} & -\frac{10}{3} \\ -1 & 1 & -2 \end{pmatrix} \begin{pmatrix} y_1 \\ y_2 \\ y_3 \end{pmatrix}.$$

Thus the required inverse is

$$\begin{pmatrix} 1 & -\frac{2}{3} & \frac{5}{3} \\ -1 & \frac{4}{3} & -\frac{10}{3} \\ -1 & 1 & -2 \end{pmatrix}.$$

The product of this with the original matrix should be $\mathbf{I}$.

PROBLEMS

1. Explain why in Manipulation 3 we do not allow multiplications by the scalar 0, whereas in Manipulation 1 we do.

2. Find all solutions of the following systems by reducing the augmented matrix to echelon form. Express your answers in the form:

one vector plus all linear combinations of some other vectors (see Examples 2 and 3).

(a) $$\begin{aligned} 2x + 3y &= 1, \\ x - y &= 2. \end{aligned}$$

(b) $$\begin{aligned} 2x + 3y + 4z &= 0, \\ x + y + z &= 0, \\ 4x + 5y + 6z &= 0. \end{aligned}$$

(c) $$\begin{aligned} 2x + 3y + 4z &= 0, \\ x + y + z &= 1, \\ 4x + 5y + 6z &= 1. \end{aligned}$$

(d) $$\begin{aligned} a_{11}x_1 + a_{12}x_2 &= b_1, \\ a_{21}x_1 + a_{22}x_2 &= b_2, \end{aligned}$$
assuming $a_{11}a_{22} - a_{12}a_{21} \neq 0$.

(e) $$\begin{aligned} 2x_1 + 3x_3 + 4x_4 + x_5 + 5x_6 &= 0, \\ x_1 + x_2 + x_3 + x_4 + x_6 &= 1, \\ x_1 - x_2 + x_3 - x_4 + x_6 &= 1, \\ 3x_1 + x_2 - x_3 + 2x_4 &= 0, \\ x_2 + x_3 - x_6 &= 1. \end{aligned}$$

3. Find the kernels of the following linear mappings by reducing the appropriate matrices to echelon form:

(a) $f(x, y, z) = [2x + y, x - y - z, y + z]$;

(b) $f(x_1, x_2, x_3) = [x_1 + x_2, x_2 + x_3]$;

(c) $f(x_1, x_2, x_3, x_4) = [2x_1 + x_2, x_1 - x_3, x_1 + x_2 + x_4]$;

(d) the linear mapping whose matrix is

$$\begin{pmatrix} 4 & 0 & 6 & 3 \\ 2 & 1 & 1 & 0 \\ 4 & 4 & -2 & -3 \end{pmatrix}.$$

4. Find the inverses of the following matrices:

(a) $\begin{pmatrix} 1 & 3 \\ 4 & 2 \end{pmatrix}$; (b) $\begin{pmatrix} a & 0 \\ b & c \end{pmatrix}$; (c) $\begin{pmatrix} 4 & 6 \\ 6 & 9 \end{pmatrix}$;

(d) $\begin{pmatrix} 1 & 0 & 1 \\ 2 & 1 & 1 \\ -3 & 4 & 5 \end{pmatrix}$; (e) $\begin{pmatrix} a_{11} & a_{12} \\ a_{21} & a_{22} \end{pmatrix}$;

(f) $\begin{pmatrix} 1 & 2 & 3 & 4 \\ 2 & 3 & 4 & 5 \\ 1 & 0 & 0 & 1 \\ 2 & 1 & 2 & 3 \end{pmatrix}$; (g) $\begin{pmatrix} 1 & 0 & 0 & 0 \\ a & 1 & 0 & 0 \\ b & d & 1 & 0 \\ c & e & f & 1 \end{pmatrix}$.

5. Consider the following planes:

(P_1) through the origin and the points (1, 1, 2) and (2, 1, 1);

(P_2) through the origin, perpendicular to $\mathbf{i} + \mathbf{j} + \mathbf{k}$;

(P_3) through (1, 1, 1), perpendicular to $\mathbf{i} + \mathbf{j}$;

(P_4) through (1, 0, 0) and (0, 1, 1), and perpendicular to the xy-plane.

Find the intersection of P_1, P_2, P_3, and P_4; of P_1, P_2, and P_3; of P_1 and P_2.

6. Find the point of intersection (if any) of the following pairs of lines:

(a) first line: $2x + 3y + 4z = 1, \quad x + y + z = 0,$
second line: $x + y - 2z = 0, \quad x - y = -2;$

(b) first line: same as in part (a)
second line: $x + 2y - z = 1, \quad x + z = 1;$

(c) first line: $2x + 3y + az = 4, \quad x - y = 1,$
second line: $3x + y + z = 5, \quad x - y - z = -1.$

In part (c), for what values of a do the lines intersect and for what values of a are they skew? Show that the remaining possibility, that they are parallel, cannot happen.

7. Find an equation of the plane through the points (1, 2, 1), (3, 0, 5), (−2, 2, 2) by the second (nongeometric) method of Example 4, Chapter 2, Section 1.

8. Find the equation of the hyperplane $Ax_1 + Bx_2 + Cx_3 + Dx_4 + E = 0$ through the four points (1, 1, 2, 0), (2, 0, 1, −1), (0, 0, 2, 3), (−1, 1, 1, 2). Use the second method in Example 4, Chapter 2, Section 1.

9. Find all vectors $\mathbf{u}$ such that $f(\mathbf{u}) = \mathbf{v}$, where the matrix of f is

$$\begin{pmatrix} 2 & 3 & 4 \\ 1 & 2 & 3 \\ -2 & -1 & 0 \end{pmatrix}$$

and $\mathbf{v} = \mathbf{i} + \mathbf{j} + \mathbf{k}$. Do the same if $\mathbf{v} = \mathbf{i} + \mathbf{j} - \mathbf{k}$.

10. Show that if $\mathbf{u} = \mathbf{u}_0$ is one vector satisfying the equation $f(\mathbf{u}) = \mathbf{v}$ ($\mathbf{v}$ is a given vector, f is a linear mapping), then all vectors satisfying this equation are given by $\mathbf{u} = \mathbf{u}_0$ + vector in the kernel of f.

11. Consider systems of n linear equations in n unknowns; that is, systems with square coefficient matrix. If we reduce the *coefficient* matrix and get an echelon form with *all* columns distinguished, then show that

(a) the echelon form is the identity matrix;

(b) the original system of equations has exactly one solution (it does not matter how the *augmented* matrix behaves);

(c) the situation envisioned in 1.2 cannot occur.

12. Consider systems of n linear equations in n unknowns, as in Problem 11. If such a system has exactly one solution, show that we must be in the situation described in Problem 11: Every column is distinguished in the echelon form to which the coefficient matrix is reduced.

13. If we have a system of m linear equations in n unknowns and if $m \neq n$, is it possible to have exactly one solution vector? Why?

14. Let f be a linear mapping with matrix

$$\begin{pmatrix} 1 & 2 \\ 0 & 1 \end{pmatrix}.$$

Find all vectors that are not turned by f, that is, all vectors $\mathbf{v}$ such that $f(\mathbf{v})$ is parallel to $\mathbf{v}$.

15. Let f be as in Problem 14 and let λ be an unspecified scalar. Show that for most values of λ, the kernel of $f - \lambda I$ consists of the zero vector alone. For what values of λ is this not the case, and for these λ's what is the kernel of $f - \lambda I$?

16. Show that Problem 15 is just a reformulation of Problem 14, since the statement $f(\mathbf{v}) = \lambda\mathbf{v}$ is the same as the statement $\mathbf{v}$ is in the kernel of $f - \lambda I$.

17. Same as Problems 14 and 15, but with f having the matrix

$$\begin{pmatrix} 2 & 3 \\ 3 & 2 \end{pmatrix}.$$

2. CLOSER ANALYSIS OF THE SOLUTIONS AND THE SOLUTION METHOD

2.1 Definition

If in a system of equations $\sum_q a_{pq}x_q = b_p$ as in Eq. (1.1) the b_p's are all zero, the system is called *homogeneous*. If in a nonhomogeneous system we replace all the b_p's by zero, we get a new system, which is called the *corresponding homogeneous system.*

2.2 Theorem

To get all solutions of a nonhomogeneous system, add one solution (any one solution will do) to all solutions of the corresponding homogeneous system.

Proof. This fact is visible in the solutions to all problems in the preceding section, since these solutions appeared as one vector, say $\mathbf{x}_0$, plus all linear combinations of some other vectors $\mathbf{v}_1, \ldots, \mathbf{v}_t$. Since $\mathbf{0}$ is a linear combination of $\mathbf{v}_1, \ldots, \mathbf{v}_t$, $\mathbf{x}_0 + \mathbf{0} = \mathbf{x}_0$ is indeed one solution of the original system. All solutions equal this solution $\mathbf{x}_0$ plus linear combinations of $\mathbf{v}_1, \ldots, \mathbf{v}_t$. A proof of 2.2 can then be constructed by showing that these linear combinations of $\mathbf{v}_1, \ldots, \mathbf{v}_t$ are exactly the solutions of the corresponding homogeneous system. However, we prefer the following elegant proof.

The solutions of the system of equations are exactly the vectors $\mathbf{x}$ that satisfy $f(\mathbf{x}) = \mathbf{b}$, where f is the linear mapping whose matrix is the matrix of coefficients and $\mathbf{b}$ is the vector whose components form the last column of the augmented matrix (see Eqs. (1.1) and (1.3) for display of notations). Choose one solution vector $\mathbf{x}_0$ (any one will do, but fix it), so $f(\mathbf{x}_0) = \mathbf{b}$. If $\mathbf{x}$ is any other solution vector, that is, if $f(\mathbf{x}) = \mathbf{b}$, then since f is linear, $f(\mathbf{x} - \mathbf{x}_0) = f(\mathbf{x}) - f(\mathbf{x}_0) = \mathbf{b} - \mathbf{b} = \mathbf{0}$. This says that the vector $\mathbf{y} = \mathbf{x} - \mathbf{x}_0$ satisfies the equation $f(\mathbf{y}) = \mathbf{0}$ and $\mathbf{x} = \mathbf{x}_0 + \mathbf{y}$ is the sum of $\mathbf{x}_0$ and a solution of the corresponding homogeneous system. Conversely, if $\mathbf{y}$ is a solution of the corresponding homogeneous system then $\mathbf{x}_0 + \mathbf{y}$ is a solution of the original system, because $f(\mathbf{x}_0 + \mathbf{y}) = f(\mathbf{x}_0) + f(\mathbf{y}) = \mathbf{b} + \mathbf{0} = \mathbf{b}$.

This completes the proof. What we have shown is that the set of vectors obtained by adding $\mathbf{x}_0$ to all the solution vectors of the homogeneous system is the same as the set of all solution vectors of the nonhomogeneous system (we showed that $\mathbf{x}$ is in one of these sets if and only if it is in the other, which is how to show that two sets are the same). This deserves two comments.

First, we have established a one-to-one correspondence between solutions of the nonhomogeneous system (if there are any at all, so that we have one to use for $\mathbf{x}_0$) and the solutions of the corresponding homogeneous system. The correspondence associates to each solution $\mathbf{y}$ of the homogeneous system the solution $\mathbf{x}_0 + \mathbf{y}$ of the nonhomogeneous system. The two systems have the "same number" of solutions in a rather strong sense.

Second, we can visualize geometrically the correspondence in the preceding comment. For example, suppose the system we are discussing is just one equation in three unknowns. The solutions of the nonhomogeneous system are the position vectors of points on a certain plane $a_{11}x_1 + a_{12}x_2 + a_{13}x_3 = b_1$; the solutions of the homogeneous system are the position vectors of the points on the parallel plane through the origin, $a_{11}x_1 + a_{12}x_2 + a_{13}x_3 = 0$. The position vectors of points on the first plane can indeed be obtained by adding one fixed one of them to all the position vectors on the plane through the origin (see Fig. 4.1). If instead we have a system of two equations in three unknowns, the solutions are the position vectors of points on the intersection of two planes; the solutions of the homogeneous system are the position vectors of the points on the intersection of the two planes through the origin that are parallel to the original two planes. If the original two planes do intersect, then the sum of the position vector of a point on the line of intersection plus all position vectors of the intersection of the planes through the origin will again give all the position vectors of the points of intersection of the original planes.

The question of how many solutions a system has is intimately associated with the following number which we encountered in the preceding section. We give this number a name.

2.3 Definition

The *rank* of a matrix is the number of nonzero rows in an associated echelon form. This is the same as the number of distinguished columns in the echelon form.

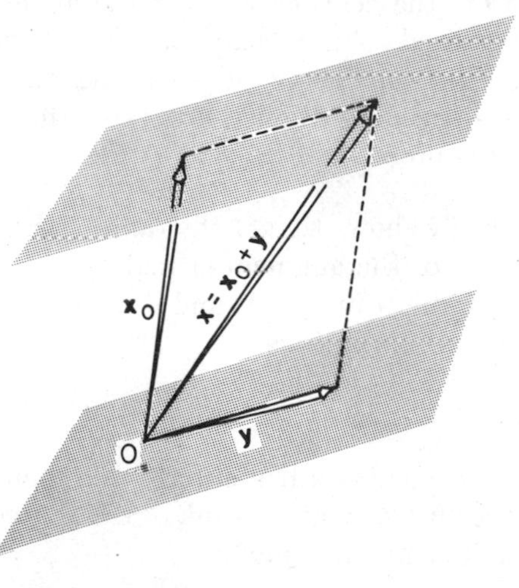

Figure 4.1

This definition has a flaw: It is conceivable that different sequences of manipulations on the same matrix might lead to two different echelon forms, and even two different ranks. It is conceivable, but it is not true; the echelon form is unique. As we have no use for this fact, we do not stop to prove it. We do need the fact that one matrix has only one rank, however, and we prove this fact in the next chapter by showing that the rank of a matrix **A** is equal to "the dimension of the space of all linear combinations of the rows of **A**," which obviously depends only on **A** and not on any possible ambiguities in a reduction process. A more sophisticated treatment of linear algebra would have started with Chapter 5, defined rank as the dimension mentioned above, and used our present definition as a computational scheme for finding the rank.

2.4 The rank of an m by n matrix is $\leq m$ and $\leq n$.

2.5 Theorem

A system of equations has a solution if and only if the rank of the coefficient matrix equals the rank of the augmented matrix.

Proof. According to 1.2, we get no solution if and only if the last distinguished column is the last column in the echelon form of the augmented matrix. Since the echelon form is constructed by improving columns beginning at the left, in the process of reducing the augmented matrix to echelon form we will also be reducing the coefficient matrix to echelon form; if we drop the last column of the former echelon form, we get the latter echelon form. If this last column is distinguished, then the echelon form of the augmented matrix has one more distinguished column than does the echelon form of the coefficient matrix, and conversely. In this case, the rank of the augmented matrix is one more than the rank of the coefficient matrix. This completes the proof.

The proof actually shows that if the rank of the coefficient matrix is r, then the rank of the augmented matrix of the same system of equations will be either r or $r + 1$, and it is exactly in the latter case that the system has no solutions.

2.6 Theorem

A system of linear equations has exactly one solution if and only if the rank of the coefficient matrix, rank of the augmented matrix, and number of unknowns are all equal.

Proof. By 2.5, equality of the two ranks means that there is at least one solution. If these ranks also equal the number of unknowns, then there are as many distinguished columns as unknowns, so all the unknowns are in Group 1. This means we can solve for each of the unknowns; none of the unknowns may be chosen arbitrarily.

2.7 Theorem

Let **A** be a square n by n matrix and let f be the corresponding linear function from R^n to R^n. Then the following statements about **A**, about f, and about the systems of linear equations with coefficient matrix **A** are all equivalent:

(a) The rank of **A** is n.

(b) The echelon form to which **A** can be reduced is the identity matrix.

(c) Every system of equations with **A** as coefficient matrix has one and only one solution.

(d) **A** and f are invertible.

(e) **A** and f have one-sided inverses: there is a matrix **B** such that $\mathbf{BA} = \mathbf{I}$, and a function g such that $g \circ f = I$.

(f) The kernel of f consists of **0** alone.

(In Chapter 6, we add a seventh condition: the determinant of **A** or of f is not zero.)

Proof. We show (a) implies (b) which implies (c) which implies (d) which implies (e) which implies (f) which implies (a) again. This will show all are equivalent.

If **A** has rank n, the corresponding echelon form has n distinguished columns; since it has only n columns all together, these columns are exactly $\mathbf{i}_1, \ldots, \mathbf{i}_n$, that is, the echelon form is **I**.

To show that (b) implies (c), consider any system of equations with **A** as coefficient matrix. The augmented matrix is then **A** with an extra column adjoined on the right edge. Since **A** can be reduced to the n by n identity matrix, its rank is n. The rank of the augmented matrix must be either n or $n + 1$, but since it has only n rows, it cannot have rank $n + 1$. Hence, the hypotheses of 2.6 are satisfied, and the system has exactly one solution.

To prove (d), it is enough to show that f is invertible (then **A** is, too). The hypothesis is now (c), which, translated as in Eqs. (1.1) and (1.3), asserts that for every vector **b** there is one and only one vector **x** with the property $f(\mathbf{x}) = \mathbf{b}$. This then specifies a function associating to every **b** an **x**, and this function is f^{-1} (see Chapter 3, Section 5).

Of course, if f has an inverse g such that $f \circ g = g \circ f = I$, then this g is also a one-sided inverse, that is, $g \circ f = I$.

If $g \circ f = I$ and **x** is in the kernel of f, then $f(\mathbf{x}) = \mathbf{0}$ and so $g(f(\mathbf{x})) = g(\mathbf{0}) = \mathbf{0}$; on the other hand, $g(f(\mathbf{x})) = (g \circ f)(\mathbf{x}) = I(\mathbf{x}) = \mathbf{x}$; these two equalities together imply $\mathbf{x} = \mathbf{0}$.

Finally, we prove (f) implies (a) by looking at the contrapositive form (see the Appendix). Suppose **A** has rank r different from n, and consider the system of homogeneous equations with coefficient matrix **A**. On reducing to echelon form, there will be some columns that are not distinguished (in fact $n - r$ of them, besides the column of zero constants), hence there will be some unknowns that can be chosen arbitrarily; this means that there will be infinitely many solutions; in particular, there will be solutions other than **0**. This says that the kernel of f (which is the same as the set of solutions of these homogeneous equations) does not consist of **0** alone. Hence if (f) is true, then **A** must have rank n. This completes the proof of 2.7.

The reader should notice what we have proved even more closely than he would just from examining the proof. For example, we proved 2.7(d) implies 2.7(e), which is both trivial and dull. But the end result was that 2.7(d) and 2.7(e) are equivalent, so 2.7(e) implies 2.7(d), which is more interesting: if we want to prove that a linear function has an inverse, it is enough to check that it has a one-sided inverse.

Another detail of this kind that is worth noting explicitly is that 2.7(f) implies 2.7(c): If the homogeneous system with coefficient matrix **A** has only the zero solution, then every system (homogeneous or not) with coefficient matrix **A** has a unique solution.

There are two more conditions, each of which is equivalent to 2.7(a)–(f): "The image of f is all of R^n," and "f has a one-sided inverse g in the sense that $f \circ g = I$." It is most convenient to postpone until Chapter 5 (Problem 4, Section 3) the proofs that these are also equivalent to the invertibility of f.

Example 1. For what numbers λ will the matrix

$$\begin{pmatrix} 2-\lambda & 3 \\ 3 & 2-\lambda \end{pmatrix}$$

be invertible? This is equivalent to asking for what numbers λ the matrix has rank 2. We reduce toward echelon form:

$$\begin{pmatrix} 2-\lambda & 3 \\ 3 & 2-\lambda \end{pmatrix} \to \begin{pmatrix} 3 & 2-\lambda \\ 2-\lambda & 3 \end{pmatrix}$$

$$\to \begin{pmatrix} 1 & \dfrac{(2-\lambda)}{3} \\ 2-\lambda & 3 \end{pmatrix} \to \begin{pmatrix} 1 & \dfrac{(2-\lambda)}{3} \\ 0 & 3 - \dfrac{(2-\lambda)^2}{3} \end{pmatrix}.$$

This will reduce to **I** if and only if $3 - (2-\lambda)^2/3 \neq 0$. But $3 - (2-\lambda)^2/3 = 0$ exactly when $2 - \lambda = \pm 3$, so the answer to our question is: all numbers except 5 and -1.

This example is one of an important type that we will deal with more thoroughly in Chapters 6 and 7. The present example comes up, for instance, when we ask which vectors are left unturned (though possibly stretched) by the mapping f whose matrix is

$$\begin{pmatrix} 2 & 3 \\ 3 & 2 \end{pmatrix},$$

because this question involves finding all vectors $\mathbf{x}$ such that $f(\mathbf{x}) = \lambda\mathbf{x}$. This is the same as asking for those $\mathbf{x}$ for which $(f - \lambda I)(\mathbf{x}) = \mathbf{0}$; we are asking for the kernel of $f - \lambda I$. The answer will be "the zero vector and no other" when $f - \lambda I$ is invertible (see 2.7). If we use the results of Example 2.1, we know that a full answer to the question consists of the kernels of $f - 5I$ and $f + I$. (Compare Problems 14–17, Section 1.)

We conclude with two theorems that we will need in the next two chapters. We also get one corollary that is useful immediately.

2.8 Theorem

A homogeneous system of m linear equations in more than m unknowns always has infinitely many solutions. In particular, it has an infinite number of nonzero solutions.

Proof. A look at echelon forms shows that the number of unknowns in Group 1 is at most equal to the number m of equations (actually it is exactly equal to the rank of the matrix of coefficients). In the present circumstances, this means that there is at least one unknown in Group 2. Since the system is homogeneous, solutions do exist (for example, $[0, \ldots, 0]$), and we get infinitely many solutions by choosing arbitrary values for the Group 2 unknowns.

Note the geometrical content of this theorem: If f is a linear mapping from R^n to R^m and if $m < n$, then f has a kernel consisting of more than just the zero vector; if we try to stuff a high-dimensional space into a lower-dimensional space, some vectors are going to get squeezed into $\mathbf{0}$.

2.9 If a matrix $\mathbf{B}$ is obtained from a matrix $\mathbf{A}$ by a sequence of Manipulations 1, 2, and 3, then $\mathbf{B} = \mathbf{PA}$ for some invertible matrix $\mathbf{P}$. This matrix $\mathbf{P}$ is the result of applying the same sequence of manipulations to the identity matrix. This $\mathbf{P}$ is a product of *elementary matrices,* where an elementary matrix is defined as the result of applying one of Manipulations 1, 2, or 3 to the identity matrix.

Proof. It is sufficient to show (1) that if $\mathbf{A}'$ is the result of applying one of Manipulations 1, 2, or 3 to $\mathbf{A}$, and $\mathbf{I}'$ is the result of applying the same manipulation to the m by m identity matrix (here m is the number of rows in $\mathbf{A}$), then $\mathbf{A}' = \mathbf{I}'\mathbf{A}$; and (2) that each $\mathbf{I}'$ is invertible. For then if $\mathbf{B}$ is obtained from $\mathbf{A}$ by a succession of manipulations, repeated use of (1) will say

$$\mathbf{B} = \mathbf{I}'_k(\mathbf{I}'_{k-1}(\cdots\mathbf{I}'_1\mathbf{A})) = \mathbf{PA},$$

with

$$\mathbf{P} = \mathbf{I}'_k\mathbf{I}'_{k-1}\cdots\mathbf{I}'_1;$$

and if each $\mathbf{I}'$ is invertible, so is $\mathbf{P}$; and once we know that for every $\mathbf{A}$, $\mathbf{PA} = \mathbf{I}'_k\cdots\mathbf{I}'_1\mathbf{A}$ is the result of applying this sequence of manipulations to $\mathbf{A}$, we may take $\mathbf{A} = \mathbf{I}$ and find that $\mathbf{I}'_k\cdots\mathbf{I}'_1\mathbf{I} = \mathbf{P}$ is the result of applying this sequence of manipulations to $\mathbf{I}$.

We prove (1) for Manipulation 1 and leave the proofs for Manipulations 2 and 3 to the reader. Let $\mathbf{A}'$ be the result of adding a times the qth row of $\mathbf{A}$ to the pth row, and let $\mathbf{I}'$ be the result of applying the same manipulation to $\mathbf{I}$. We show $\mathbf{A}' = \mathbf{I}'\mathbf{A}$: The rows of $\mathbf{I}$ are $\mathbf{i}_1, \ldots, \mathbf{i}_n$; let the columns of $\mathbf{A}$ be called $\mathbf{v}_1, \ldots, \mathbf{v}_n$. If $r \neq p$, the rth row of $\mathbf{A}'$ is the same as the rth row of $\mathbf{A}$, namely, $[\mathbf{i}_r \cdot \mathbf{v}_1, \ldots, \mathbf{i}_r \cdot \mathbf{v}_n]$; similarly, the rth row of $\mathbf{I}'$ is $\mathbf{i}_r$, so the rth row of $\mathbf{I}'\mathbf{A}$ is $[\mathbf{i}_r \cdot \mathbf{v}_1, \ldots, \mathbf{i}_r \cdot \mathbf{v}_n]$. Thus $\mathbf{A}'$ and $\mathbf{I}'\mathbf{A}$ have all rows equal except possibly the pth. The pth row of $\mathbf{A}'$ is

$$\begin{aligned}
&[\mathbf{i}_p \cdot \mathbf{v}_1, \ldots, \mathbf{i}_p \cdot \mathbf{v}_n] + a[\mathbf{i}_q \cdot \mathbf{v}_1, \ldots, \mathbf{i}_q \cdot \mathbf{v}_n] \\
&\quad = [\mathbf{i}_p \cdot \mathbf{v}_1 + a\mathbf{i}_q \cdot \mathbf{v}_1, \ldots, \mathbf{i}_p \cdot \mathbf{v}_n + a\mathbf{i}_q \cdot \mathbf{v}_n] \\
&\quad = [(\mathbf{i}_p + a\mathbf{i}_q) \cdot \mathbf{v}_1, \ldots, (\mathbf{i}_p + a\mathbf{i}_q) \cdot \mathbf{v}_n];
\end{aligned}$$

but $\mathbf{i}_p + a\mathbf{i}_q$ is the pth row of $\mathbf{I}'$, so the pth rows of $\mathbf{A}'$ and of $\mathbf{I}'\mathbf{A}$ also coincide.

As for (2), we can show each elementary matrix is invertible by direct computation, or by indirection, as follows: If $\mathbf{I}'_1$ is the result of adding a times the qth row of $\mathbf{I}$ to the pth and $\mathbf{I}'_2$ is the result of adding $-a$ times the qth row of $\mathbf{I}$ to the pth, then by (1), $\mathbf{I}'_2(\mathbf{I}'_1)$ is the result of adding $-a$ times the qth row of $\mathbf{I}'_1$ to the pth row of $\mathbf{I}'_1$—which results in the matrix $\mathbf{I}$ again. Thus $\mathbf{I}'_2\mathbf{I}'_1 = \mathbf{I}$ and, similarly, $\mathbf{I}'_1\mathbf{I}'_2 = \mathbf{I}$, so that $\mathbf{I}'_1$ is invertible and its inverse is $\mathbf{I}'_2$. If $\mathbf{I}'_3$ is the result of interchanging two rows of $\mathbf{I}$, the proof is even easier, since $\mathbf{I}'_3\mathbf{I}'_3 = \mathbf{I}$ and $\mathbf{I}'_3$ is its own inverse. The proof for an $\mathbf{I}'$ produced by Manipulation 3 is similar.

2.10 If a matrix $\mathbf{A}$ is invertible and is carried into echelon form by a sequence of manipulations, then this same sequence of manipulations when applied to the identity matrix leads to the inverse $\mathbf{A}^{-1}$.

Proof. We use 2.9: If a sequence of manipulations reduces $\mathbf{A}$ to echelon form, then the result of applying this same sequence to $\mathbf{I}$ is a matrix $\mathbf{P}$ with $\mathbf{PA}$ = the echelon form. But 2.7(b) asserts that the echelon form is $\mathbf{I}$, so $\mathbf{PA} = \mathbf{I}$. Multiply this equation on the right by $\mathbf{A}^{-1}$ to get $\mathbf{PAA}^{-1} = \mathbf{A}^{-1}$, whence $\mathbf{P} = \mathbf{A}^{-1}$.

The following example illustrates a handy way of executing the strategy for finding inverses proposed in 2.10.

Example 2. To find the inverse of

$$\mathbf{A} = \begin{pmatrix} 1 & 2 \\ 3 & 4 \end{pmatrix}$$

we write the identity matrix next to $\mathbf{A}$ to get

$$\begin{pmatrix} 1 & 2 & 1 & 0 \\ 3 & 4 & 0 & 1 \end{pmatrix}$$

and then use manipulations that will carry the first two of these columns (that is, $\mathbf{A}$) to echelon form:

$$\begin{pmatrix} 1 & 2 & 1 & 0 \\ 3 & 4 & 0 & 1 \end{pmatrix} \to \begin{pmatrix} 1 & 2 & 1 & 0 \\ 0 & -2 & -3 & 1 \end{pmatrix}$$
$$\to \begin{pmatrix} 1 & 2 & 1 & 0 \\ 0 & 1 & \frac{3}{2} & -\frac{1}{2} \end{pmatrix} \to \begin{pmatrix} 1 & 0 & -2 & 1 \\ 0 & 1 & \frac{3}{2} & -\frac{1}{2} \end{pmatrix}.$$

The last two columns then form $\mathbf{A}^{-1}$;

$$\mathbf{A}^{-1} = \begin{pmatrix} -2 & 1 \\ \frac{3}{2} & -\frac{1}{2} \end{pmatrix}.$$

This technique is not very different from that of Example 4, Section 1.

PROBLEMS

1. Repeat Problem 4, Section 1 using the method of 2.10.

2. Show that if the rank of a matrix is 0, then the matrix is the zero matrix (all entries are 0).

3. Compute the ranks of the following matrices:

$$\begin{pmatrix} 0 & 1 \\ 0 & 0 \end{pmatrix}; \quad \begin{pmatrix} 1 & 2 \\ 3 & 4 \end{pmatrix}; \quad \begin{pmatrix} 1 & 2 \\ 4 & 8 \end{pmatrix};$$

$$\begin{pmatrix} 1 & 2 & 3 \\ 4 & 5 & 6 \\ 7 & 8 & 9 \end{pmatrix}; \quad \begin{pmatrix} 1 & 2 & 4 \\ 1 & 3 & 9 \\ 1 & 4 & 16 \end{pmatrix}; \quad \begin{pmatrix} 1 & a & a^2 \\ 1 & b & b^2 \\ 1 & c & c^2 \end{pmatrix};$$

$$\begin{pmatrix} 1 & 3 & 0 & 6 \\ 2 & 4 & 0 & 7 \\ 2 & 2 & 1 & 1 \end{pmatrix}; \quad \begin{pmatrix} a \\ b \\ c \end{pmatrix} (d \quad e \quad f) \text{ (matrix product)};$$

$$\begin{pmatrix} 4 & 3 & 1 & 2 & 5 \\ 2 & -1 & 1 & 0 & -1 \\ 3 & -4 & 2 & -1 & -5 \\ 11 & 2 & 4 & 3 & 5 \end{pmatrix}.$$

4. By computing ranks, tell which of the following systems have solutions. Tell which have unique solutions.

(a) $$\begin{aligned} 2x + 3y + 4z &= 5, \\ x + y + z &= 2, \\ 3x + 5y + 7z &= 8. \end{aligned}$$

(b) $$\begin{aligned} 2x + 3y + 4z &= 5, \\ x + y + z &= 2, \\ 3x + 5y + 7z &= 7. \end{aligned}$$

(c) $$\begin{aligned} 2x + 3y + 4z &= a, \\ x + y + z &= b, \\ 3x + 5y + 7z &= c. \end{aligned}$$

(d) $$\begin{aligned} 2x + 3y + 4z &= 0, \\ x + y + z &= 0. \end{aligned}$$

(e) $$\begin{aligned} 2x + 3y + 4z &= a, \\ x + y + z &= b. \end{aligned}$$

(f) $$\begin{aligned} 2x + 2y + 2z &= 0, \\ x + y + z &= 0. \end{aligned}$$

(g) $$\begin{aligned} 2x + 2y + 2z &= a, \\ x + y + z &= b. \end{aligned}$$

(h) $$\begin{aligned} z + y &= 0, \\ x + 2y &= 0, \\ 2x + y &= 0. \end{aligned}$$

(i) $x + y = 1,$ $x + 2y = 1,$ $2x + y = 1.$

(j) $x + y = 1,$ $x + 2y = 1,$ $2x + y = a.$

5. Each of the following columns gives data on a system of linear equations. Tell whether these data predict for the system (α) no solutions, (β) one solution, (γ) infinitely many solutions, (δ) the data are impossible (contradictory), or (ϵ) the data are insufficient to determine the number of solutions.

Number of equations	4	4	5	5	5	5
Number of unknowns	3	3	3	3	4	4
Rank of coefficient matrix	3	3	3	3	3	3
Rank of augmented matrix	3	4	4	2	3	4

6. Why does the number of equations not enter into the statement of 2.6? Give examples of systems of equations with exactly one solution but with more equations than unknowns.

7. Argue that the number of distinguished columns in an echelon form is the same as the number of the last nonzero row; that is, if r is the former number, argue that the rth row is not zero but that the $(r + 1)$st, $(r + 2)$nd, ..., nth rows have only zero entries.

8. For what numbers λ will each of the following matrices be invertible?

$$\begin{pmatrix} 1-\lambda & 2 \\ 2 & 1-\lambda \end{pmatrix}; \quad \begin{pmatrix} 1 & 1 \\ 0 & 1 \end{pmatrix} - \lambda \mathbf{I};$$

$$\begin{pmatrix} 1-\lambda & 2 & 3 \\ 0 & 1-\lambda & 4 \\ 0 & 0 & 1-\lambda \end{pmatrix}; \quad \begin{pmatrix} 1-\lambda & 1 & 1 \\ 0 & 2-\lambda & 3 \\ 0 & 3 & 2-\lambda \end{pmatrix}.$$

9. Use the results and the calculations of Problem 8 to find all vectors that are not turned by mappings with each of the following matrices:

$$\begin{pmatrix} 1 & 2 \\ 2 & 1 \end{pmatrix}; \quad \begin{pmatrix} 1 & 1 \\ 0 & 1 \end{pmatrix}; \quad \begin{pmatrix} 1 & 2 & 3 \\ 0 & 1 & 4 \\ 0 & 0 & 1 \end{pmatrix}; \quad \begin{pmatrix} 1 & 1 & 1 \\ 0 & 2 & 3 \\ 0 & 3 & 2 \end{pmatrix}.$$

10. Show that if $\mathbf{A}$ and $\mathbf{B}$ are n by n matrices and $\mathbf{A}$ is invertible, then the rank of $\mathbf{AB}$ equals the rank of $\mathbf{B}$. Use 2.9.

11. Show that every invertible matrix is a product of elementary matrices (see 2.9).

12. Pick one point on the plane $2x + 3y + 4z = 6$, for example, the point $(0, 2, 0)$. Show that all the position vectors of points on this plane are expressible as $[0, 2, 0]$ + position vectors of points on the plane $2x + 3y + 4z = 0$. Do *not* use 2.2. Repeat, using a different point on the plane.

13. Consider the line of intersection of the planes $2x + 3y + 4z = 6$ and $x + y + z = 2$.

(a) Find the point on this line where $z = 0$.

(b) Find a vector parallel to the line.

(c) Use parts (a) and (b) to find parametric equations of the line.

(d) Use the methods of this chapter to solve the given pair of equations, and compare with part (c).

(e) Find parametric equations of the line of intersection of $2x + 3y + 4z = 0$ and $x + y + z = 0$. Verify 2.2.

CHAPTER

5

Dimension

1. SUBSPACES

1.1 Definition

A *subspace* of R^n is a nonempty set S of vectors in R^n with the properties

(a) whenever $\mathbf{x}$ and $\mathbf{y}$ are vectors in S, so is $\mathbf{x} + \mathbf{y}$;

(b) whenever $\mathbf{x}$ is a vector in S, so is $a\mathbf{x}$ for every scalar a.

Example 1. The set of all vectors in R^n is a subspace of R^n.

Example 2. The set of all vectors in a plane through the origin in R^3 is a subspace because $\mathbf{x} + \mathbf{y}$ is in the same plane as $\mathbf{x}$ and $\mathbf{y}$, and so is $a\mathbf{x}$.

Example 3. The set of all vectors on a line through the origin in R^3 is a subspace of R^3.

Example 4. The set consisting of $\mathbf{0}$ alone is a subspace of R^n because $\mathbf{0} + \mathbf{0} = \mathbf{0}$ and $a\mathbf{0} = \mathbf{0}$.

Nonexample 5. (a) The set of all position vectors of points on a plane not through the origin is not a subspace, because $\mathbf{0}$ is not in this set; but every subspace must contain $\mathbf{0}$: if $\mathbf{x}$ is one vector in the subspace, then so is $0\mathbf{x}$ by 1.1(a), and $0\mathbf{x} = \mathbf{0}$.

(b) The set of all position vectors of the points on a parabola is not a subspace. We leave to the reader the arguments that 1.1(a) fails. In fact, both 1.1(a) and 1.1(b) fail, but as soon as one of them fails, the set of vectors in question is not a subspace.

Example 6. If $\mathbf{u}$ and $\mathbf{v}$ are two fixed vectors in R^n, then the set of all linear combinations of $\mathbf{u}$ and $\mathbf{v}$ is a subspace of R^n, because, for any scalars b, c, b', c',

$$(b\mathbf{u} + c\mathbf{v}) + (b'\mathbf{u} + c'\mathbf{v}) = (b + b')\mathbf{u} + (c + c')\mathbf{v},$$

which is a linear combination of $\mathbf{u}$ and $\mathbf{v}$, and

$$a(b\mathbf{u} + c\mathbf{v}) = (ab)\mathbf{u} + (ac)\mathbf{v},$$

which is also a linear combination of $\mathbf{u}$ and $\mathbf{v}$. (If $n = 3$, this is a repeat of Example 2. Why?)

Example 7. If $\mathbf{u}_1, \ldots, \mathbf{u}_r$ are fixed vectors in R^n, the set of all linear combinations of $\mathbf{u}_1, \ldots, \mathbf{u}_r$ is a subspace of R^n. The proof is left to the reader.

Example 8. The kernel of a linear mapping $f\colon R^n \to R^m$ is a subspace of R^n because if $f(\mathbf{x}) = \mathbf{0}$ and $f(\mathbf{y}) = \mathbf{0}$, then

$$f(\mathbf{x} + \mathbf{y}) = f(\mathbf{x}) + f(\mathbf{y}) = \mathbf{0} + \mathbf{0} = \mathbf{0}$$

and

$$f(a\mathbf{x}) = a\mathbf{0} = \mathbf{0},$$

so that $\mathbf{x} + \mathbf{y}$ and $a\mathbf{x}$ are also in the kernel of f.

Example 9. The image of a linear mapping $f\colon R^n \to R^m$ is a subspace of R^m because if $f(\mathbf{x})$ and $f(\mathbf{y})$ are any two vectors in the image of f (all vectors in the image are so expressible), then

$$f(\mathbf{x}) + f(\mathbf{y}) = f(\mathbf{x} + \mathbf{y}),$$

which is also in the image of f, and $af(\mathbf{x})$ is, too, because it equals $f(a\mathbf{x})$.

We shall soon show that all subspaces can be described as in Example 7. Example 8 is not so described, but it can be: The kernel of f is the set of solutions of a system of homogeneous equations (whose coefficient matrix is the same as the matrix of f), and the solution methods of the preceding chapter are exactly designed to express these solutions as the set of all linear combinations of a few of the solution vectors.

We introduce some handy terminology for these ideas.

1.2 Definition

If $\mathbf{u}_1, \ldots, \mathbf{u}_r$ are vectors in R^n, the set of all linear combinations of them is called the *subspace* (of R^n) *spanned by* $\mathbf{u}_1, \ldots, \mathbf{u}_r$ (often abbreviated to the *space spanned by* $\mathbf{u}_1, \ldots, \mathbf{u}_r$). The set $\{\mathbf{u}_1, \ldots, \mathbf{u}_r\}$ is called a *spanning set* of this subspace.

PROBLEMS

1. Verify that the set of vectors in Example 7 is a subspace of R^n.

2. Consider the set of all vectors in R^3 that are perpendicular to $\mathbf{i} + \mathbf{j} + \mathbf{k}$. Is this a subspace? If so, find a spanning set.

3. Consider the set of solutions of the nonhomogeneous system

$$2x + 3y + 4z = 1,$$

$$x - y + 2z = 2.$$

Do they form a subspace of R^3? If so, find a spanning set.

4. Under what conditions will the solutions of a system of linear equations $\sum_{q=1}^{n} a_{pq}x_q = b_p$ $(p = 1, \ldots, m)$ form a subspace? Subspace of what?

5. Find a spanning set of the solution space of the system

$$2x + 3y + 4z = 0,$$

$$x - y + z = 0.$$

6. Find a spanning set of the space of solutions of

$$2x - y + 3z + t = 0,$$

$$2x + y + 3z + t = 0,$$

$$4x + 2y + z - t = 0.$$

7. Find a spanning set of the kernel of the linear mapping whose matrix is

$$\begin{pmatrix} 1 & 1 & 1 \\ 2 & 4 & 8 \\ 3 & 9 & 27 \end{pmatrix}.$$

8. What is a spanning set of the kernel of the linear mapping from R^3 to R^1 that associates to each vector its own first component?

9. Show that the space spanned by $\mathbf{u}_1, \ldots, \mathbf{u}_r$ and $\mathbf{0}$ is the same as the space spanned by $\mathbf{u}_1, \ldots, \mathbf{u}_r$.

10. Show that the space spanned by the rows of a matrix is the same as the space spanned by the rows of the corresponding echelon form.

11. Show that $\mathbf{u}_1, \ldots, \mathbf{u}_r$ and $\mathbf{v}_1, \ldots, \mathbf{v}_s$ span the same subspace of R^n if and only if both (1) each $\mathbf{u}_p$ ($p = 1, \ldots, r$) is a linear combination of $\mathbf{v}_1, \ldots, \mathbf{v}_s$ and (2) each $\mathbf{v}_q$ is a linear combination of $\mathbf{u}_1, \ldots, \mathbf{u}_r$ ($q = 1, \ldots, s$).

12. What is the space spanned by $\mathbf{i}$, $\mathbf{j}$, and $\mathbf{k}$? By $\mathbf{i} + \mathbf{j}$ and $\mathbf{i} - \mathbf{j}$? By $\mathbf{i}$, $\mathbf{i} + \mathbf{j}$, and $\mathbf{i} - \mathbf{j}$? By $\mathbf{i}$, $\mathbf{i} + \mathbf{j}$, and $\mathbf{i} + \mathbf{j} + 2\mathbf{k}$? Where possible, describe the subspace geometrically.

13. Can the space spanned by $[1, 2, 0, -1]$, $[2, 4, 0, 2]$, and $[-3, -6, 0, 3]$ be spanned by fewer than three vectors?

14. If $\mathbf{u}$, $\mathbf{v}$, and $\mathbf{w}$ are noncoplanar vectors in R^3, what is the space spanned by them?

15. If $\mathbf{u}$, $\mathbf{v}$, and $\mathbf{w}$ are coplanar vectors in R^3 (but not all parallel), what is the space spanned by them?

16. If $\mathbf{u}$ and $\mathbf{v}$ are two vectors in R^3 that are not parallel, what is the space they span?

17. If $\mathbf{u}$ and $\mathbf{v}$ are parallel vectors in R^3 (but not both $\mathbf{0}$), what is the space they span?

18. If $\mathbf{u}$ is a nonzero vector in R^3, what is the space it spans? What if $\mathbf{u} = \mathbf{0}$?

19. Consider the subspace of R^4 spanned by the vectors $[-1, 1, -3, -3]$, $[7, 2, 6, 0]$, $[-1, 4, -8, -10]$, $[4, 2, 2, -2]$. Find another spanning set of this subspace with the smallest possible number of vectors in it. Show that the spanning set you have is really the smallest possible. You can use Problem 10.

20. Show that if a set of vectors in a subspace S contains a spanning set of S, then it is also a spanning set of S.

21. Consider the space spanned by [1, 2, 0], [2, 1, 0], and [1, 0, 0]. Can it be spanned by fewer than three vectors?

22. Can the space spanned by [1, 0, 0], [0, 1, 0], and [0, 0, 1] be spanned by fewer than three vectors?

23. Can the space spanned by [1, 2, 3, 4], [5, 6, 7, 8], and [9, 10, 11, 12] be spanned by fewer than three vectors?

24. Show that if $\{\mathbf{u}_1, \ldots, \mathbf{u}_r\}$ is a spanning set of a certain subspace, then $\{\mathbf{u}_2, \ldots, \mathbf{u}_r\}$ spans the same subspace (that is, $\mathbf{u}_1$ is redundant) if and only if $\mathbf{u}_1$ is a linear combination of $\mathbf{u}_2, \ldots, \mathbf{u}_r$.

25. If $\mathbf{u}_1, \ldots, \mathbf{u}_r$ is any set of vectors in R^n (it could even be an infinite set; for example, the set of all vectors along a line) and if V is the set of all vectors that are orthogonal to every one of the $\mathbf{u}_p$, show that V is a subspace of R^n.

2. DIMENSION

2.1 Definition

If a subspace S of R^n is spanned by r vectors, but cannot be spanned by fewer than r vectors, we say that S has *dimension* r, or is *r-dimensional*. Any spanning set consisting of exactly r vectors is called a *basis* of S. We make the convention that the subspace consisting of $\mathbf{0}$ alone has dimension zero and that its only basis is the empty set consisting of no vectors.

Example 1. The set of vectors in a plane through the origin in R^3 is a two-dimensional subspace of R^3 because it can be spanned by two vectors (Problem 16, Section 1) but not by one (Problem 18, Section 1).

2.2 R^3 is three-dimensional because it can be spanned by $\mathbf{i}$, $\mathbf{j}$, and $\mathbf{k}$ but cannot be spanned by any two vectors (two vectors span at most a plane).

In 2.8 we shall prove that the dimension of R^n is n for all n. To do this, we first devise a technique for computing dimensions.

We need some way of telling whether or not a subspace can be spanned by one vector, by two vectors, and so on. This amounts to the question: Suppose we know a set of vectors spanning the subspace; how can we tell if they form a basis? That is, how can we tell if the number of them is really the dimension of the subspace? Two criteria are now given.

2.3 Theorem

The following conditions on a set of vectors $\{\mathbf{u}_1, \ldots, \mathbf{u}_r\}$ are equivalent to each other:

(a) $\mathbf{u}_1, \ldots, \mathbf{u}_r$ form a basis of the space they span.

(b) No one of the $\mathbf{u}$'s is a linear combination of the others.

(c) If $a_1\mathbf{u}_1 + \cdots + a_r\mathbf{u}_r = \mathbf{0}$, then $a_1 = \cdots = a_r = 0$.

Proof. We show (a) implies (b), (b) implies (c), and (c) implies (a). This will complete a circle that shows that each one of (a), (b), and (c) implies the others. In fact, in all three proofs we prove the contrapositive form (see Appendix).

First suppose (b) fails—one of the $\mathbf{u}$'s, say $\mathbf{u}_1$, is a linear combination of the others: $\mathbf{u}_1 = a_2\mathbf{u}_2 + \cdots + a_r\mathbf{u}_r$. Then every linear combination of $\mathbf{u}_1, \ldots, \mathbf{u}_r$ is a linear combination of $\mathbf{u}_2, \ldots, \mathbf{u}_r$ because

$$\sum_{p=1}^{r} b_p\mathbf{u}_p = b_1(a_2\mathbf{u}_2 + \cdots + a_r\mathbf{u}_r) + b_2\mathbf{u}_2 + \cdots + b_r\mathbf{u}_r$$

and we may multiply out the parenthesis and collect coefficients of the $\mathbf{u}$'s. Thus every vector in the space spanned by $\mathbf{u}_1, \ldots, \mathbf{u}_r$ is also in the space spanned by $\mathbf{u}_2, \ldots, \mathbf{u}_r$ (for the last part of this argument, compare Problems 11 and 24, Section 1 or Problem 28, Chapter 1, Section 5). This says that the former space is spanned by $r - 1$ vectors, so $\mathbf{u}_1, \ldots, \mathbf{u}_r$ cannot form a basis. We have proved that (a) fails, so (a) implies (b).

Next we show that if (c) fails, then (b) fails. The failure of (c) means that there exist some $a_1, \ldots, a_r$ not all zero such that $\sum a_p\mathbf{u}_p = \mathbf{0}$ (see the Appendix). Suppose, for example, a_1 is not zero (we leave it to the reader to supply the analogous proof when it is some other a_p that is known to be nonzero). Then we can solve $\sum a_p\mathbf{u}_p = \mathbf{0}$ for $\mathbf{u}_1$, getting $\mathbf{u}_1 = \sum_{p=2}^{r} (-a_p/a_1)\mathbf{u}_p$, so that (b) fails. Thus (b) implies (c).

Finally, we assume (a) fails and prove that (c) fails. Let S denote the subspace spanned by $\mathbf{u}_1, \ldots, \mathbf{u}_r$. If (a) fails, then S can be spanned by $r - 1$ vectors, say $\mathbf{v}_1, \ldots, \mathbf{v}_{r-1}$; all linear combinations of $\mathbf{u}_1, \ldots, \mathbf{u}_r$ will be linear combinations of $\mathbf{v}_1, \ldots, \mathbf{v}_{r-1}$. In particular, $\mathbf{u}_1, \ldots, \mathbf{u}_r$

will be linear combinations of $\mathbf{v}_1, \ldots, \mathbf{v}_{r-1}$:

$$\mathbf{u}_p = \sum_{q=1}^{r-1} c_{pq}\mathbf{v}_q \qquad (p = 1, \ldots, r),$$

where $c_{11}, \ldots, c_{r,r-1}$ are scalars.

We aim to produce scalars $a_1, \ldots, a_r$, not all zero, such that $\sum a_p\mathbf{u}_p = \mathbf{0}$. To do this it is enough to find a's satisfying

$$\sum_p a_p c_{pq} = 0 \qquad (q = 1, \ldots, r-1), \tag{2.1}$$

$$[a_1, \ldots, a_r] \neq [0, \ldots, 0]$$

because then

$$\begin{aligned} \sum_p a_p\mathbf{u}_p &= \sum_p a_p\Big(\sum_q c_{pq}\mathbf{v}_q\Big) \\ &= \sum_q \Big(\sum_p a_p c_{pq}\Big)\mathbf{v}_q \\ &= 0\mathbf{v}_1 + \cdots + 0\mathbf{v}_{r-1} \\ &= \mathbf{0}. \end{aligned}$$

But Eq. (2.1) is a homogeneous system of $r - 1$ equations in r unknowns $[a_1, \ldots, a_r]$, so there is always a nonzero solution by Chapter 4, Theorem 2.8.

2.4 Definition

Vectors $\mathbf{u}_1, \ldots, \mathbf{u}_r$ in R^n are said to be *linearly independent,* or just *independent,* if they satisfy one (and hence all) of the conditions 2.3(a), (b), and (c).

The moral of all this is *to find the dimension of a subspace, find an independent spanning set.* The number of vectors in this set is the dimension.

This condition of independence is important. In n-space it is a substitute for some very simple geometric conditions on vectors in 3-space (see 2.5, 2.6, and 2.7). It is logically a little intricate because it can be thought of in two ways, 2.3(b) and 2.3(c), and because each way involves several logical connectives. Therefore we pause to examine the condition from a few other viewpoints. We begin with special cases.

2.5 Two vectors in 3-space are independent if and only if they are not parallel, because 2.3(b) says neither one may be a scalar times the other, and Chapter 1, 4.4 says this means the vectors are not parallel.

2.6 Three vectors in 3-space are independent if and only if they are not coplanar, because 2.3(b) again says that no one may be in the plane spanned by the other two. It is not enough that every two of these three vectors be independent; three coplanar vectors such as **i**, **j**, and **i** + **j** can easily have the property that no two of them are parallel, yet the fact that they are coplanar spoils their independence.

2.7 One vector is independent if and only if it is not zero.

Condition 2.3(c) is clear enough: $a\mathbf{u} = \mathbf{0}$ only if $a = 0$; this means $\mathbf{u} \neq \mathbf{0}$. (We can also use 2.3(b) if we define what we mean by the space spanned by an empty set of vectors; it is the set consisting of $\mathbf{0}$ alone.)

Example 2. Do the vectors $[2, 4, -1]$, $[1, 0, 2]$, $[3, 7, 0]$ form an independent set? 2.3(c) says we must see whether

$$a_1[2, 4, -1] + a_2[1, 0, 2] + a_3[3, 7, 0] = [0, 0, 0] \qquad (2.2)$$

implies $a_1 = a_2 = a_3 = 0$. We find *all* numbers a_1, a_2, a_3 satisfying Eq. (2.2); this means that we find all a_1, a_2, a_3 satisfying

$$[2a_1 + a_2 + 3a_3, 4a_1 + 7a_3, -a_1 + 2a_2] = [0, 0, 0];$$

that is,

$$\begin{aligned} 2a_1 + a_2 + 3a_3 &= 0, \\ 4a_1 \qquad + 7a_3 &= 0, \\ -a_1 + 2a_2 \qquad &= 0. \end{aligned}$$

As in Chapter 4, we do this by reducing the augmented matrix to echelon form

$$\begin{pmatrix} 2 & 1 & 3 & 0 \\ 4 & 0 & 7 & 0 \\ -1 & 2 & 0 & 0 \end{pmatrix} \to \begin{pmatrix} -1 & 2 & 0 & 0 \\ 4 & 0 & 7 & 0 \\ 2 & 1 & 3 & 0 \end{pmatrix}$$

$$\to \begin{pmatrix} 1 & -2 & 0 & 0 \\ 0 & 8 & 7 & 0 \\ 0 & 5 & 3 & 0 \end{pmatrix} \to \begin{pmatrix} 1 & -2 & 0 & 0 \\ 0 & 1 & \frac{7}{8} & 0 \\ 0 & 0 & -\frac{11}{8} & 0 \end{pmatrix}$$

$$\to \begin{pmatrix} 1 & 0 & 0 & 0 \\ 0 & 1 & 0 & 0 \\ 0 & 0 & 1 & 0 \end{pmatrix}$$

so a_1, a_2, a_3 satisfy Eq. (2.2) if and only if they are a solution of

$$\begin{aligned} a_1 \qquad\qquad &= 0, \\ a_2 \qquad &= 0, \\ a_3 &= 0. \end{aligned}$$

This implies that $a_1 = a_2 = a_3 = 0$; the original set of three vectors is an independent set.

The calculation in Example 2 gives a general, computational scheme for detecting independent sets of vectors:

2.8 Given vectors $\mathbf{v}_1, \ldots, \mathbf{v}_r$ in R^n, construct the n by r matrix whose columns are the given vectors. Reduce to echelon form. Then the $\mathbf{v}$'s are independent if and only if all the columns in the echelon form are distinguished, that is, if and only if the original matrix (or the echelon form) has rank r.

Proof. All the columns in the echelon form are distinguished if and only if the homogeneous system of equations with this matrix of coefficients has only the zero solution. But as in Example 2, this system of scalar equations is the system obtained by writing the components of the vector equation

$$a_1\mathbf{v}_1 + \cdots + a_r\mathbf{v}_r = 0$$

(unknowns $a_1, \ldots, a_r$). This vector equation has only the solution $a_1 = a_2 = \cdots = a_r = 0$ if and only if $\mathbf{v}_1, \mathbf{v}_2, \ldots, \mathbf{v}_r$ are independent (2.3(c)).

2.9 R^n is n-dimensional.

Proof. A spanning set is $\mathbf{i}_1, \mathbf{i}_2, \ldots, \mathbf{i}_n$. But this is in fact an independent spanning set, because if

$$a_1\mathbf{i}_1 + \cdots + a_n\mathbf{i}_n = 0,$$

then

$$a_1[1, 0, \ldots, 0] + a_2[0, 1, 0, \ldots, 0] + \cdots + a_n[0, 0, \ldots, 0, 1] = \mathbf{0},$$

which means

$$[a_1, a_2, \ldots, a_n] = [0, 0, \ldots, 0]$$

and so

$$a_1 = a_2 = \cdots = a_n = 0.$$

Then 2.3 asserts that the n vectors $\mathbf{i}_1, \ldots, \mathbf{i}_n$ are independent, that they form a basis of R^n, and that the dimension of R^n is n. See also Problem 2 and the next section.

The proof of 2.9 also gives us a little more intuition about the concept of a basis. Vectors that form a basis of a subspace are to be thought of as describing a coordinate system in the subspace, just as the vectors $\mathbf{i}_1, \ldots, \mathbf{i}_n$ describe a coordinate system in R^n. In Problem 13 we give a further reinforcement of this analogy between "basis" and "coordinate system."

To develop a little more familiarity with independent sets, let us list a few things they must never contain.

2.10 An independent set never contains $\mathbf{0}$, because $\mathbf{0}$ is a linear combination of every set of vectors, so 2.3(b) could not hold.

An independent set never contains two equal vectors, nor two vectors that are scalar multiples of one another, because once again 2.3(b) could not hold.

Most generally, an independent set cannot contain any set of vectors that is not itself an independent set. For, suppose $\{\mathbf{u}_1, \ldots, \mathbf{u}_r\}$ is a set of vectors and the set $\{\mathbf{u}_1, \ldots, \mathbf{u}_s\}$ is not independent for some $s \leq r$. Then $a_1\mathbf{u}_1 + \cdots + a_s\mathbf{u}_s = \mathbf{0}$ while the coefficients $a_1, \ldots, a_s$ are not all zero. But then

$$a_1\mathbf{u}_1 + \cdots + a_2\mathbf{u}_s + 0\mathbf{u}_{s+1} + \cdots + 0\mathbf{u}_r = 0,$$

and these coefficients are not all zero, either.

Next, a few comments on the relationship between 2.3(b) and 2.3(c) are appropriate. Although both are concerned with linear combinations, note that 2.3(b) says nothing about whether or not the coefficients are nonzero, whereas the condition that the coefficients in 2.3(c) are not all zero is the heart of that version of independence. Perhaps the difference can be remembered best if we give a very short direct proof that 2.3(c) implies 2.3(b), again in the contrapositive form: If 2.3(b) does not hold, then one of the $\mathbf{u}$'s, say $\mathbf{u}_1$, is a linear combination of the others: $\mathbf{u}_1 = b_2\mathbf{u}_2 + \cdots + b_r\mathbf{u}_r$. It does not matter here how many or how few of the b's are zero, since we subtract $\mathbf{u}_1$ from both sides and get $(-1)\mathbf{u}_1 + b_2\mathbf{u}_2 + \cdots + b_r\mathbf{u}_r = \mathbf{0}$, and these coefficients are not all zero, because the first is -1.

Notice also that in the examples, 2.3(b) was used mostly in discussing generalities like 2.10. When it comes to specific computations like the proof of 2.9, 2.3(c) is preferable, partly because it is more symmetric (it does not single out "one of the **u**'s"), but mainly because the computation is more direct; it is not necessary to consider the r separate cases where $\mathbf{u}_1$ is a linear combination of $\mathbf{u}_2, \ldots, \mathbf{u}_r$, where $\mathbf{u}_2$ is a linear combination of $\mathbf{u}_1, \mathbf{u}_3, \ldots, \mathbf{u}_r$, and so on.

By now the reader will be aware that it is about as important to recognize the negation of 2.3(c) as it is to recognize 2.3(c) itself (see the Appendix): If a set of vectors is not independent we say it is a *dependent* set. A set is dependent if and only if there are scalars $a_1, \ldots, a_r$ that are not all zero, but are such that $a_1\mathbf{u}_1 + \cdots + a_r\mathbf{u}_r = \mathbf{0}$.

Then 2.5–2.7 can also read: Two vectors in 3-space are dependent if and only if they are parallel, three vectors in 3-space are dependent if and only if they are coplanar, and one vector (in any space) is dependent if and only if it is zero.

It is something of a pity that the important circumstance—the normal and usual circumstance—namely, independence, should carry the name with the negative prefix, whereas the degenerate special case should be described by the positive word, dependence. This is also true in the political uses of these words, however, and it is true in the special cases in 2.5–2.7, where the unusual circumstances are described by the words "parallel," "coplanar," and "zero."

2.11 Theorem

If T is an m-dimensional space, then every set of more than m vectors in T will always be dependent.

Proof. This is almost a repetition of the proof that 2.3(c) implies 2.3(a). Let $\mathbf{v}_1, \ldots, \mathbf{v}_m$ be a basis of T and let $\mathbf{u}_1, \ldots, \mathbf{u}_n$ be any n vectors in T and assume $n > m$. Then each $\mathbf{u}$ is a linear combination of the $\mathbf{v}$'s:

$$\mathbf{u}_p = c_{p1}\mathbf{v}_1 + \cdots + c_{pm}\mathbf{v}_m \qquad (p = 1, \ldots, n)$$

for some scalars c_{pq}. By Theorem 2.8, Chapter 4 again, we can find numbers $a_1, \ldots, a_n$ not all zero such that $\sum_{p=1}^{n} a_p c_{pq} = 0$ for all q, so that $\sum_p a_p \mathbf{u}_p = \mathbf{0}$, that is, $\mathbf{u}_1, \ldots, \mathbf{u}_n$ are dependent.

2.12 Four or more vectors in R^3 are never independent. More than n vectors in R^n are never independent.

Proof. Use 2.11 and 2.9. Note that 2.12, when written out completely, becomes a repetition of Theorem 2.8, Chapter 4.

2.13 Theorem

If S and T are subspaces of R^n and S is contained in T (we say that S is a subspace of T), then the dimension of S is $\leq$ the dimension of T.

Proof. Let the dimension of T be m and the dimension of S be r. Take any basis of $\mathbf{u}_1, \ldots, \mathbf{u}_r$ of S. Then $\mathbf{u}_1, \ldots, \mathbf{u}_r$ are also vectors in T and, since they form a basis of S, they are independent. By 2.11, there cannot be more than m of them, that is, $r \leq m$. See Problem 11, Section 4 for more information.

PROBLEMS

1. Repeat Problems 16–28, Chapter 1, Section 5, rephrasing them in terms of subspaces, spanning sets, and independent sets.

2. Show that if vectors $\mathbf{u}_1, \ldots, \mathbf{u}_n$ are orthogonal ($\mathbf{u}_p \cdot \mathbf{u}_q = 0$ whenever $p \neq q$) and nonzero (each $\mathbf{u}_p \neq \mathbf{0}$), then they are independent. Show by counterexample that the converse is false.

3. Let f be a linear mapping from R^n to R^m and let $\mathbf{u}_1, \ldots, \mathbf{u}_r$ be vectors in R^n. Show that if $f(\mathbf{u}_1), \ldots, f(\mathbf{u}_r)$ are independent, then so are $\mathbf{u}_1, \ldots, \mathbf{u}_r$, but not conversely (the converse just requires a counterexample).

4. Find the dimension of the subspace of R^n spanned by each of the following sets. Which sets are bases of the subspaces they span?

(a) $[1, 2, 3], [4, 5, 6], [7, 8, 9]$;

(b) $[1, 2, 3], [1, 0, 1], [0, 0, 2]$;

(c) $[1, 0, 0, 0], [0, 1, 0, 0], [1, 2, 0, 1], [0, 0, 0, 1]$;

(d) $[1, 0, 1, 2], [2, 0, 2, 3], [0, 1, -1, 1], [1, 1, 1, 1]$.

5. Find bases of the subspaces of R^n in Problem 2.4. In each case find a basis by discarding some of the given vectors.

6. Find the dimensions of the image and the kernel of the linear mapping whose matrix is

$$\begin{pmatrix} 2 & 1 & 3 & 4 \\ 0 & 0 & 1 & 2 \end{pmatrix}.$$

7. The set of vectors $[1, 0, 0, *, *, *]$, $[0, 1, 0, *, *, *]$, $[0, 0, 1, *, *, *]$ is independent no matter what numbers are in the positions indicated by $*$'s.

8. Let $\mathbf{A}$ be a matrix in echelon form. The methods of Chapter 4, Section 1 express the solutions of the homogeneous system $\mathbf{Ax} = \mathbf{0}$ as the space spanned by certain vectors. Show that these vectors are independent.

9. Use Problem 8 to conclude that the solutions of a homogeneous system of m equations in n unknowns form a subspace of R^n of dimension $n - r$ where r is the rank of the coefficient matrix.

10. Which of the following sets of vectors are independent? In each case find a basis of the space they span.

(a) $[1, 0, 1], [2, 1, 3], [-1, 1, 1]$;

(b) $\mathbf{i}, \mathbf{i} + \mathbf{j}, \mathbf{i} + \mathbf{j} + \mathbf{k}$;

(c) $\mathbf{i} + \mathbf{j} + \mathbf{k}, 2\mathbf{i} - \mathbf{j} + \mathbf{k}, -\mathbf{i} + 5\mathbf{j} + \mathbf{k}$;

(d) $[1, 1, 0, 1], [2, 0, 0, 1], [3, 1, -1, 1]$;

(e) $[1, 1, 0, 1], [2, 2, 0, 2], [3, 1, -1, 1]$;

(f) $[1, 0, 0, 0], [0, 1, 0, 0], [a, b, 0, 0]$.

11. Find a vector in R^4 which, together with the vectors in Problem 10(d), form a basis of R^4.

12. Find a so that the following vectors are dependent:

$$[1, 1, a]; \qquad [2, 0, 3]; \qquad [1, -1, 1].$$

13. Prove that a set of vectors $\mathbf{u}_1, \ldots, \mathbf{u}_r$ is a basis of a certain subspace if and only if every vector in the subspace can be written in *exactly* one way as a linear combination of $\mathbf{u}_1, \ldots, \mathbf{u}_r$.

14. Write out 2.13 in enough detail to show that it is a repetition of Theorem 2.8, Chapter 4.

3. ORTHOGONALITY AND DIMENSION

One last connotation of dimension that may help link it to your intuitive concept:

3.1 A subspace of R^n is r-dimensional if and only if it has some spanning set consisting of r nonzero, mutually orthogonal vectors.

For example, R^3 is three-dimensional because it can be spanned by **i**, **j**, and **k**, each of which is perpendicular to the other two. Any plane through the origin in R^3 (more accurately, the space of vectors in such a plane) is two-dimensional because you can find two perpendicular vectors in it, and their linear combinations will give all the vectors in the plane.

Proof. If: Let $\mathbf{u}_1, \ldots, \mathbf{u}_r$ be r mutually orthogonal vectors spanning the subspace. If we show they are independent, then they form a basis of the subspace, and the subspace has dimension r. But nonzero orthogonal vectors are always independent: Suppose $a_1\mathbf{u}_1 + \cdots + a_r\mathbf{u}_r = \mathbf{0}$. Dot the equation with $\mathbf{u}_p$ (any p) and get $a_1(\mathbf{u}_1 \cdot \mathbf{u}_p) + \cdots + a_p(\mathbf{u}_p \cdot \mathbf{u}_p) + \cdots + a_r(\mathbf{u}_r \cdot \mathbf{u}_p) = 0$. But $\mathbf{u}_1 \cdot \mathbf{u}_p, \ldots, \mathbf{u}_r \cdot \mathbf{u}_p$ are all zero because of the orthogonality of the **u**'s except for $\mathbf{u}_p \cdot \mathbf{u}_p$, which is 1. Thus the last equation says $a_p = 0$. This holds for every p, so all the a's are 0. This proves $\mathbf{u}_1, \ldots, \mathbf{u}_r$ are independent.

Only if: Suppose a space has dimension r. Then it has a basis of r vectors, say $\mathbf{v}_1, \cdots, \mathbf{v}_r$. We must show that it has another spanning set (which will also be a basis, incidentally) of orthogonal vectors. We do this by what is known as the *Gram-Schmidt process*. The geometric motivation is easy. To find a new, orthogonal basis $\mathbf{u}_1, \ldots, \mathbf{u}_r$ we start by taking $\mathbf{u}_1 = \mathbf{v}_1$. Now take $\mathbf{u}_2$ in the plane of $\mathbf{v}_1$ and $\mathbf{v}_2$ but orthogonal to $\mathbf{u}_1$. How to do this in the abstract, multidimensional context? Simply subtract from $\mathbf{v}_2$ the projection of $\mathbf{v}_2$ on $\mathbf{u}_1$: According to Chapter 1, 6.5, in R^3 the formula for the (scalar) *component* of $\mathbf{v}_2$ along $\mathbf{u}_1$ should be $(\mathbf{v}_2 \cdot \mathbf{u}_1) \, \| \mathbf{u}_1 \|^{-1}$. The (vector) *projection* of $\mathbf{v}_2$ along $\mathbf{u}_1$ should be this scalar times a unit vector in the direction of $\mathbf{u}_1$; the unit vector is $\| \mathbf{u}_1 \|^{-1}\mathbf{u}_1$. Multiplying, we get a formula for the projection of $\mathbf{v}_2$ along $\mathbf{u}_1$, namely $(\mathbf{v}_2 \cdot \mathbf{u}_1) \, \| \mathbf{u}_1 \|^{-2}\mathbf{u}_1$, or better $(\mathbf{v}_2 \cdot \mathbf{u}_1)(\mathbf{u}_1 \cdot \mathbf{u}_1)^{-1}\mathbf{u}_1$. Strictly speaking this deduction pretended that we were in R^3. Nevertheless, whatever vector space we are in, we can define

$$\mathbf{u}_2 = \mathbf{v}_2 - (\mathbf{v}_2 \cdot \mathbf{u}_1)(\mathbf{u}_1 \cdot \mathbf{u}_1)^{-1}\mathbf{u}_1$$

and we see that $\mathbf{u}_2$ is in fact orthogonal to $\mathbf{u}_1$ because $\mathbf{u}_2 \cdot \mathbf{u}_1 = \mathbf{v}_2 \cdot \mathbf{u}_1 - (\mathbf{v}_2 \cdot \mathbf{u}_1)(\mathbf{u}_1 \cdot \mathbf{u}_1)^{-1}(\mathbf{u}_1 \cdot \mathbf{u}_1) = 0$. Furthermore, $\mathbf{v}_1$ and $\mathbf{v}_2$ are linear combinations of $\mathbf{u}_1$ and $\mathbf{u}_2$, as well as vice versa, so the space spanned by $\mathbf{v}_1$ and $\mathbf{v}_2$ is the same as the space spanned by $\mathbf{u}_1$ and $\mathbf{u}_2$.

We continue in this way. Define $\mathbf{u}_3$ to be the result of subtracting from $\mathbf{v}_3$ its projections on $\mathbf{u}_1$ and $\mathbf{u}_2$:

$$\mathbf{u}_3 = \mathbf{v}_3 - (\mathbf{v}_3 \cdot \mathbf{u}_1)(\mathbf{u}_1 \cdot \mathbf{u}_1)^{-1}\mathbf{u}_1 - (\mathbf{v}_3 \cdot \mathbf{u}_2)(\mathbf{u}_2 \cdot \mathbf{u}_2)^{-1}\mathbf{u}_2.$$

It is then easy to check that $\mathbf{u}_3$ is orthogonal to $\mathbf{u}_1$ and $\mathbf{u}_2$, and that $\mathbf{u}_1$, $\mathbf{u}_2$ and $\mathbf{u}_3$ span the same space as $\mathbf{v}_1$, $\mathbf{v}_2$, and $\mathbf{v}_3$. The process continues until we arrive at r orthogonal vectors $\mathbf{u}_1, \ldots, \mathbf{u}_r$, which span the same space as $\mathbf{v}_1, \ldots, \mathbf{v}_r$.

Example 1. Find an orthogonal basis of the subspace of R^4 spanned by $\mathbf{v}_1 = [1, 1, 3, 0]$, $\mathbf{v}_2 = [2, 0, 1, -1]$, and $\mathbf{v}_3 = [1, -2, 0, 0]$. Solution: Use the Gram-Schmidt process. Let $\mathbf{u}_1 = [1, 1, 3, 0]$. Then $\mathbf{v}_2 \cdot \mathbf{u}_1 = 5$ and $\mathbf{u}_1 \cdot \mathbf{u}_1 = 11$, so we set $\mathbf{u}_2 = [2, 0, 1, -1] - (5/11)[1, 1, 3, 0] = (1/11)[17, -5, -4, -11]$. Next $\mathbf{v}_3 \cdot \mathbf{u}_1 = -1$, $\mathbf{v}_3 \cdot \mathbf{u}_2 = 27/11$, and $\mathbf{u}_1 \cdot \mathbf{u}_1 = 11$ as before and $\mathbf{u}_2 \cdot \mathbf{u}_2 = 451/(11)^2$, so $\mathbf{u}_3 = [1, -2, 0, 0] - (1/11)[1, 1, 3, 0] - (27 \times 11/451)(1/11)[17, -5, -4, -11] = (1/451)[33, -726, 231, 297]$.

P

3.2 Corollary

Every subspace of R^n has an orthonormal basis, that is, a basis consisting of vectors of length one and each orthogonal to all the others.

Proof. 3.1 converts any basis into an orthogonal basis. Then multiply each of these basis vectors $\mathbf{v}$ by a suitable scalar (namely $\|\mathbf{v}\|^{-1}$) to get new basis vectors of length one. They will still be orthogonal.

PROBLEMS

1. Use the Gram-Schmidt process to find an orthogonal basis of the subspace of R^4 spanned by $[1, 0, 1, 0]$, $[2, 3, 2, 3]$, and $[0, -1, 2, -1]$.

2. Use the Gram-Schmidt process to find an orthogonal basis of the subspace of R^4 spanned by $[1, 1, 0, 0]$, $[-1, 2, 1, 0]$, and $[1, -1, 2, 0]$. Then find another answer more quickly, without Gram-Schmidt.

3. Find two perpendicular vectors in the plane $2x + 4y + 3z = 0$ by using the techniques of Chapter 4 to find a spanning set of the solution space of the equation, then the Gram-Schmidt process to find an orthonormal spanning set.

Then solve the original problem more quickly, using cross products in 3-space.

4. The set of vectors $[x, y, z, t]$ which satisfy both $x + y + t = 0$ and $2x + 4y + 3z = 0$ form a two-dimensional subspace of R^4. Use the techniques of Chapter 4 to find a spanning set of this subspace, then Gram-Schmidt to find an orthogonal spanning set.

5. Use the Gram-Schmidt process on the *dependent* set of vectors $\mathbf{i} + 2\mathbf{j} + \mathbf{k}, \mathbf{i} + \mathbf{j}, \mathbf{i} - \mathbf{j} - 2\mathbf{k}$. Can you predict before you start, from geometrical considerations, that you will get a zero vector?

6. Is this always true: Vectors $\mathbf{v}_1, \ldots, \mathbf{v}_r$ are dependent if and only if the Gram-Schmidt process yields at least one zero vector?

7. The Gram-Schmidt process is an algorithm; it gives the same $\mathbf{v}_1, \ldots, \mathbf{v}_r$ every time, provided you start with the same vectors $\mathbf{u}_1, \ldots, \mathbf{u}_r$. Does this mean that there is only one set of mutually orthogonal vectors that span the same space as $\mathbf{v}_1, \ldots, \mathbf{v}_r$? How can you get others? However, see the next problem.

8. Prove that the Gram-Schmidt process gives the only orthogonal vectors $\mathbf{u}_1, \ldots, \mathbf{u}_r$ (except for scalar multiples) with the property that $\mathbf{u}_1, \ldots, \mathbf{u}_p$ span the same space as $\mathbf{v}_1, \ldots, \mathbf{v}_p$ for every $p = 1, 2, \ldots, r$.

9. Prove what was said to be easy in the proof of 3.1:

$$\mathbf{v}_3 - (\mathbf{v}_3 \cdot \mathbf{u}_1)(\mathbf{u}_1 \cdot \mathbf{u}_1)^{-1}\mathbf{u}_1 - (\mathbf{v}_3 \cdot \mathbf{u}_2)(\mathbf{u}_2 \cdot \mathbf{u}_2)^{-1}\mathbf{u}_2$$

is orthogonal to both $\mathbf{u}_1$ and $\mathbf{u}_2$, assuming that $\mathbf{u}_1$ and $\mathbf{u}_2$ are already orthogonal to each other.

Similarly, show that if $\mathbf{u}_1, \ldots, \mathbf{u}_{p-1}$ are orthogonal to one another, then $\mathbf{v}_p - (\mathbf{v}_p \cdot \mathbf{u}_1)(\mathbf{u}_1 \cdot \mathbf{u}_1)^{-1}\mathbf{u}_1 - (\mathbf{v}_p \cdot \mathbf{u}_2)(\mathbf{u}_2 \cdot \mathbf{u}_2)^{-1}\mathbf{u}_2 - \cdots - (\mathbf{v}_p \cdot \mathbf{u}_{p-1})^{-1}\mathbf{u}_{p-1}$ is orthogonal to $\mathbf{u}_1$, and to $\mathbf{u}_2, \ldots,$ and to $\mathbf{u}_{p-1}$.

4. RANK

In Chapter 4, Section 2 we gave a definition of rank of a matrix that depended on reduction to echelon form; it was not clear that different reduction procedures must necessarily lead to the same ranks. In 4.2, we shall correct this defect by showing that the rank of a matrix coincides with the dimension of certain subspaces that have nothing to do with any reduction process; the ranks of a matrix computed by two different reduction processes must both be equal to this dimension, and therefore must be equal to each other.

4.1 Definition

If $\mathbf{A}$ is an m by n matrix, then the *row space* of $\mathbf{A}$ is the subspace of R^n spanned by the rows of $\mathbf{A}$. The *column space* of $\mathbf{A}$ is the subspace of R^m spanned by the columns of $\mathbf{A}$.

4.2 Theorem

Let **A** be an m by n matrix and let f be the corresponding linear mapping from R^n to R^m. Then

the rank of **A** (or the rank of f) =

the dimension of the row space of **A** =

the dimension of the column space of **A** =

the dimension of the image of f =

n minus the dimension of the kernel of f (or of the solution space of **A**).

We carry out the proof in a sequence of lemmas, always using the notation in Theorem 4.2.

4.3 The dimension of the image of f equals the dimension of the column space of **A**.

Proof. According to Chapter 3, 6.2 the image of f *is* the column space of **A**, so of course they have the same dimension.

4.4 The dimension of the image of f = n minus the dimension of the kernel of f.

Proof. Let s denote the dimension of the kernel of f and let $\{\mathbf{u}_1, \ldots, \mathbf{u}_s\}$ be a basis of this kernel. This, then, is an independent set in R^n. If it spans R^n, then the kernel of f is R^n, which means that $f(\mathbf{v}) = \mathbf{0}$ for every vector $\mathbf{v}$ in R^n. Then the image of f consists of the zero vector alone and 4.4 is true: The dimension of the image of f is 0 and the dimension of the kernel of f is n.

If $\{\mathbf{u}_1, \ldots, \mathbf{u}_s\}$ does not span all of R^n, we pick a vector that is not a linear combination of them, and call this vector $\mathbf{u}_{s+1}$. We prove that $\{\mathbf{u}_1, \ldots, \mathbf{u}_s, \mathbf{u}_{s+1}\}$ is also an independent set by assuming

$$a_1\mathbf{u}_1 + \cdots + a_s\mathbf{u}_s + a_{s+1}\mathbf{u}_{s+1} = 0 \tag{4.1}$$

and showing

$$a_1 = \cdots = a_s = a_{s+1} = 0.$$

First, $a_{s+1} = 0$ because if not, we can solve for $\mathbf{u}_{s+1}$,

$$\mathbf{u}_{s+1} = -\left(\frac{a_1}{a_{s+1}}\right)\mathbf{u}_1 - \cdots - \left(\frac{a_s}{a_{s+1}}\right)\mathbf{u}_s,$$

so that $\mathbf{u}_{s+1}$ is a linear combination of $\mathbf{u}_1, \ldots, \mathbf{u}_s$, contrary to the way we picked it in the first place. But if $a_{s+1} = 0$, Eq. (4.1) becomes

$$a_1\mathbf{u}_1 + \cdots + a_s\mathbf{u}_s = \mathbf{0},$$

which by the independence of $\mathbf{u}_1, \ldots, \mathbf{u}_s$ implies that $a_1 = \cdots = a_s = 0$. This shows all the coefficients of Eq. (4.1) are zero.

If the set $\{\mathbf{u}_1, \ldots, \mathbf{u}_s, \mathbf{u}_{s+1}\}$ spans R^n, we stop. If not, there is a vector $\mathbf{u}_{s+2}$ that is not a linear combination of $\mathbf{u}_1, \ldots, \mathbf{u}_s, \mathbf{u}_{s+1}$ and the same argument again shows that $\{\mathbf{u}_1, \ldots, \mathbf{u}_s, \mathbf{u}_{s+1}, \mathbf{u}_{s+2}\}$ is an independent set. If this set spans R^n, we stop. If not, we pick $\mathbf{u}_{s+3}$, and so on. This process of picking new vectors cannot continue forever, because each step gives an independent set with one more vector in it than in the previous step—and we know that in R^n no independent set can have more than n vectors in it. Thus the process must stop after at most n steps. But the process stops only when we have an independent set spanning R^n. Then we know there are n vectors in this set by 2.9. We started with a basis of the kernel of f and we picked $n - s$ new vectors to get a basis

$$\{\mathbf{u}_1, \ldots, \mathbf{u}_s, \mathbf{u}_{s+1}, \ldots, \mathbf{u}_n\}$$

of R^n.

Now apply f to all these $\mathbf{u}$'s. Since $\mathbf{u}_1, \ldots, \mathbf{u}_s$ are in the kernel of f, we know $f(\mathbf{u}_1) = \cdots = f(\mathbf{u}_s) = \mathbf{0}$. We also know that $f(\mathbf{u}_{s+1}), \ldots, f(\mathbf{u}_n)$ are not zero because, by the way we picked the new $\mathbf{u}$'s, none of these $\mathbf{u}$'s is in the kernel. We now prove more; we prove that $f(\mathbf{u}_{s+1}), \ldots, f(\mathbf{u}_n)$ form a basis of the image of f.

First, they span the image; that is, every vector in the image of f is a linear combination of $f(\mathbf{u}_{s+1}), \cdots, f(\mathbf{u}_n)$: every vector in the image is $f(\mathbf{v})$ for some $\mathbf{v}$ in R^n. Since $\{\mathbf{u}_1, \ldots, \mathbf{u}_s, \mathbf{u}_{s+1}, \ldots, \mathbf{u}_n\}$ spans R^n, we can write $\mathbf{v} = \sum_{p=1}^n b_p\mathbf{u}_p$ and so $f(\mathbf{v}) = \sum b_p f(\mathbf{u}_p)$ because f is linear. However, as we remarked earlier,

$$f(\mathbf{u}_1) = \cdots = f(\mathbf{u}_s) = \mathbf{0}, \qquad \text{so} \quad f(\mathbf{v}) = \sum_{p=s+1}^{n} b_p f(\mathbf{u}_p)$$

as desired.

Second, $\{f(\mathbf{u}_{s+1}), \ldots, f(\mathbf{u}_n)\}$ is an independent set: Suppose a linear combination of them vanishes,

$$a_{s+1} f(\mathbf{u}_{s+1}) + \cdots + a_n f(\mathbf{u}_n) = \mathbf{0};$$

we must show that all the a's are zero. Since f is linear, our supposition implies

$$f(a_{s+1}\mathbf{u}_{s+1} + \cdots + a_n\mathbf{u}_n) = \mathbf{0},$$

which says that $a_{s+1}\mathbf{u}_{s+1} + \cdots + a_n\mathbf{u}_n$ is in the kernel of f. But this kernel is spanned by $\{\mathbf{u}_1, \ldots, \mathbf{u}_s\}$, so $a_{s+1}\mathbf{u}_{s+1} + \cdots + a_n\mathbf{u}_n$ is a linear combination of $\mathbf{u}_1, \ldots, \mathbf{u}_s$, say equal to $a_1\mathbf{u}_1 + \cdots + a_s\mathbf{u}_s$. Therefore

$$a_1\mathbf{u}_1 + \cdots + a_s\mathbf{u}_s - a_{s+1}\mathbf{u}_{s+1} - \cdots - a_n\mathbf{u}_n = \mathbf{0}.$$

The independence of $\{\mathbf{u}_1, \ldots, \mathbf{u}_n\}$ then shows that all the a's are zero; in particular, $a_{s+1}, \ldots, a_n$ are all zero, which is what was needed to prove the independence of $\{f(\mathbf{u}_{s+1}), \ldots, f(\mathbf{u}_n)\}$.

The result of all this is that the image of f has a basis $\{f(\mathbf{u}_{s+1}), \ldots, f(\mathbf{u}_n)\}$, and so has dimension $n - s$. Since s is the dimension of the kernel, we have proved 4.4.

4.5 The rank of $\mathbf{A} = n$ minus the dimension of the kernel of f.

Proof. Let $\mathbf{E}$ be the echelon form to which $\mathbf{A}$ can be reduced by the process in Chapter 4. Then the rank r of $\mathbf{A}$ is the number of distinguished columns in $\mathbf{E}$. These distinguished column vectors are the vectors $\mathbf{i}_1, \ldots, \mathbf{i}_r$ in R^m, so of course they are independent. But every vector in the column space of $\mathbf{E}$ has zero for its $(r + 1)$st, $\ldots$, mth components because the $(r + 1)$st, $\ldots$, mth rows of $\mathbf{E}$ are all zero; in other words, every vector in the column space of $\mathbf{E}$ is a linear combination of $\mathbf{i}_1, \ldots, \mathbf{i}_r$. This shows that the dimension of the column space of $\mathbf{E}$ is r.

If g is the linear function whose matrix is $\mathbf{E}$, then by 4.3 and 4.4, we have $r = n -$ dimension of the kernel of g. Now the kernel of g is the same as the kernel of f by Chapter 4, 1.3. Therefore, the dimension of the kernel of $g =$ the dimension of the kernel of f and so the rank of $\mathbf{A} = r = n$ minus the dimension of the kernel of f.

4.6 The rank of $\mathbf{A} =$ the dimension of the row space of $\mathbf{A}$.

Proof. We again use the notations introduced in the proof of 4.5. Now $\mathbf{E}$ has exactly r nonzero rows (this was one version of the definition of rank). Call them $\mathbf{v}_1, \ldots, \mathbf{v}_r$. Then $\mathbf{v}_1$ has a 1 in one component where all the other $\mathbf{v}$'s have the component 0 (this 1 and these 0's form the first distinguished column in $\mathbf{E}$), so $a_1\mathbf{v}_1 + \cdots + a_r\mathbf{v}_r$ has this component equal to a_1. A similar argument using $\mathbf{v}_2$ in place of $\mathbf{v}_1$ shows that another of the components of $a_1\mathbf{v}_1 + \cdots + a_r\mathbf{v}_r$ equals a_2, and so on. Hence if $a_1\mathbf{v}_1 + \cdots + a_r\mathbf{v}_r = \mathbf{0}$, then $a_1 = 0$ and $a_2 = 0$ and $\cdots$ $a_r = 0$. This proves that the r nonzero rows of $\mathbf{E}$ are independent, so the rank of $\mathbf{A} =$ the dimension of the row space of $\mathbf{E}$.

It will be enough to show now that the row space of $\mathbf{A}$ is the same as the row space of $\mathbf{E}$, for then these two row spaces will have the same dimension, and 4.6 will be proved. We get from **A** to **E** by a sequence of manipulations; we claim that each of these manipulations, although it changes the matrix, does not change the row space. If we start with a matrix with rows $\mathbf{u}_1, \ldots, \mathbf{u}_m$ and apply one manipulation to get a matrix with rows $\mathbf{v}_1, \ldots, \mathbf{v}_m$, then each of the **v**'s is a linear combination of the **u**'s (in fact, all but two of the **v**'s equal the corresponding **u**'s), so that any linear combination of the **v**'s is also a linear combination of the **u**'s; that is, any vector in the row space of the second matrix is also in the row space of the first. Since we can also apply one manipulation to the second matrix to get back to the first, the same result says that every vector in the row space of the first matrix is in the row space of the second. This shows that the two row spaces are the same. Thus, after a sequence of manipulations, we still end up with a matrix having the same row space as the original matrix.

Combining 4.3, 4.4, 4.5, and 4.6 proves Theorem 4.2.

Theorem 4.2 really consists of two distinct parts. First, it gives three invariant versions of the concept of rank: the dimension of the row space, of the column space, and of the image. These versions are invariant in the sense that they do not depend on the choice of a reduction method, as did the definition of rank in Chapter 4. We might well have used one of these three versions as the definition of rank, instead of the definition we gave in Chapter 4. The reader would be well advised to make such a substitution of definitions in his mind now. In particular, we now define the *rank of a linear mapping* as the dimension of its image or, equivalently, as the rank of its matrix.

Second, the theorem states that all these various versions of rank equal n minus the dimension of the kernel. This is the quantitative result promised in Chapter 3: "the bigger the kernel, the smaller the image."

In the course of the proof of 4.6 we showed that manipulations on the rows of a matrix do not change the row space. The same proof, of course, shows that if we perform manipulations on the columns of a matrix (add a multiple of one column to another, interchange two columns, multiply a column by a nonzero scalar), we will not change the column space; then we will not change the dimension of the column space, either; and, by 4.2, we will not change the rank. This is a handy corollary of 4.2, which is useful in computing ranks.

4.7 If we make free use of Manipulations 1, 2, and 3 (from Chapter 4, Section 1) on both rows and columns, the resulting matrix will have the same rank as the original.

Caution: If the matrix we manipulate is the augmented matrix of a system of equations, and if we manipulate the *columns*, we *do* change the solutions. By manipulating columns, we can find the rank, which determines the dimension of the space of solutions (assuming we have a homogeneous system, so that the solutions do form a subspace), but we cannot find the solutions themselves this way.

Next we investigate the ranks of composites and sums of linear mappings. To do this, we need a lemma.

4.8 Let f be a linear mapping from R^n to R^m and let V be a subspace of R^n. Then $f(V)$, the set of all vectors $f(\mathbf{v})$ with $\mathbf{v}$ ranging over V, is a subspace of R^m and the dimension of $f(V)$ is $\leq$ the dimension of V. If f is invertible, then these two dimensions are equal.

Proof. Suppose $\{\mathbf{u}_1, \ldots, \mathbf{u}_r\}$ is a basis of V, so that every vector in V is of the form $\mathbf{v} = \sum a_p\mathbf{u}_p$ for some scalars $a_1, \ldots, a_r$. Then $f(V)$ consists of all

$$f(\mathbf{v}) = f(\sum a_p\mathbf{u}_p) = \sum a_p f(\mathbf{u}_p);$$

thus $f(V)$ is the space spanned by $\{f(\mathbf{u}_1), \ldots, f(\mathbf{u}_r)\}$. This proves that it is a subspace of R^m and that its dimension is at most r. It is not usually true that $\{f(\mathbf{u}_1), \ldots, f(\mathbf{u}_r)\}$ is a basis of $f(V)$. If f is invertible, we apply to f^{-1} what we just proved: for every subspace W of R^m the dimension of $f^{-1}(W)$ is $\leq$ the dimension of W; take $W = f(V)$, so that $f^{-1}(W) = V$, and find that the dimension of V is $\leq$ the dimension of $f(V)$, which is the opposite of the foregoing inequality. Hence, $f(V)$ and V have the same dimension.

4.9 Let f and g be linear mappings such that $f \circ g$ is defined. Then the rank of $f \circ g$ is $\leq$ the rank of g. If f is invertible, the equality holds. Similarly, the rank of $f \circ g$ is $\leq$ the rank of f. If g is invertible, this equality holds.

Proof. Suppose g maps R^n to R^m. Then the rank of $f \circ g$ is the dimension of $f \circ g(R^n) = f(g(R^n))$. Use 4.8 with $V = g(R^n)$; then the rank of $f \circ g$ = dimension of $f(V) \leq$ dimension of V = dimension of image of g = rank of g. If f is invertible, 4.8 says that the lone inequality in this string is an equality, so all the items in the string are equal.

For the other half of 4.9, since $g(R^n)$ is contained in R^m, $f(g(R^n))$ is contained in $f(R^m)$, the image of f. Thus the dimension of $f(g(R^n))$ is $\leq$ the dimension of the image of f, by 2.13, and the rank of $f \circ g$ is $\leq$

the rank of f. If g is invertible, then the image of g is all of R^m, so this inequality becomes an equality.

4.10 Let f and g be linear mappings such that $f + g$ is defined. Then the rank of $f + g$ is $\leq$ the rank of $f +$ the rank of g.

Proof. Suppose f and g map R^n into R^m. Then the image of $f + g$ consists of all $f(\mathbf{v}) + g(\mathbf{v})$ as $\mathbf{v}$ ranges over R^n. If $\{\mathbf{u}_1, \ldots, \mathbf{u}_r\}$ is a basis of $f(R^n)$ and $\{\mathbf{u}_1', \ldots, \mathbf{u}_s'\}$ is a basis of $g(R^n)$, then $f(\mathbf{v})$ is a linear combination of $\mathbf{u}_1, \ldots, \mathbf{u}_r$ and $g(\mathbf{v})$ is a linear combination of $\mathbf{u}_1', \ldots, \mathbf{u}_s'$, so that $f(\mathbf{v}) + g(\mathbf{v})$ is a linear combination of $\mathbf{u}_1, \ldots, \mathbf{u}_r, \mathbf{u}_1', \ldots, \mathbf{u}_s'$. Thus the image of $f + g$ has a spanning set consisting of $r + s$ vectors, so the dimension of this image is $\leq r + s$. But $r =$ rank of f and $s =$ rank of g.

Of course these theorems translate immediately to theorems on matrices:

4.11 If **A** and **B** are two matrices such that **AB** is defined, then the rank of **AB** is $\leq$ the rank of **B**; if **A** is invertible, equality holds. Moreover, the rank of **AB** is $\leq$ the rank of **A**; if **B** is invertible, equality holds.

4.12 If **A** and **B** are matrices such that **A** + **B** is defined, then the rank of **A** + **B** is $\leq$ the rank of **A** + the rank of **B**.

Here is a theorem that we promised in Chapter 3:

4.13 Every invertible matrix is square. If f is a linear function from R^n to R^m and f is invertible, then $m = n$.

Proof. It should be clear that the two statements in 4.13 are equivalent to each other. We prove the second. If there is a function g that maps R^m to R^n such that $g \circ f = I_n$ ($=$ the identity map on R^n), then 4.9 asserts that n, which is the rank of I_n, is $\leq$ the rank of f. But since f maps R^n into R^m, the rank of f cannot be larger than m (you can give both a geometric and a matrix argument for this). Hence, $n \leq m$. Similarly, since $f \circ g = I_m$, we get $m \leq$ the rank of $g \leq n$. Therefore $m = n$.

Example 1. As in Chapter 4, we see again that the rank of an invertible linear mapping from R^n to R^n is n (see the proof of 4.13).

Example 2. The rank of any rotation of 3-space is 3, since rotations are invertible. This checks with other evidence, too: If f is a rotation, then $f(\mathbf{i})$, $f(\mathbf{j})$, and $f(\mathbf{k})$ are mutually orthogonal. By Problem 2, Section 2 or the proof of 3.1, these vectors are independent, so the dimension of the image of f is 3.

Example 3. Projection of 3-space onto the x-axis is a linear mapping of rank 1, since its image is the set of vectors along the x-axis, which is a one-dimensional subspace of R^3. Check this against the matrix of this projection.

Example 4. If neither f nor g is invertible, we cannot expect equalities or any other refinements in 4.9 and 4.10. For example, if f is projection of 3-space onto the x-axis and g is projection onto the y-axis, then $f \circ g = 0$, so the rank of $f \circ g$ is strictly less than both the rank of f and the rank of g. Similarly, if $f = -g$ and neither mapping is zero, then the rank of $f + g$ is $<$ the rank of $f +$ the rank of g.

Example 5. 4.9 says that if we multiply an m by 1 matrix by a 1 by n matrix, we get an m by n matrix of rank 1 or less. We can prove a kind of converse: Every m by n matrix of rank 1 is a product of an m by 1 matrix by a 1 by n matrix.

Proof. If $\mathbf{A}$ has rank 1, then the space spanned by its rows can be spanned by one of the rows. This row is the 1 by n matrix we are looking for; call it $\mathbf{B}$. Each row of $\mathbf{A}$ is a scalar multiple of $\mathbf{B}$, say the pth row is $a_p\mathbf{B}$. Then

$$\mathbf{A} = \begin{pmatrix} a_1 \\ \vdots \\ a_m \end{pmatrix} \mathbf{B}.$$

PROBLEMS

1. Use Problem 3, Section 2 to give another proof that the dimension of $f(V) \leq$ dimension of V. Use this version of the dimension of a subspace: the maximum number of vectors in an independent set in the subspace.

2. If $\mathbf{A}$ is an m by 2 matrix and $\mathbf{B}$ is a 2 by n matrix, show that the m by n matrix $\mathbf{AB}$ has rank ≤ 2.

Conversely, given any m by n matrix of rank 2, show that it is expressible as $\mathbf{AB}$ for some m by 2 matrix $\mathbf{A}$ and 2 by n matrix $\mathbf{B}$.

3. Let f be a linear mapping from R^3 to R^3. Show that

(a) If the rank of f is 0, then the kernel and the image of f are R^3 and $\mathbf{0}$, respectively;

(b) If the rank of f is 1, then the kernel and the image of f are a plane and a line (both through the origin), respectively;

(c) If the rank of f is 2, then the kernel and the image of f are a line and a plane (through the origin), respectively;

(d) If the rank of f is 3, then the kernel and the image of f are $\mathbf{0}$ and R^3, respectively.

Give examples of each of cases (a)–(d).

4. Compare Problem 3 with the results of Chapter 4:

(a) If the rank of a 3 by 3 matrix is zero, then the solutions of the homogeneous system of equations with matrix $\mathbf{A}$ comprise all vectors in R^3. The only vector $\mathbf{v}$ for which the nonhomogeneous system $\mathbf{Ax} = \mathbf{v}$ has a solution, $\mathbf{x}$, is $\mathbf{v} = \mathbf{0}$.

(b) If the rank of $\mathbf{A}$ is one, then the solutions of the homogeneous system of equations with matrix $\mathbf{A}$ are all scalars times a single nonzero vector. The set of vectors $\mathbf{v}$ for which $\mathbf{Ax} = \mathbf{v}$ has a solution are all linear combinations of two nonparallel vectors.

(c) Write the matrix and linear equation analogs of Problem 3(c) and (d), and prove them using Chapter 4.

5. Let f be a linear mapping from R^n to R^n. Use 4.2 to show that the following conditions are equivalent:

(a) f is invertible;

(b) the kernel of f is $\mathbf{0}$;

(c) the image of f is R^n;

(d) $g \circ f$ = identity, for some g mapping R^n to R^n;

(e) $f \circ h$ = identity, for some h mapping R^n to R^n.

If these conditions hold, show that g and h in parts (d) and (e) must be equal (use the associative law on $g \circ f \circ h$) so that they are both f^{-1}.

6. Write the matrix analogs of Problems 5(a)–(e). Which of these theorems have we already proved in Chapters 3 and 4?

7. Find an example of nonsquare matrices $\mathbf{A}$ and $\mathbf{B}$ with $\mathbf{AB} = \mathbf{I}$. Show that 4.11 and 4.13 or both prevent $\mathbf{BA}$ from equaling $\mathbf{I}$.

8. In a system of m linear equations in n unknowns, if the augmented matrix has rank strictly less than m, show that one of the equations is redundant; you get the same solutions if you omit this equation. (Show that one row is a linear combination of the others by arguing the dimension of the row space.)

9. Use the argument in the proof of 4.4 to prove that if T is a subspace of R^n and $\{\mathbf{u}_1, \ldots, \mathbf{u}_s\}$ is an independent set in T, then we can pick $\mathbf{u}_{s+1}, \ldots, \mathbf{u}_m$ in T so that

$$\{\mathbf{u}_1, \ldots, \mathbf{u}_s, \mathbf{u}_{s+1}, \ldots, \mathbf{u}_m\}$$

is a basis of T.

10. As a corollary of Problem 9, prove that if $\{\mathbf{u}_1, \ldots, \mathbf{u}_s\}$ is an independent set of vectors in a subspace T and if s is the dimension of T, then this set is a basis of T.

11. As a corollary of Problem 10, prove a kind of supplement to 2.13: If S and T are subspaces of R^n, if S is contained in T, and if the dimensions of S and T are equal, then S and T are the same.

12. (a) Let S and S' be two subspaces of R^n and let $S \cap S'$ denote their intersection, the set of all vectors that lie in both S and S'. Show that $S \cap S'$ is a subspace of R^n.

(b) Let $S + S'$ denote the set of all vectors $\mathbf{v} + \mathbf{w}$ when $\mathbf{v}$ ranges over S and $\mathbf{w}$ ranges over S'. Show that $S + S'$ is a subspace of R^n and contains both S and S'.

(c) Let $\{\mathbf{u}_1, \ldots, \mathbf{u}_r\}$ be a basis of $S \cap S'$; use Problem 9 to extend it to a basis $\{\mathbf{u}_1, \ldots, \mathbf{u}_r, \mathbf{u}_{r+1}, \ldots, \mathbf{u}_s\}$ of S and again to a basis $\{\mathbf{u}_1, \ldots, \mathbf{u}_r, \mathbf{u}'_{r+1}, \ldots, \mathbf{u}'_t\}$ of S'. Prove that $\{\mathbf{u}_1, \ldots, \mathbf{u}_r, \mathbf{u}_{r+1}, \ldots, \mathbf{u}_s, \mathbf{u}'_{r+1}, \ldots, \mathbf{u}_t\}$ is a basis of $S + S'$.

(d) Deduce

$$\dim(S + S') = \dim S + \dim S' - \dim(S \cap S').$$

(e) Use part (d) to argue that two planes through the origin in R^3 must intersect in more points than just the origin, but in R^4 this need not be true.

13. If S is a subspace of R^n, the *orthogonal complement* of S is the set $S^\perp$ consisting of all vectors that are orthogonal to all the vectors in S. For example, if S is the set of all vectors in the xy-plane in R^3, its orthogonal complement $S^\perp$ is the set of all vectors along the z-axis.

(a) Show that $S^\perp$ is a subspace of R^n.

(b) Show that if the dimension of S is r, then the dimension of $S^\perp$ is $n - r$. (Hint: Take a basis of S, and let f be the linear mapping whose matrix has these basis vectors as its rows. By 4.6, the rank of f is r, and by 4.5, the dimension of the kernel of f is $n - r$. Then prove that the kernel of f is $S^\perp$.)

(c) Show that $S \cap S^\perp$ consists of $\mathbf{0}$ alone.

(d) Use Problems 12(d), 13(c), and 11 to show $S + S^\perp = R^n$.

(e) Show that the orthogonal complement of $S^\perp$ is S.

14. Let S and T be any two subspaces of R^n with the property that $S \cap T = \mathbf{0}$ and $S + T = R^n$. Show that every vector in R^n is uniquely expressible as a sum of a vector in S and a vector in T. In particular, every vector in R^n is uniquely expressible as a sum of a vector in S and a vector in $S^\perp$.

CHAPTER

6

Determinants and Transposes

1. COMPUTATION OF DETERMINANTS

If **A** is a square matrix, the determinant of **A**, written det **A**, is a number. If **A** is 1 by 1, det **A** is the one entry in **A**. If **A** is 2 by 2, det **A** is plus or minus the area of the parallelogram two of whose sides are the vectors that are the rows of **A**. If **A** is 3 by 3, det **A** is plus or minus the volume of the parallelepiped three of whose edges are the vectors that are the rows of **A**. In higher dimensions, the determinant will be defined so that it has four of the fundamental properties of areas and volumes: see 1.1–1.4.

For those readers who by now automatically think about linear functions when we mention matrices, the determinant of a linear function can be defined as the determinant of its matrix. We have more to say of this in 3.11 and in Problem 8, Chapter 8, Section 3.

We run into an interesting logical exercise in axiomatics here which can be skipped in your first reading of Sections 1 and 2. We take 1.1–1.4 as axioms or postulates for a determinant function, but we do not know in advance that there is any function that satisfies these four conditions (we must prove its existence) and, even if we knew there were such a function, we would not know but that there were many such, all satisfying 1.1–1.4 (we must prove uniqueness). The strategy we use is reminiscent of the way you first solved equations: If $2x + 3 = 7$, then $2x = 4$, then $x = 2$, which proves uniqueness of the solution (any solution must equal 2); then a check verifies that 2 indeed has the property $2(2) + 3 = 7$, so a solution exists. By analogy, we let det be any function satisfying 1.1–1.4 and we deduce from that a scheme for computing det $\mathbf{A}$ that will show that there is only one such function (see 1.6 and the arguments following it). In other words, we shall have demonstrated the uniqueness of the determinant function. We leave the existence proof to Section 2, especially 2.2 and the discussion following it. The net result will be that for each $n = 1, 2, \ldots$, there is one and only one function associating to each n by n matrix a scalar in such a way that 1.1–1.4 hold.

The nonvanishing of this determinant will turn out to be a necessary and sufficient condition for the rows of a matrix to be independent or, equivalently, for the matrix to be invertible. In 3-space this makes good geometric sense; three vectors are dependent if and only if the parallelepiped they generate is degenerate or collapsed and so has a zero volume. Moreover, in 2.6 we shall be able to use determinants to give an explicit formula for $\mathbf{A}^{-1}$. The reader should be warned, however, that this determinant test for invertibility and the determinant formula for the inverse are valuable theoretical results, but are computationally almost useless for any but very small or very special matrices.

If the rows of $\mathbf{A}$ are the vectors $\mathbf{u}_1, \ldots, \mathbf{u}_n$, we shall also denote det $\mathbf{A}$ by $\det(\mathbf{u}_1, \ldots, \mathbf{u}_n)$. The properties we demand from the determinant are as follows.

1.1 If we add a multiple of one row of $\mathbf{A}$ to another row, we get another matrix with the same determinant:

$$\det(\mathbf{u}_1, \ldots, \mathbf{u}_{p-1}, \mathbf{u}_p + a\mathbf{u}_q, \mathbf{u}_{p+1}, \ldots, \mathbf{u}_n) = \det(\mathbf{u}_1, \mathbf{u}_2, \ldots, \mathbf{u}_n).$$

1.2 If we interchange two rows of $\mathbf{A}$, we get another matrix whose determinant is minus that of $\mathbf{A}$:

$$\det(\mathbf{u}_1, \ldots, \mathbf{u}_p, \ldots, \mathbf{u}_q, \ldots, \mathbf{u}_n) = -\det(\mathbf{u}_1, \ldots, \mathbf{u}_q, \ldots, \mathbf{u}_p, \ldots, \mathbf{u}_n).$$

1.3 If we multiply one row of $\mathbf{A}$ by a scalar, the determinant of the new matrix is this scalar times det $\mathbf{A}$:

$$\det(\mathbf{u}_1, \ldots, a\mathbf{u}_p, \ldots, \mathbf{u}_n) = a \det(\mathbf{u}_1, \ldots, \mathbf{u}_p, \ldots, \mathbf{u}_n).$$

We may also read 1.3 as saying that a scalar may be "factored out" of one row.

Note that we are specifying what should happen to det $\mathbf{A}$ on applying Manipulations 1, 2, 3 of Chapter 4, except that 1.3 allows multiplication by a zero scalar. We impose one further demand.

1.4 The determinant of the identity matrix is 1.

First we check that in 3-space the volume of a parallelepiped satisfies these demands. More accurately, let $d(\mathbf{u}, \mathbf{v}, \mathbf{w})$ denote plus or minus the volume of the parallelepiped three of whose edges are $\mathbf{u}$, $\mathbf{v}$, and $\mathbf{w}$; plus if $\mathbf{u}$, $\mathbf{v}$, $\mathbf{w}$ form a right-handed system, minus if they form a left-handed system. (Of course $\mathbf{u}$, $\mathbf{v}$, and $\mathbf{w}$ are not necessarily perpendicular to each other, but the definition of right-handedness and left-handedness in Chapter 1 still works.) Then this function d satisfies the conditions 1.1–1.4 for a determinant function. The last three are the easiest to check.

Verification of 1.2. Interchanging two of $\mathbf{u}$, $\mathbf{v}$, and $\mathbf{w}$ changes a right-handed system to a left-handed one and vice versa, but leaves the parallelepiped alone, so it will change the sign of $d(\mathbf{u}, \mathbf{v}, \mathbf{w})$.

Verification of 1.3. Lengthening one edge of a parallelepiped by a factor a leaves the base alone (say the base is the parallelogram generated by the other two edges), but multiplies the altitude by a, hence multiplies the volume by a—if a is positive. If a is negative, the volume will be multiplied by $|a|$, that is, by $-a$, but multiplying by a negative scalar reverses the direction of one of the vectors, which changes the handedness of the system, resulting in a d of opposite sign. The net result is to multiply by $-a$ and by -1, that is, to multiply by a.

Verification of 1.4. The rows of the identity matrix are $\mathbf{i}$, $\mathbf{j}$, $\mathbf{k}$ in that order. The parallelepiped is rectangular with all three dimensions equal to 1, so its volume is 1. Furthermore, $\mathbf{i}$, $\mathbf{j}$, $\mathbf{k}$ is a right-handed system, so $d(\mathbf{i}, \mathbf{j}, \mathbf{k}) = 1$.

Verification of 1.1. Verifying 1.1 is not much more difficult (refer

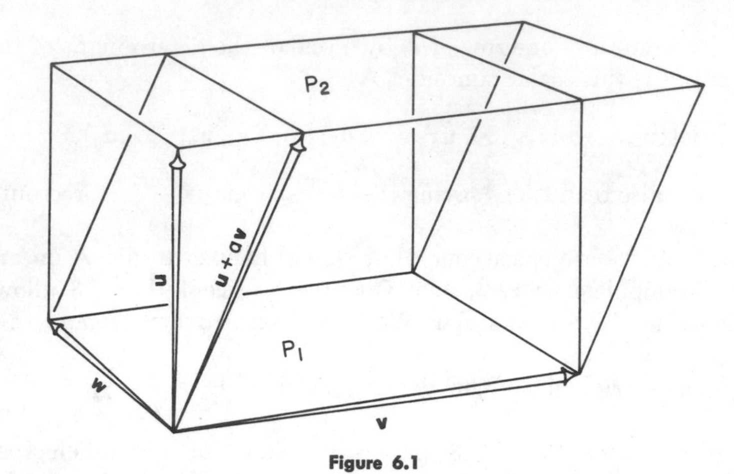

Figure 6.1

to Fig. 6.1). Let P_1 be the plane containing **v** and **w** and P_2 be the plane parallel to P_1 through the head of **u**. Then $d(\mathbf{u}, \mathbf{v}, \mathbf{w})$ is the area of the parallelogram generated by **v** and **w** (the base) multiplied by the perpendicular distances between the planes P_1 and P_2 (the altitude), then multiplied by ± 1. Now consider $d(\mathbf{u} + a\mathbf{v}, \mathbf{v}, \mathbf{w})$. Since $\mathbf{u} + a\mathbf{v}$ is computed by placing at the head of **u** a vector parallel to $a\mathbf{v}$ (hence, parallel to the planes P_1 and P_2, we well), the head of $\mathbf{u} + a\mathbf{v}$ will turn out to lie in the plane P_2 so the base times altitude computation gives $d(\mathbf{u} + a\mathbf{v}, \mathbf{v}, \mathbf{w})$ = the area of this same parallelogram times the distance between the same planes P_1 and P_2, times the same ± 1 (since the handedness will not change as long as the vector $\mathbf{u} + a\mathbf{v}$ stays on the same side of the plane P_1, as it does). We have proved 1.1 for the case where we add a multiple of **v** to **u**; clearly, the same argument can be carried out when we add a multiple of any one of **u**, **v**, **w** to any other one. Compare also Problem 9, Chapter 1, Section 7, on areas.

In view of 7.12 in Chapter 1, and assuming that in 3-space there is only one determinant function, we have shown that $\det(\mathbf{u}, \mathbf{v}, \mathbf{w}) = (\mathbf{u} \times \mathbf{v}) \cdot \mathbf{w}$. This, of course, holds only in 3-space, since the cross product is defined only there.

We now deduce a few properties of the determinant function, assuming it satisfies 1.1–1.4. We shall end up showing that there is only one possible function, and shall have an algorithmic technique for computing it.

1.5 If one row of $\mathbf{A}$ is $\mathbf{0}$, then $\det \mathbf{A} = 0$.

Proof.

$$\begin{aligned}\det(\mathbf{u}_1, \ldots, \mathbf{u}_{p-1}, \mathbf{0}, \mathbf{u}_{p+1}, \ldots, \mathbf{u}_n) &= \det(\mathbf{u}_1, \ldots, \mathbf{u}_{p-1}, 0\mathbf{0}, \mathbf{u}_{p+1}, \ldots, \mathbf{u}_n) \\ &= 0 \det(\mathbf{u}_1, \ldots, \mathbf{0}, \ldots, \mathbf{u}_n) \qquad \text{(by 1.3)} \\ &= 0.\end{aligned}$$

1.6 Theorem

Let $\mathbf{A}$ be a square n by n matrix. Then the following four conditions are equivalent:

det $\mathbf{A} = 0$;

the rank of $\mathbf{A}$ is less than n;

the rows of $\mathbf{A}$ are dependent;

$\mathbf{A}$ is not invertible.

Or, what amounts to the same thing (see the Appendix), the following four conditions are equivalent:

det $\mathbf{A} \neq 0$;

the rank of $\mathbf{A}$ is n;

the rows of $\mathbf{A}$ are independent;

$\mathbf{A}$ is invertible.

Proof. We already know that $\mathbf{A}$ has rank less than n if and only if the rows of $\mathbf{A}$ are dependent, and that this is also equivalent to $\mathbf{A}$ being not invertible. We need only show that these conditions are equivalent to det $\mathbf{A} = 0$.

As in Chapter 4 we reduce $\mathbf{A}$ to echelon form $\mathbf{B}$ by a sequence of manipulations 1, 2, and 3. According to 1.1, 1.2, and 1.3, det $\mathbf{A} = c \det \mathbf{B}$ for some nonzero number c (c is the product of ± 1's from manipulations 2, and of nonzero scalars from manipulations 3), so det $\mathbf{A} = 0$ if and only det $\mathbf{B} = 0$. Moreover, we know the rank of $\mathbf{A}$ is less than n if and only if not all columns of $\mathbf{B}$ are distinguished, that is, $\mathbf{B} \neq \mathbf{I}$. So we have reduced the proof of "det $\mathbf{A} = 0$ if and only if rank of $\mathbf{A}$ is less than n" to a problem on the echelon form: "det $\mathbf{B} = 0$ if and only if $\mathbf{B} \neq \mathbf{I}$." The "if" part is the contrapositive (see the Appendix) of 1.4. The "only if" part reads, "if $\mathbf{B} \neq \mathbf{I}$, then det $\mathbf{B} = 0$"; but an n by n echelon form that is not $\mathbf{I}$ has a row of zeros, so has determinant 0 by 1.5.

The proof of 1.6 tells how to compute determinants: We reduce to echelon form, keeping track of how many times we interchange rows (the determinant changes sign each time) and of the nonzero scalars by which we multiply rows (the determinant is multiplied by each such scalar). When we arrive at the echelon form, if there is a row of zeros, we know the original determinant was 0. If not, the echelon form will be exactly the identity matrix, and we know its determinant by 1.4.

This computation process also shows that there is only one determinant function satisfying the demands 1.1–1.4. If we have two determinant functions, they must both associate to **A** the number 0 if the echelon form **B** has a row of zeros, and the number c (= product of ± 1's and nonzero scalars, as earlier) if the echelon form **B** is the identity matrix. In short, two determinant functions have to associate the same number to the matrix **A** for every **A**; they must be the same function.

Example 1. Find det **A** if

$$\mathbf{A} = \begin{pmatrix} 1 & 2 & 3 & 4 \\ 1 & 0 & 0 & 2 \\ 2 & 1 & 3 & 5 \\ -1 & 1 & 1 & 1 \end{pmatrix}.$$

$$\det \mathbf{A} = -\det \begin{pmatrix} 1 & 0 & 0 & 2 \\ 1 & 2 & 3 & 4 \\ 2 & 1 & 3 & 5 \\ -1 & 1 & 1 & 1 \end{pmatrix} = -\det \begin{pmatrix} 1 & 0 & 0 & 2 \\ 0 & 2 & 3 & 2 \\ 0 & 1 & 3 & 1 \\ 0 & 1 & 1 & 3 \end{pmatrix}$$

$$= \det \begin{pmatrix} 1 & 0 & 0 & 2 \\ 0 & 1 & 1 & 3 \\ 0 & 1 & 3 & 1 \\ 0 & 2 & 3 & 2 \end{pmatrix} = \det \begin{pmatrix} 1 & 0 & 0 & 2 \\ 0 & 1 & 1 & 3 \\ 0 & 0 & 2 & -2 \\ 0 & 0 & 1 & -4 \end{pmatrix}$$

$$= 2 \det \begin{pmatrix} 1 & 0 & 0 & 2 \\ 0 & 1 & 1 & 3 \\ 0 & 0 & 1 & -1 \\ 0 & 0 & 1 & -4 \end{pmatrix} = 2 \det \begin{pmatrix} 1 & 0 & 0 & 2 \\ 0 & 1 & 1 & 3 \\ 0 & 0 & 1 & -1 \\ 0 & 0 & 0 & -3 \end{pmatrix}$$

$$= -6 \det \begin{pmatrix} 1 & 0 & 0 & 2 \\ 0 & 1 & 1 & 3 \\ 0 & 0 & 1 & -1 \\ 0 & 0 & 0 & 1 \end{pmatrix} = -6 \det I = -6.$$

Example 2.

$$\det \begin{pmatrix} a & b \\ c & d \end{pmatrix} = ad - bc$$

because, if $a \neq 0$, then

$$\det\begin{pmatrix} a & b \\ c & d \end{pmatrix} = a \det\begin{pmatrix} 1 & a^{-1}b \\ c & d \end{pmatrix} = a \det\begin{pmatrix} 1 & a^{-1}b \\ 0 & d - ca^{-1}b \end{pmatrix}$$

$$= a(d - ca^{-1}b) \det\begin{pmatrix} 1 & a^{-1}b \\ 0 & 1 \end{pmatrix}$$

$$= a(d - ca^{-1}b) \det \mathbf{I} = ad - bc.$$

We leave it to the reader to give a similar proof when $b \neq 0$. If neither a nor b is different from zero, the asserted formula is true by 1.5. Therefore the formula is always true.

1.7 $\det(\mathbf{u} + \mathbf{u}', \mathbf{u}_2, \ldots, \mathbf{u}_n)$

$$= \det(\mathbf{u}, \mathbf{u}_2, \ldots, \mathbf{u}_n) + \det(\mathbf{u}', \mathbf{u}_2, \ldots, \mathbf{u}_n).$$

Proof. First, suppose $\mathbf{u} = \sum_{p=2}^{n} a_p\mathbf{u}_p$. Then by repeated use of 1.1 (subtract $a_2\mathbf{u}_2$ from $\mathbf{u} + \mathbf{u}'$, $a_3\mathbf{u}_3$ from the result, and so on) we find

$$\det(\mathbf{u} + \mathbf{u}', \mathbf{u}_2, \ldots, \mathbf{u}_n) = \det(\mathbf{u}', \mathbf{u}_2, \ldots, \mathbf{u}_n).$$

In particular, if $\mathbf{u}' = \mathbf{0}$, we get $\det(\mathbf{u}, \mathbf{u}_2, \ldots, \mathbf{u}_n) = 0$ by 1.5. These two facts together prove 1.7 when $\mathbf{u}$ is a linear combination of $\mathbf{u}_2, \ldots, \mathbf{u}_n$. Now suppose $\mathbf{u}$ is not a linear combination of $\mathbf{u}_2, \ldots, \mathbf{u}_n$. If $\{\mathbf{u}_2, \ldots, \mathbf{u}_n\}$ is a dependent set, then so are $\{\mathbf{u} + \mathbf{u}', \mathbf{u}_2, \ldots, \mathbf{u}_n\}$, $\{\mathbf{u}, \mathbf{u}_2, \ldots, \mathbf{u}_n\}$ and $\{\mathbf{u}', \mathbf{u}_2, \ldots, \mathbf{u}_n\}$, so all three determinants in 1.7 are zero, by 1.6, and the statement is true. If $\{\mathbf{u}_2, \ldots, \mathbf{u}_n\}$ is an independent set and $\mathbf{u}$ is not a linear combination of $\mathbf{u}_2, \ldots, \mathbf{u}_n$, then $\{\mathbf{u}, \mathbf{u}_2, \ldots, \mathbf{u}_n\}$ is an independent set (why?), hence, it is a basis of R^n. This means that $\mathbf{u}'$ is a linear combination $\mathbf{u}' = a\mathbf{u} + \sum_{p=2}^{n} a_p\mathbf{u}_p$. Then

$$\det(\mathbf{u} + \mathbf{u}', \mathbf{u}_2, \ldots, \mathbf{u}_n) = \det\Big(\sum_{p=2}^{n} a_p\mathbf{u}_p + (1 + a)\mathbf{u}, \mathbf{u}_2, \ldots, \mathbf{u}_n\Big)$$

and again our first vector is a sum with the first summand a linear combination of $\mathbf{u}_2, \ldots, \mathbf{u}_n$. Therefore the determinant is

$$\det((1 + a)\mathbf{u}, \mathbf{u}_2, \ldots, \mathbf{u}_n)$$
$$= (1 + a) \det(\mathbf{u}, \mathbf{u}_2, \ldots, \mathbf{u}_n)$$
$$= \det(\mathbf{u}, \mathbf{u}_2, \ldots, \mathbf{u}_n) + a \det(\mathbf{u}, \mathbf{u}_2, \ldots, \mathbf{u}_n)$$
$$= \det(\mathbf{u}, \mathbf{u}_2, \ldots, \mathbf{u}_n) + \det(a\mathbf{u}, \mathbf{u}_2, \ldots, \mathbf{u}_n)$$
$$= \det(\mathbf{u}, \mathbf{u}_2, \ldots, \mathbf{u}_n) + \det\Big(a\mathbf{u} + \sum_{p=2}^{n} a_p\mathbf{u}_p, \mathbf{u}_2, \ldots, \mathbf{u}_n\Big)$$
$$= \det(\mathbf{u}, \mathbf{u}_2, \ldots, \mathbf{u}_n) + \det(\mathbf{u}', \mathbf{u}_2, \cdots, \mathbf{u}_n).$$

1.8 $\det(\mathbf{AB}) = (\det \mathbf{A})(\det \mathbf{B})$.

Proof. First, suppose $\mathbf{A}$ is an elementary matrix (Chapter 4, 2.9) obtained from the identity matrix $\mathbf{I}$ by adding a times the pth row to the qth. By 1.1, $\det \mathbf{A} = \det \mathbf{I} = 1$ and $\det(\mathbf{AB}) = \det \mathbf{B}$ since $\mathbf{AB}$ is obtained by adding a times the pth row of $\mathbf{B}$ to the qth. Thus 1.8 is true in this case. Similarly, we prove 1.8 when $\mathbf{A}$ is an elementary matrix corresponding to manipulations of type 2 or 3. Now every n by n matrix $\mathbf{A}$ of rank n has $\mathbf{I}$ as its echelon form, so is obtainable from $\mathbf{I}$ by multiplying by a sequence of elementary matrices: $\mathbf{A} = \mathbf{P}_1\mathbf{P}_2\cdots\mathbf{P}_s\mathbf{I}$. Since each $\mathbf{P}_p$ is elementary, we now know

$$\begin{aligned}\det(\mathbf{AB}) &= \det(\mathbf{P}_1(\mathbf{P}_2\cdots\mathbf{P}_s\mathbf{B}))\\ &= (\det \mathbf{P}_1)(\det \mathbf{P}_2\cdots\mathbf{P}_s\mathbf{B})\\ &= (\det \mathbf{P}_1)(\det \mathbf{P}_2)(\det \mathbf{P}_3\cdots\mathbf{P}_s\mathbf{B})\\ &= \cdots = \left(\prod_{p=1}^{s} \det \mathbf{P}_p\right)(\det \mathbf{B}).\end{aligned}$$

(The notation $\prod_{p=1}^{s}$ is like the notation $\sum_{p=1}^{s}$ but with products replacing sums, which explains the capital pi replacing the capital sigma; $\prod_{p=1}^{s} \det \mathbf{P}_p$ means $(\det \mathbf{P}_1)(\det \mathbf{P}_2)\cdots(\det \mathbf{P}_s)$.) In particular, if $\mathbf{B} = \mathbf{I}$, we get $\det \mathbf{A} = \prod_{p=1}^{s} \det \mathbf{P}_p$. Putting these last two equations together, we get 1.8 for all $\mathbf{A}$ of rank n. Finally, if $\mathbf{A}$ has rank less than n, so has $\mathbf{AB}$, according to 4.11 in Chapter 5; then 1.6 says that both sides of 1.8 are zero. This completes the proof.

There is one last computational result you should be aware of, which we shall prove in 3.10: Manipulations on the columns of a matrix produce the same effect on the determinant as the same manipulations on the rows.

PROBLEMS

1. Compute the determinants of the following matrices. Tell which of the matrices are invertible.

(a) $\begin{pmatrix}1 & 0 & 0\\ 0 & 1 & 0\\ 0 & 0 & 1\end{pmatrix}$; (b) $\begin{pmatrix}1 & 0 & 0\\ 0 & 2 & 0\\ 0 & 0 & 3\end{pmatrix}$; (c) $\begin{pmatrix}1 & 2 & 3\\ 4 & 5 & 6\\ 7 & 8 & 9\end{pmatrix}$;

(d) $\begin{pmatrix}1 & 1 & 2\\ 2 & 2 & 4\\ 3 & 1 & 0\end{pmatrix}$; (e) $\begin{pmatrix}1 & 0 & 1\\ 2 & 1 & 2\\ 3 & 4 & 1\end{pmatrix}$;

(f) $\begin{pmatrix} 1 & 0 & 1 & 1 & 3 \\ 2 & 1 & -1 & 1 & 2 \\ 5 & 4 & 0 & -2 & 1 \\ 1 & -1 & 0 & 0 & 2 \\ 0 & 1 & 2 & -1 & -1 \end{pmatrix}$; (g) $\begin{pmatrix} 1 & 0 & 0 & 0 \\ 0 & 1 & 0 & 0 \\ a & b & p & q \\ c & d & r & s \end{pmatrix}$;

(h) $\begin{pmatrix} 1 & 0 & a & c \\ 0 & 1 & b & d \\ 0 & 0 & p & r \\ 0 & 0 & q & s \end{pmatrix}$.

2. Prove that the determinant of a diagonal matrix is the product of the diagonal entries.

3. Prove that the determinant of a triangular matrix ($a_{pq} = 0$ whenever $p < q$) is the product of the diagonal entries. Hint: If all the diagonal entries are nonzero, use 1.1 to reduce to Problem 2. If one of the diagonal entries is zero, prove that the rank is less than n and so the determinant is zero.

4. Use determinants to check whether the following sets of vectors are dependent or independent.

$$[1, 2], [2, 1]; \qquad [1, 2, 3], [4, 5, 6], [7, 8, 9];$$
$$[1, 0, 0], [0, 1, 0], [0, 0, 1].$$

5. Find all numbers x such that the vectors $[x, 1, 1]$, $[1, x, 0]$, $[1, 1, 2]$ are dependent. Use determinants.

6. If

$$\mathbf{A} = \begin{pmatrix} a & 0 & 0 & \cdots & 0 \\ 0 & * & * & \cdots & * \\ 0 & * & * & \cdots & * \\ \vdots & \vdots & \vdots & & \vdots \\ 0 & * & * & \cdots & * \end{pmatrix}$$

is a square n by n matrix where the asterisks comprise an $(n - 1)$ by $(n - 1)$ matrix that we shall call $\mathbf{B}$, show that $\det \mathbf{A} = a \det \mathbf{B}$. (Argue that to reduce $\mathbf{A}$ to echelon form it suffices to reduce $\mathbf{B}$ to echelon form.)

7. (a) Suppose an n by n matrix $\mathbf{U}$ can be partitioned into blocks

$$\mathbf{U} = \begin{pmatrix} \mathbf{A} & \mathbf{O} \\ \mathbf{O} & \mathbf{B} \end{pmatrix},$$

where $\mathbf{A}$ is an r by r matrix, $\mathbf{B}$ is an $n - r$ by $n - r$ matrix, and the $\mathbf{O}$'s are of suitable size to fill out the n by n matrix (one is r by $n - r$

and the other is $n - r$ by r). Show $\det \mathbf{U} = \det \mathbf{A} \det \mathbf{B}$ by generalizing the argument in Problem 6.

(b) Suppose a similar partitioning of an n by n matrix $\mathbf{V}$ results in

$$\mathbf{V} = \begin{pmatrix} \mathbf{I} & \mathbf{O} \\ \mathbf{C} & \mathbf{B} \end{pmatrix}$$

Show $\det \mathbf{V} = \det \mathbf{B}$.

(c) If a matrix partitions into

$$\begin{pmatrix} \mathbf{A} & \mathbf{O} \\ \mathbf{C} & \mathbf{B} \end{pmatrix}$$

show that its determinant is $\det \mathbf{A} \det \mathbf{B}$. Hint: show

$$\begin{pmatrix} \mathbf{A} & \mathbf{O} \\ \mathbf{C} & \mathbf{B} \end{pmatrix} = \begin{pmatrix} \mathbf{A} & \mathbf{O} \\ \mathbf{O} & \mathbf{I} \end{pmatrix} \begin{pmatrix} \mathbf{I} & \mathbf{O} \\ \mathbf{C} & \mathbf{B} \end{pmatrix}$$

and use 1.8.

8. If $\mathbf{A}$ is invertible, show $\det(\mathbf{A}^{-1}) = (\det \mathbf{A})^{-1}$.

9. (a) If $\mathbf{A}^2 = \mathbf{A}$ but $\mathbf{A} \neq \mathbf{I}$, show $\det \mathbf{A} = 0$. (Show that if $\mathbf{A}^2 = \mathbf{A}$ and $\mathbf{A}^{-1}$ exists, then $\mathbf{A} = \mathbf{I}$.)

(b) If $\mathbf{A}$ is the 3 by 3 matrix that represents projection on a fixed plane in R^3 (for every $\mathbf{v}$ in R^3, $\mathbf{Av}$ is the vector projection of $\mathbf{v}$ in the given plane) then $\det \mathbf{A} = 0$. Prove this in two ways: show $\mathbf{A}^2 = \mathbf{A}$ by geometry, and use (a); and show rank $\mathbf{A} < 3$ by finding the dimension of the image of the corresponding linear mapping.

10. (a) If one column of $\mathbf{A}$ is $\mathbf{0}$, show $\det \mathbf{A} = 0$. (Use 1.6.)

(b) If one column of $\mathbf{A}$ is a scalar times another, then $\det \mathbf{A} = 0$.

(c) $\det \mathbf{A} = 0$ if and only if the columns of $\mathbf{A}$ are dependent. (Use Chapter 5, 4.2.)

11. If $\mathbf{A}$ is an n by n matrix and a is a scalar, show

$$\det(a\mathbf{A}) = a^n \det \mathbf{A}.$$

In particular, $\det(-\mathbf{A}) = \det \mathbf{A}$ or $-\det \mathbf{A}$ according as n is even or odd.

12. There is no general theorem about the determinant of a sum of matrices. Give examples to show

$$\det(\mathbf{A} + \mathbf{B}) \neq \det \mathbf{A} + \det \mathbf{B}.$$

13. As a corollary of Theorem 1.6, prove that a system of n linear equations in n unknowns has a unique solution if and only if the determinant of its coefficient matrix is not zero.

2. EXPLICIT FORMULAS

2.1 If **A** is a 2 by 2 matrix (a_{pq}), where $p, q = 1, 2$, then

$$\det \mathbf{A} = a_{11}a_{22} - a_{12}a_{21}.$$

If **A** is a 3 by 3 matrix (a_{pq}), where $p, q = 1, 2, 3$, then

$$\det \mathbf{A} = a_{11}a_{22}a_{33} + a_{12}a_{23}a_{31} + a_{13}a_{21}a_{32}$$
$$- a_{13}a_{22}a_{31} - a_{12}a_{21}a_{33} - a_{11}a_{23}a_{32}.$$

The first of these formulas is proved in Example 2, Section 1. Similar techniques can show that the second formula is also true.

You can remember these two formulas by multiplying the entries on each arrow in the following diagrams and adding the terms from

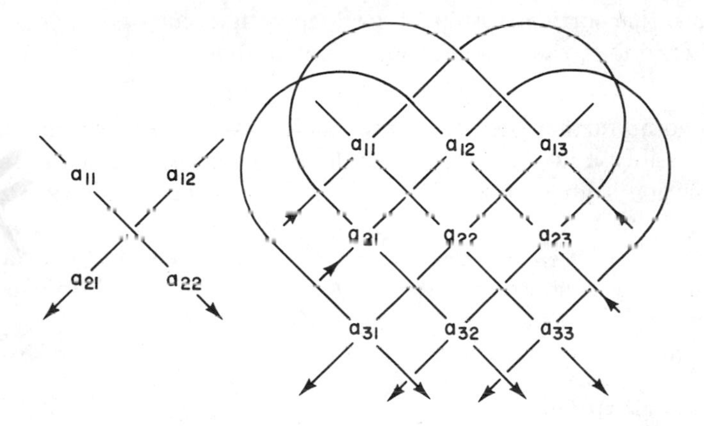

arrows with heads pointing down right and subtracting the terms from arrows with heads pointing down left. Note that the analogue of the 3 by 3 diagram does not work for 2 by 2's: The result of

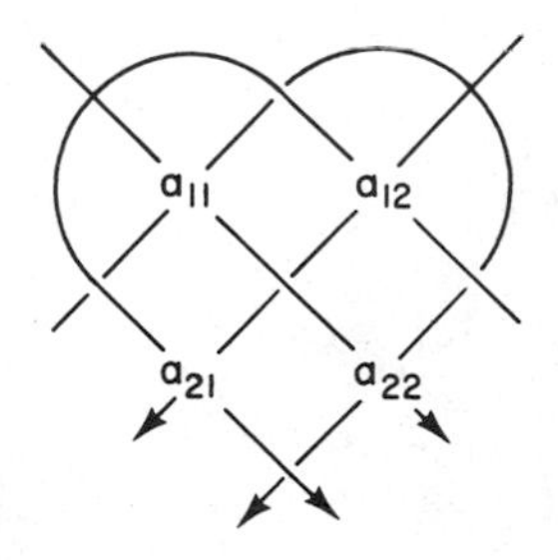

is always zero. There is no analogous diagram for computing 4 by 4 or larger matrices. There is a formula for the determinant of an n by n matrix that generalizes 2.1:

2.2 If $\mathbf{A}$ is an n by n matrix (a_{pq}), $p, q = 1, 2, \ldots, n$, then det $\mathbf{A}$ is the sum of all possible terms

$$\sum \pm a_{1p_1}a_{2p_2}a_{3p_3}\cdots a_{np_n},$$

where $p_1, p_2, \ldots, p_n$ are the numbers from 1 to n in some order; in other words, each term is a product of entries, one out of each row such that no two are out of the same column. Then all possible such products are added with a suitable $\pm$ sign in front of each. This sign is determined by looking at the sequence of numbers denoted by $p_1, p_2, \ldots, p_n$; this sequence is a permutation of the numbers 1, 2, $\ldots$, n and can be returned to this natural order by repeatedly interchanging pairs of numbers. If you need an even number of interchanges, this particular product is added with a plus sign; if you need an odd number, the product takes a minus sign.

We go no further with this formula. It has $n!$ terms, which makes it impossibly unwieldy even for a 10 by 10 matrix. A more useful formulation is given in 2.4. It is an inductive formula, allowing us to compute the determinant of an n by n matrix by computing n determinants, each $(n - 1)$ by $(n - 1)$; each of these in turn can be computed by computing $n - 1$ $(n - 2)$ by $(n - 2)$'s, and so on. Computationally, this is still horrendous for large n.

2.3 Definition

If $\mathbf{A} = (a_{pq})$ is an n by n matrix, its *pq-minor* M_{pq} is the determinant of the $(n - 1)$ by $(n - 1)$ matrix obtained from $\mathbf{A}$ by striking out the pth row and the qth column. The *pq-cofactor* A_{pq} is $(-1)^{p+q}M_{pq}$. Note that these cofactors are numbers.

2.4 Expansion by minors

$$\begin{aligned}\det \mathbf{A} &= a_{11}A_{11} + a_{12}A_{12} + \cdots + a_{1n}A_{1n}\\ &= a_{11}M_{11} - a_{12}M_{12} + \cdots \pm a_{1n}M_{1n}.\end{aligned}$$

Similarly, we may use the pth row instead of the first:

$$\det \mathbf{A} = \sum_{q=1}^{n} a_{pq}A_{pq} \qquad (p \text{ fixed})$$

or any column:

$$\det \mathbf{A} = \sum_{p=1}^{n} a_{pq}A_{pq} \qquad (q \text{ fixed}).$$

We shall not give a complete proof of either 2.2 or 2.4 here. Since we know there is at most one function of matrices that satisfies 1.1–1.4, all we would have to do is verify that the formula 2.2 does in fact give such a function, and similarly for the formula in 2.4. To verify this for 2.2 would take us too far afield into a study of permutations. We refer the interested reader to any book on determinants. The proof that 2.4 gives a function satisfying 1.1–1.4 is sketched in Problem 10.

We should note that these omitted proofs constitute the proof of the existence of a determinant that we promised at the outset in Section 1. In fact, we have exhibited such a function in 2.2 and in 2.4 because these satisfy the conditions 1.1–1.4.

Cofactors have one essential useful property, which can be phrased in several ways (see 2.3–2.7). We begin with a system of n linear equations in n unknowns:

$$\sum_{q=1}^{n} a_{pq}x_q = b_p \qquad (p = 1, \ldots, n).$$

We call the coefficient matrix $\mathbf{A}$. If $n = 2$, we can easily take a linear combination of the two equations that contains only x_1 and not x_2—and then solve for x_1. For larger n, we should be able to take a linear combination of the n equations that contains only x_1 and neither x_2 nor $x_3, \ldots,$ nor x_n, and then solve for x_1. The coefficients $c_1, \ldots, c_n$ of this linear combination need to have the properties

$$c_1a_{11} + c_2a_{21} + c_na_{n1} \neq 0$$

and

$$\begin{aligned} c_1a_{12} + c_2a_{22} + \cdots + c_na_{n2} &= 0, \\ &\vdots \\ c_1a_{1n} + c_2a_{2n} + \cdots + c_na_{nn} &= 0; \end{aligned}$$

that is, the vector $[c_1, \ldots, c_n]$ must be orthogonal to all the columns of **A** except the first. We assert that $c_p = A_{p1}$ will do.

2.5 The n-tuple $\mathbf{A}_1 = [A_{11}, A_{21}, \ldots, A_{n1}]$ is orthogonal to all columns of **A** except the first, and the dot product of $\mathbf{A}_1$ with the first column of **A** is det **A**.

Proof. $\sum a_{p1}A_{p1} = \det \mathbf{A}$ from 2.4, proving the last statement of 2.5. The dot product of $\mathbf{A}_1$ with the second column of **A**, $\sum a_{p2}A_{p1}$, can also be thought of as a determinant, namely, that of the matrix whose last $n - 1$ columns are the same as in **A**, so that its $p1$-cofactor is the same as A_{p1} for all p, but whose first column contains all the a_{p2}'s. But then the first and second columns of this matrix are both

$$[a_{12}, a_{22}, \ldots, a_{n2}],$$

and a matrix with two equal columns has determinant zero. The remaining dot products vanish similarly. If we single out the qth column instead of the first, we get the following.

2.6 The n-tuple $\mathbf{A}_q = [A_{1q}, A_{2q}, \ldots, A_{nq}]$ is orthogonal to all the columns of **A** except the qth, and the dot product of $\mathbf{A}_q$ with the qth column of **A** is det **A**.

This motivates the following definition.

2.7 Definition

The *adjoint* of an n by n matrix **A** is the matrix whose entry in the qth row, pth column is the pq-cofactor A_{pq} of **A**. It is the transpose of the matrix made of the cofactors of **A** in their natural order. Its rows are the vectors $\mathbf{A}_q$ of 2.6.

2.8 Theorem

If **A** is an n by n matrix and **B** is its adjoint, then $\mathbf{BA} = (\det \mathbf{A})\mathbf{I}$. If **A** is invertible, then $\mathbf{A}^{-1}$ is $(\det \mathbf{A})^{-1}$ times the adjoint of **A**.

Proof. This is merely a restatement of 2.6.

2.9 Cramer's Rule

Consider a system of n equations in n unknowns whose coefficient matrix **A** is invertible:

$$\sum_{q=1}^{n} a_{pq}x_q = b_p \qquad (p = 1, \ldots, n).$$

Then the system has a unique solution, given by the formulas

$$x_q = \frac{\det \mathbf{B}_q}{\det \mathbf{A}}$$

where $\mathbf{B}_q$ is the matrix obtained from **A** by replacing the qth column by

$$\begin{pmatrix} b_1 \\ \vdots \\ b_n \end{pmatrix}$$

Proof. If we used the cofactors $A_{1q}, \ldots, A_{nq}$ as coefficients in taking a linear combination of the equations (as promised earlier), 2.6 says that we get

$$\left(\sum_{p=1}^{n} A_{pq}a_{pq}\right)x_q = \sum_{p=1}^{n} A_{pq}b_p.$$

Both those sums are determinants; the first is det **A**, by 2.4 and the second is det $\mathbf{B}_q$, by the same argument as was used in 2.5.

Example 1. Let

$$\mathbf{A} = \begin{pmatrix} 1 & 0 & 2 & 0 \\ 0 & 3 & 0 & 4 \\ 5 & 0 & 6 & 0 \\ 0 & 7 & 0 & 8 \end{pmatrix}$$

and compute det **A** and $\mathbf{A}^{-1}$ by 2.4 and 2.8.

$$A_{11} = \det\begin{pmatrix} 3 & 0 & 4 \\ 0 & 6 & 0 \\ 7 & 0 & 8 \end{pmatrix}; \qquad A_{12} = -\det\begin{pmatrix} 0 & 0 & 4 \\ 5 & 6 & 0 \\ 0 & 0 & 8 \end{pmatrix};$$

$$A_{13} = \det\begin{pmatrix} 0 & 3 & 4 \\ 5 & 0 & 0 \\ 0 & 7 & 8 \end{pmatrix}; \qquad A_{14} = -\det\begin{pmatrix} 6 & 3 & 0 \\ 5 & 0 & 6 \\ 0 & 7 & 0 \end{pmatrix}.$$

Hence,

$$\det \mathbf{A} = 1A_{11} + 0A_{12} + 2A_{13} + 0A_{14} = A_{11} + 2A_{13}.$$

Compute A_{11} similarly:

$$A_{11} = 3 \det \begin{pmatrix} 6 & 0 \\ 0 & 8 \end{pmatrix} - 0 \det \begin{pmatrix} 0 & 0 \\ 7 & 8 \end{pmatrix} + 4 \det \begin{pmatrix} 0 & 6 \\ 7 & 0 \end{pmatrix}$$

$$= 3(48) + 4(-42) = -24.$$

And

$$A_{13} = 0 \det \begin{pmatrix} 0 & 0 \\ 7 & 8 \end{pmatrix} - 3 \det \begin{pmatrix} 5 & 0 \\ 0 & 8 \end{pmatrix} + 4 \det \begin{pmatrix} 5 & 0 \\ 0 & 7 \end{pmatrix}$$

$$= (-3)(40) + 4(35) = 20.$$

Hence

$$\det \mathbf{A} = -24 + 2(20) = 16.$$

Check this by reducing $\mathbf{A}$ to echelon form as in Section 1.

Since $\det \mathbf{A} \neq 0$, $\mathbf{A}$ is invertible. To use 2.8 we need to know A_{pq} for all p and q. We have just computed A_{11} and A_{13}; we leave it to the reader to compute the rest. You should find that

$$\begin{array}{llll} A_{11} = -24, & A_{12} = 0, & A_{13} = 20, & A_{14} = 0; \\ A_{21} = 0, & A_{22} = -32, & A_{23} = 0, & A_{24} = 28; \\ A_{31} = 8, & A_{32} = 0, & A_{33} = -4, & A_{34} = 0; \\ A_{41} = 0, & A_{42} = 16, & A_{43} = 0, & A_{44} = -12. \end{array}$$

Then 2.8 says

$$\mathbf{A}^{-1} = \begin{pmatrix} -\frac{24}{16} & 0 & \frac{8}{16} & 0 \\ 0 & -\frac{32}{16} & 0 & \frac{16}{16} \\ \frac{20}{16} & 0 & -\frac{4}{16} & 0 \\ 0 & \frac{28}{16} & 0 & -\frac{12}{16} \end{pmatrix}$$

$$= \begin{pmatrix} -\frac{3}{2} & 0 & \frac{1}{2} & 0 \\ 0 & -2 & 0 & 1 \\ \frac{5}{4} & 0 & -\frac{1}{4} & 0 \\ 0 & \frac{7}{4} & 0 & -\frac{3}{4} \end{pmatrix}.$$

Example 2. Solve the following system of equations, using Cramer's rule:

$$2x + 3y + z = 1, \qquad x + y + z = 2, \qquad 4x + y + z = 3.$$

The answer is

$$x = \frac{\det\begin{pmatrix}1 & 3 & 1\\ 2 & 1 & 1\\ 3 & 1 & 1\end{pmatrix}}{\det\begin{pmatrix}2 & 3 & 1\\ 1 & 1 & 1\\ 4 & 1 & 1\end{pmatrix}} = \frac{2}{6} = \frac{1}{3};$$

$$y = \frac{\det\begin{pmatrix}2 & 1 & 1\\ 1 & 2 & 1\\ 4 & 3 & 1\end{pmatrix}}{\det\begin{pmatrix}2 & 3 & 1\\ 1 & 1 & 1\\ 4 & 1 & 1\end{pmatrix}} = -\frac{4}{6} = -\frac{2}{3};$$

$$z = \frac{\det\begin{pmatrix}2 & 3 & 1\\ 1 & 1 & 2\\ 4 & 1 & 3\end{pmatrix}}{\det\begin{pmatrix}2 & 3 & 1\\ 1 & 1 & 1\\ 4 & 1 & 1\end{pmatrix}} = \frac{14}{6} = \frac{7}{3}.$$

Check this against the solution method of Chapter 4. Note that Cramer's rule is longer and (therefore?) more susceptible to error.

PROBLEMS

1. Show that 2.2 does reduce to 2.1 when $n = 2$ or 3.

2. How many terms are there in the formula 2.2 for det **A** if **A** is 2 by 2, 3 by 3, 4 by 4, 10 by 10? Write out all the terms for the determinant of an arbitrary 4 by 4 matrix.

3. If **A** is a 3 by 3 matrix, show that the vector $\mathbf{A}_1$ in 2.5 is the cross product of the last two columns of **A**, and that $\mathbf{A}_2$ and $\mathbf{A}_3$ in 2.6 are also cross products of columns of **A** (of which columns)?

4. If $\mathbf{u}_1, \mathbf{u}_2, \mathbf{u}_3$ are vectors in R^3, show explicitly that

$$\det(\mathbf{u}_1, \mathbf{u}_2, \mathbf{u}_3) = (\mathbf{u}_1 \times \mathbf{u}_2) \cdot \mathbf{u}_3.$$

5. Compute the determinants in Problem 1, Section 1 by expanding by minors as in 2.4.

6. Use 2.8 to compute the inverses of those matrices in Problem 1, Section 1 that are invertible.

7. Show that the determinant of every rotation in R^2 is 1.

8. Solve by Cramer's rule:

(a)
$$\begin{aligned} x + y + z &= 1, \\ 2x - y &= 2, \\ x + 2y - z &= 0. \end{aligned}$$

(b)
$$\begin{aligned} 2x + y &= 1, \\ x + 2y - 3z &= 0, \\ x + y + 2z + t &= 0, \\ -x - y + z + 2t &= 1. \end{aligned}$$

(c)
$$\begin{aligned} x_1 &= 2, \\ 2x_1 + x_2 &= 1, \\ 2x_1 + x_3 &= 0, \\ x_2 + x_3 + x_4 &= 1, \\ x_1 + 2x_2 + 3x_3 + 4x_4 + 5x_5 &= 6. \end{aligned}$$

(d)
$$\begin{pmatrix} 2 & 1 & -2 \\ -2 & 2 & -1 \\ 1 & 2 & 2 \end{pmatrix} \begin{pmatrix} x \\ y \\ z \end{pmatrix} = \begin{pmatrix} a \\ b \\ c \end{pmatrix} \qquad \text{(solve for } x, y, z\text{)}.$$

9. Use Cramer's rule to solve as many as possible of the systems of equations in Problem 2, Chapter 4, Section 1.

10. Prove that there is one and only one determinant function and it is computable by expansion by minors on any row.

(Expansion by minors on a column we postpone until the end of Section 3.) Follow the following pattern.

(a) For 1 by 1 matrices, show that there is a determinant function, namely, $d_1(\mathbf{A})$ = the entry in $\mathbf{A}$. This proves that for 1 by 1 matrices there is one and only one determinant.

(b) For 2 by 2 matrices show that the formula in 2.1 is a determinant function. That is, if

$$\mathbf{A} = \begin{pmatrix} a_{11} & a_{12} \\ a_{21} & a_{22} \end{pmatrix},$$

define $d_2(\mathbf{A}) = a_{11}A_{11} + a_{12}A_{12}$ (where A_{11} and A_{12} are determinants of 1 by 1 matrices, and so are computable by part (a)) and then show that d_2 satisfies 1.1–1.4. This proves the existence of a determinant function (namely, d_2) for 2 by 2 matrices.

(c) For 2 by 2 matrices, define $d_2'(\mathbf{A})$ as the expansion by minors on the second row: $d_2'(\mathbf{A}) = a_{21}A_{21} + a_{22}A_{22}$. Show that $d_2(\mathbf{A}) = d_2'(\mathbf{A})$ for every $\mathbf{A}$. This will complete the proof of most of 2.4 for 2 by 2 matrices.

(d) Assume the theorem is true for $(n-1)$ by $(n-1)$ matrices. If $\mathbf{A}$ is an n by n matrix, define $d_n(\mathbf{A})$ by the formula in 2.4:

$$d_n(\mathbf{A}) = a_{11}A_{11} + a_{12}A_{12} + \cdots + a_{1n}A_{1n}.$$

(According to our assumption, each A_{1q}, which is $\pm$ the determinant of an $(n-1)$ by $(n-1)$ matrix, is well defined and computable by minors.) Then show that d_n satisfies 1.4.

(e) Following the program in part (d), show that d_n satisfies 1.3 by studying two cases: where the first row is multiplied by a scalar (what happens to a_{1q}? to A_{1q}?) and where some other row is multiplied by a scalar.

(f) Show that d_n defined in part (d) satisfies 1.2 by studying two cases: where neither of the two rows interchanged is the first row (What happens to a_{1q}? to A_{1q}?) and where the first row is interchanged with another row. In the latter case, you will have to expand A_{1q} by minors on the row that is being moved.

(g) Show that d_n defined in part (d) satisfies 1.1. The pattern of proof is similar to that in part (f).

(h) Prove the theorem for all matrices by induction on the size of the matrix. This has been done in parts (a), (b), and (c) for 1 by 1 and 2 by 2 matrices. The induction step has nearly been completed in parts (d), (e), (f), and (g). All you have to do is show that expansion by minors on the qth row $(q \neq 1)$ gives the same result as d_n (compare part (c)).

3. TRANSPOSES

3.1 Definition

If $\mathbf{A}$ is a matrix, the *transpose* of $\mathbf{A}$, written ${}^t\mathbf{A}$ (or sometimes $\mathbf{A}'$) is the matrix whose p, q-entry (entry in the pth row, qth column) is the same as the q, p-entry in $\mathbf{A}$. In other words, ${}^t\mathbf{A}$ is obtained from

A by flipping the matrix over the main diagonal; if **A** is m by n, then ${}^t\mathbf{A}$ is n by m; the rows of **A** are the columns of ${}^t\mathbf{A}$.

Example 1.

$$ {}^t\begin{pmatrix} 1 & 2 & 3 \\ 4 & 5 & 6 \\ 7 & 8 & 9 \end{pmatrix} = \begin{pmatrix} 1 & 4 & 7 \\ 2 & 5 & 8 \\ 3 & 6 & 9 \end{pmatrix}; \qquad {}^t(1 \quad 2) = \begin{pmatrix} 1 \\ 2 \end{pmatrix}. $$

The transpose of every diagonal matrix is the same matrix. In particular, ${}^t\mathbf{I} = \mathbf{I}$.

3.2 Definition

If f is a linear mapping from R^n to R^m and its matrix is **A**, then ${}^tf$, the *transpose* of f, is the linear mapping from R^m to R^n whose matrix is ${}^t\mathbf{A}$.

3.3 (a) If f is a linear mapping from R^n to R^m, then for all vectors $\mathbf{v}$ in R^n and all vectors $\mathbf{w}$ in R^m,

$$ f(\mathbf{v}) \cdot \mathbf{w} = \mathbf{v} \cdot {}^tf(\mathbf{w}). $$

(b) Conversely, if f is a linear mapping from R^n to R^m and g is a linear mapping from R^m to R^n such that $f(\mathbf{v}) \cdot \mathbf{w} = \mathbf{v} \cdot g(\mathbf{w})$ for all $\mathbf{v}$ in R^n and all $\mathbf{w}$ in R^m, then $g = {}^tf$.

Proof. Suppose the matrix of f is (a_{pq}). This means that a_{pq} is the pth component of $f(\mathbf{i}_q)$, or $a_{pq} = f(\mathbf{i}_q) \cdot \mathbf{i}_p$. We first prove 3.3 for the special case where $\mathbf{v} = \mathbf{i}_q$ and $\mathbf{w} = \mathbf{i}_p$. Here $f(\mathbf{i}_q) \cdot \mathbf{i}_p = a_{pq}$ while $\mathbf{i}_q \cdot {}^tf(\mathbf{i}_p) =$ the qth component of ${}^tf(\mathbf{i}_p) =$ the entry in the qth row, pth column of the matrix of ${}^tf =$ the entry in the pth row, qth column of the matrix of $f = a_{pq}$.

Now for the general case, write $\mathbf{v} = \sum_q x_q\mathbf{i}_q$ and $\mathbf{w} = \sum_p y_p\mathbf{i}_p$, and compute

$$
\begin{aligned}
f(\mathbf{v}) \cdot \mathbf{w} &= f\Big(\sum_q x_q\mathbf{i}_q\Big) \cdot \mathbf{w} \\
&= \sum_q x_q f(\mathbf{i}_q) \cdot \mathbf{w} \\
&= \sum_q x_q f(\mathbf{i}_q) \cdot \sum_p y_p\mathbf{i}_p \\
&= \sum_{p,q} x_q y_p [f(\mathbf{i}_q) \cdot \mathbf{i}_p] \\
&= \sum_{p,q} x_q y_p [\mathbf{i}_q \cdot {}^tf(\mathbf{i}_p)] \\
&= \sum_q x_q\mathbf{i}_q \cdot {}^tf\Big(\sum_p y_p\mathbf{i}_p\Big) \\
&= \mathbf{v} \cdot {}^tf(\mathbf{w}).
\end{aligned}
$$

For the converse, we need an elementary lemma on dot products:

If $\mathbf{v} \cdot \mathbf{x} = \mathbf{v} \cdot \mathbf{y}$ for all $\mathbf{v}$, then $\mathbf{x} = \mathbf{y}$. (Because $\mathbf{v} \cdot \mathbf{x} = \mathbf{v} \cdot \mathbf{y}$ says $\mathbf{v} \cdot (\mathbf{x} - \mathbf{y}) = 0$ for all $\mathbf{v}$; take $\mathbf{v} = \mathbf{x} - \mathbf{y}$ to get $\|\mathbf{x} - \mathbf{y}\|^2 = 0$, $\mathbf{x} - \mathbf{y} = 0$.)

We now know, for all $\mathbf{v}$ and $\mathbf{w}$, that $f(\mathbf{v}) \cdot \mathbf{w} = \mathbf{v} \cdot {}^tf(\mathbf{w})$ by part (a), and $f(\mathbf{v}) \cdot \mathbf{w} = \mathbf{v} \cdot g(\mathbf{w})$ by hypothesis. The elementary lemma, with $\mathbf{x} = {}^tf(\mathbf{w})$ and $\mathbf{y} = g(\mathbf{w})$, implies ${}^tf(\mathbf{w}) = g(\mathbf{w})$ for all $\mathbf{w}$, which means ${}^tf = g$.

3.4 If f and g are linear functions and $f \circ g$ is defined, then ${}^t(f \circ g) = {}^tg \circ {}^tf$. If $\mathbf{A}$ and $\mathbf{B}$ are matrices and $\mathbf{AB}$ is defined, then ${}^t(\mathbf{AB}) = {}^t\mathbf{B}\,{}^t\mathbf{A}$.

Proof. By 3.3(b) we need only show for all $\mathbf{v}$ and $\mathbf{w}$,

$$(f \circ g)(\mathbf{v}) \cdot \mathbf{w} = \mathbf{v} \cdot ({}^tg \circ {}^tf)(\mathbf{w}).$$

The left-hand side is $f(g(\mathbf{v})) \cdot \mathbf{w} = g(\mathbf{v}) \cdot {}^tf(\mathbf{w})$ by 3.3(a), where $g(\mathbf{v})$ replaces $\mathbf{v}$. A second use of 3.3(a) then gives $g(\mathbf{v}) \cdot {}^tf(\mathbf{w}) = \mathbf{v} \cdot {}^tg({}^tf(\mathbf{w})) = \mathbf{v} \cdot ({}^tg \circ {}^tf)(\mathbf{w})$ as desired.

If $\mathbf{A}$ and $\mathbf{B}$ are the matrices of f and g, respectively, then $\mathbf{BA}$ is the matrix of $g \circ f$, so ${}^t(\mathbf{BA})$ is the matrix of ${}^t(g \circ f)$ and ${}^t\mathbf{A}\,{}^t\mathbf{B}$ is the matrix of ${}^tf \circ {}^tg$. The equality of linear mappings we just proved then implies the equality of the matrices ${}^t(\mathbf{BA})$ and ${}^t\mathbf{A}\,{}^t\mathbf{B}$.

3.5 For all linear mappings f and g from R^n to R^m and for all scalars a and b,

$${}^t(af + bg) = a\,{}^tf + b\,{}^tg.$$

Similarly, for all m by n matrices $\mathbf{A}$ and $\mathbf{B}$,

$${}^t(a\mathbf{A} + b\mathbf{B}) = a\,{}^t\mathbf{A} + b\,{}^t\mathbf{B}.$$

We leave the proof as an exercise (Problems 2 and 3). Note that you will need 3.3 again.

3.6 If a matrix $\mathbf{A}$ is invertible, so is ${}^t\mathbf{A}$; if a linear mapping f is invertible, so is ${}^tf$. In addition, $({}^tf)^{-1} = {}^t(f^{-1})$ and $({}^t\mathbf{A})^{-1} = {}^t(\mathbf{A}^{-1})$.

Proof. If $g = f^{-1}$ so $f \circ g = g \circ f = I$, then by 3.4,

$${}^tg \circ {}^tf = {}^t(f \circ g) = {}^tI = I$$

and

$${}^tf \circ {}^tg = {}^t(g \circ f) = {}^tI = I,$$

showing that ${}^tg$ is the inverse of ${}^tf$. The proof for matrices is the same.

3.7 ${}^t({}^tf) = f, \quad {}^t({}^t\mathbf{A}) = \mathbf{A}.$

Proof. In this case the matrix proof is easier; if we flip the matrix **A** over its diagonal twice, we get **A** back again.

3.8 Converse of 3.6. If ${}^t\mathbf{A}$ is invertible, so is **A**; if ${}^tf$ is invertible, so is f.

Proof. If ${}^t\mathbf{A}$ is invertible, then by 3.6 so is ${}^t({}^t\mathbf{A})$, which is **A** by 3.7.

We now come to a pair of results, especially 3.10, that can be said to be the motivation for introducing transposes in this chapter rather than elsewhere.

3.9 For all square matrices **A**, $\det({}^t\mathbf{A}) = \det \mathbf{A}$.

Proof. If **A** is not invertible, then neither is ${}^t\mathbf{A}$, by 3.8. In this case, both determinants are zero. Hence, we need only consider invertible matrices. We first look at the elementary ones. If **A** is the elementary matrix obtained from the identity matrix by adding a times the pth row to the qth row, we know $\det \mathbf{A} = 1$. This **A** coincides with **I** except at the q, p-entry, where **A** has the entry a; hence, ${}^t\mathbf{A}$ is the same as ${}^t\mathbf{I} = \mathbf{I}$ except that its p, q-entry is a. Therefore ${}^t\mathbf{A}$ is also an elementary matrix; it is the result of adding a times the qth row of **I** to the pth. Hence, $\det {}^t\mathbf{A} = 1$, too, proving 3.9 for such elementary matrices. We leave it to the reader to carry out similar proofs for the elementary matrices obtained from **I** by Manipulations 2 and 3, Chapter 4.

Finally, if **A** is a general invertible matrix, it is a product of elementary matrices (Chapter 4, 2.9 and 2.7(b) or Problem 11, Section 2), say, $\mathbf{A} = \mathbf{A}_1\mathbf{A}_2\cdots\mathbf{A}_r$. It follows that

$${}^t\mathbf{A} = {}^t\mathbf{A}_r\,{}^t\mathbf{A}_{r-1}\cdots{}^t\mathbf{A}_2\,{}^t\mathbf{A}_1,$$

by repeated use of 3.4, and that

$$\det {}^t\mathbf{A} = (\det {}^t\mathbf{A}_r)(\det {}^t\mathbf{A}_{r-1})\cdots(\det {}^t\mathbf{A}_1),$$

by 1.8. But

$$\det {}^t\mathbf{A}_r = \det \mathbf{A}_r, \;\ldots,\; \det {}^t\mathbf{A}_1 = \det \mathbf{A}_1$$

because $\mathbf{A}_1\cdots\mathbf{A}_r$ are all elementary. This gives

$$\det({}^t\mathbf{A}) = (\det \mathbf{A}_r)(\det \mathbf{A}_{r-1})\cdots(\det \mathbf{A}_1) = \det \mathbf{A}.$$

3.10 In computing determinants, you may use Manipulations 1, 2, and 3 on the columns as well as on rows. These manipulations on columns have the same effect on the determinant as the corresponding manipulations on rows.

Proof. We give the proof for Manipulation 1. We let $\mathbf{B}$ be the result of adding a times the pth column of $\mathbf{A}$ to the qth. Now take transposes of everything. The qth row of ${}^t\mathbf{B}$ is the qth column of $\mathbf{B}$, which equals the qth column of $\mathbf{A}$ plus a times the pth column of $\mathbf{A}$; by definition of ${}^t\mathbf{A}$, this sum is also the qth row of ${}^t\mathbf{A}$ plus a times the pth row of ${}^t\mathbf{A}$. A similar argument shows that the rth row of ${}^t\mathbf{B}$ (for every $r \neq q$) is the same as the rth row of ${}^t\mathbf{A}$. In short, if Manipulation 1 on columns carries $\mathbf{A}$ to $\mathbf{B}$, then the same manipulation on rows carries ${}^t\mathbf{A}$ to ${}^t\mathbf{B}$; but we know by 1.1 that this row manipulation does not change the determinant, so

$$\det \mathbf{B} = \det {}^t\mathbf{B} = \det {}^t\mathbf{A} = \det \mathbf{A}.$$

We leave to the reader the arguments for the other two manipulations.

Example 2. Compute $\det \mathbf{A}$ if

$$\mathbf{A} = \begin{pmatrix} 1 & -1 & 0 & 0 \\ 2 & 1 & 4 & 6 \\ 3 & 1 & -1 & 0 \\ 4 & 0 & 1 & 3 \end{pmatrix}.$$

First add the first column to the second to get a first row [1, 0, 0, 0], then subtract multiples of the first row from the others to make the first column distinguished. We get

$$\det \mathbf{A} = \det \begin{pmatrix} 1 & 0 & 0 & 0 \\ 0 & 3 & 4 & 6 \\ 0 & 4 & -1 & 0 \\ 0 & 4 & 1 & 3 \end{pmatrix}.$$

Now interchange the second and third rows and the second and third columns and multiply the new second row by -1 to get

$$\det \mathbf{A} = (-1)^3 \det \begin{pmatrix} 1 & 0 & 0 & 0 \\ 0 & 1 & -4 & 0 \\ 0 & 4 & 3 & 6 \\ 0 & 1 & 4 & 3 \end{pmatrix}.$$

Add 4 times the second column to the third, then -4 times the second row to the third row and -1 times the second row to the fourth row; the result is

$$\det \mathbf{A} = -\det \begin{pmatrix} 1 & 0 & 0 & 0 \\ 0 & 1 & 0 & 0 \\ 0 & 0 & 19 & 6 \\ 0 & 0 & 8 & 3 \end{pmatrix}.$$

We leave the rest to the reader. The answer is $\det \mathbf{A} = -9$.

Note also that we can now complete the argument necessary in 2.4: Determinants can be computed by expansion by minors on any column, because $\det \mathbf{A} = \det {}^t\mathbf{A}$ and if we expand $\det {}^t\mathbf{A}$ by minors on the qth row, we get

$$\det {}^t\mathbf{A} = a_{1q}A_{1q} + a_{2q}A_{2q} + \cdots + a_{nq}A_{nq}$$

(why is the qp-cofactor in ${}^t\mathbf{A}$ the same as the pq-cofactor in $\mathbf{A}$?), which is the expansion of $\det \mathbf{A}$ by minors on the qth column.

We end this section with a geometrical reprise. In Section 1, after 1.4, we described the determinant of a 3 by 3 matrix as a "signed volume"—plus or minus the volume of the parallelepiped three of whose edges are the three row vectors in the matrix. We shall think of the determinant of an n by n matrix in similar terms—plus or minus the n-dimensional "volume" of the n-dimensional parallelotope defined by taking the rows of the matrix as the n edges issuing from the origin.

If f is a linear mapping from R^n to R^n, we agreed to define $\det f$ as the determinant of the matrix $\mathbf{A}$ of f. Since the columns of $\mathbf{A}$ (or the rows of ${}^t\mathbf{A}$) are $f(\mathbf{i}_1), \ldots, f(\mathbf{i}_n)$ and $\det \mathbf{A} = \det {}^t\mathbf{A} = \det(f(\mathbf{i}_1), \ldots, f(\mathbf{i}_n))$, the determinant of f is the "signed volume" of the parallelotope into which f deforms the basic unit parallelotope (whose edges at the origin are the standard basis vectors $\mathbf{i}_1, \ldots, \mathbf{i}_n$). The fact is (3.11) that the basic unit parallelotope need not be involved in this statement. Take any parallelotope defined by n vectors $\mathbf{u}_1, \ldots, \mathbf{u}_n$ in R^n. Suppose its signed volume is u. Then f carries this parallelotope into a new parallelotope defined by the vectors $f(\mathbf{u}_1), \ldots, f(\mathbf{u}_n)$, and the new parallelotope has signed volume u times the determinant of f. This is then a good, geometric way of looking at $\det f$, independent of matrices or of the coordinate system: it is the factor by which volumes of parallelotopes are changed when you operate with f. We can write this down and prove it as follows.

3.11 If $\mathbf{u}_1, \ldots, \mathbf{u}_n$ are vectors in R^n and f is a linear mapping of R^n

to R^n, then

$$\det(f(\mathbf{u}_1), \ldots, f(\mathbf{u}_n)) = (\det f)(\det(\mathbf{u}_1, \ldots, \mathbf{u}_n)).$$

Proof. Let $\mathbf{A}$ be the matrix of f and $\mathbf{U}$ be the matrix whose rows are $\mathbf{u}_1, \ldots, \mathbf{u}_n$, so the pth column of ${}^t\mathbf{U}$ is $\mathbf{u}_p$ for $p = 1, \ldots, n$. If we want the matrix whose rows are $f(\mathbf{u}_1), \ldots, f(\mathbf{u}_n)$, it is then the transpose of $\mathbf{A}\,{}^t\mathbf{U}$. By 3.4, this is $\mathbf{U}\,{}^t\mathbf{A}$. Therefore, $\det(f(\mathbf{u}_1), \ldots, f(\mathbf{u}_n)) = \det(\mathbf{U}\,{}^t\mathbf{A}) = \det \mathbf{U} \det \mathbf{A} = \det(\mathbf{u}_1, \ldots, \mathbf{u}_n) \det \mathbf{A}$ and, of course, $\det \mathbf{A} = \det f$.

PROBLEMS

1. Find the transposes of the following matrices:

$$\begin{pmatrix} 1 & 2 & 3 & 4 \\ 5 & 6 & 7 & 8 \\ 9 & 10 & 11 & 12 \end{pmatrix}; \quad \begin{pmatrix} 0 & a & b \\ -a & 0 & c \\ -b & -c & 0 \end{pmatrix}; \quad \begin{pmatrix} a & b & c \\ b & d & e \\ c & e & f \end{pmatrix};$$

$$\begin{pmatrix} 2 & 1 \\ 3 & 0 \end{pmatrix}; \quad (1 \quad 1 \quad 2 \quad 3).$$

2. Prove 3.5 in the style of the proof of 3.4.

3. Prove 3.5 for matrices directly, without use of linear mappings. Hint: If a_{pq} is the p, q-entry in $\mathbf{A}$ and b_{pq} is the p, q-entry in $\mathbf{B}$, what is the p, q-entry in $a\mathbf{A} + b\mathbf{B}$? The q, p-entry in ${}^t(a\mathbf{A} + b\mathbf{B})$? The q, p-entry in $a({}^t\mathbf{A}) + b({}^t\mathbf{B})$?

4. Prove 3.4 in the style of Problem 3. Hint: The p, q-entry in ${}^t(\mathbf{BA})$ is the q, p-entry in $\mathbf{BA}$, which is the dot product of the qth row of $\mathbf{B}$ by the pth row of $\mathbf{A}$.

5. Fill the gaps left in the proof of 3.9 by proving 3.9 in the case where $\mathbf{A}$ is obtained from $\mathbf{I}$ by a manipulation of type 2; of type 3.

6. Fill the gaps in the proof of 3.10 by showing that if $\mathbf{B}$ is the result of interchanging two columns of $\mathbf{A}$, then $\det \mathbf{B} = -\det \mathbf{A}$, and if $\mathbf{B}$ is obtained by multiplying one column of $\mathbf{A}$ by a scalar a, then $\det \mathbf{B} = a \det \mathbf{A}$.

7. Let $\mathbf{u}$ and $\mathbf{w}$ be vectors in R^m and let $\mathbf{U}$ and $\mathbf{W}$ be the n by 1 matrices (column vectors) with the same entries as $\mathbf{u}$ and $\mathbf{w}$, respectively. Then show that the dot product $\mathbf{u} \cdot \mathbf{w}$ is the matrix product $({}^t\mathbf{U})\mathbf{W}$. If f is a linear mapping from R^n to R^m, if $\mathbf{v}$ is any vector in R^n, and if $\mathbf{V}$ is the corresponding n by 1 matrix, show that

$$f(\mathbf{v}) \cdot \mathbf{w} = {}^t\mathbf{V}\,{}^t\mathbf{A}\mathbf{W}$$

and use this to argue once again that $f(\mathbf{v}) \cdot \mathbf{w} = \mathbf{v} \cdot {}^tf(\mathbf{w})$.

8. Give another proof of the last argument used in 3.3(b): If ϕ and ψ are linear mappings from R^n to R^m and if $\mathbf{v} \cdot \phi(\mathbf{w}) = \mathbf{v} \cdot \psi(\mathbf{w})$ for all $\mathbf{v}$ and $\mathbf{w}$, then $\phi = \psi$. Do this by proving that ϕ and ψ have the same matrix (the entries in the matrix of ϕ are expressible as $\mathbf{v} \cdot \phi(\mathbf{w})$ if the right $\mathbf{v}$ and $\mathbf{w}$ are chosen). Note, however, that the proof given in 3.3 works even for nonlinear ϕ and ψ.

9. Prove that if $\mathbf{v} \cdot \mathbf{x} = \mathbf{v} \cdot \mathbf{y}$ for every $\mathbf{v}$, then $\mathbf{x} = \mathbf{y}$ (the lemma in 3.3) by taking $\mathbf{v}$ successively to be $\mathbf{i}_1, \mathbf{i}_2, \ldots, \mathbf{i}_n$.

10. Pretend you have never heard of transposes of matrices. Show that the equation in 3.3 can be used to *define* ${}^tf$ as follows. For given f and $\mathbf{w}$ consider the function $\phi(\mathbf{v}) = f(\mathbf{v}) \cdot \mathbf{w}$

(a) Show that ϕ is a linear function from R^n to R^1.

(b) Show that every linear function ψ from R^n to R^1 is of the form $\psi(\mathbf{v}) = \mathbf{v} \cdot \mathbf{v}_0$ for some $\mathbf{v}_0$ in R^n (that is, if ψ is linear, then there is a vector $\mathbf{v}_0$ such that $\psi(\mathbf{v}) = \mathbf{v} \cdot \mathbf{v}_0$ for all $\mathbf{v}$). Show also that, given ψ, there is only one $\mathbf{v}_0$ that will do this. (See Problem 9.)

(c) Parts (a) and (b) associate to each fixed f and $\mathbf{w}$ a vector $\mathbf{v}_0$. Now hold f fixed, but let $\mathbf{w}$ vary; then $\mathbf{v}_0$ will probably vary, too; we are associating to each $\mathbf{w}$ in R^m a vector $\mathbf{v}_0$ in R^n. Show that this association is a linear function.

(d) Call the function in part (c) ${}^tf$. Directly from this definition of ${}^tf$ you should see that $f(\mathbf{v}) \cdot \mathbf{w} = \mathbf{v} \cdot {}^tf(\mathbf{w})$.

11. (a) Let $\mathbf{P}$ be the elementary matrix obtained from the n by n identity matrix by adding a times the pth row to the qth. We know that for every matrix $\mathbf{A}$ with n rows, $\mathbf{PA}$ is the matrix obtained by performing this same manipulation on $\mathbf{A}$. Compute ${}^t\mathbf{P}$ and by matrix multiplication show that, for every matrix $\mathbf{B}$ with n columns, $\mathbf{B}({}^t\mathbf{P})$ is the result of adding a times the pth column of $\mathbf{B}$ to the qth. Thus Manipulation 1 on the columns of a matrix can equally well be effected by multiplying on the right by an elementary matrix.

(b) Compute $\det({}^t\mathbf{P})$ and use it with 1.8 to argue again that Manipulation 1 on the columns of a matrix does not change the determinant.

(c) Use the same type of argument with the two other types of manipulations to give a new proof for all of 3.10.

12. Using row ranks and column ranks, prove that the rank of $\mathbf{A}$ is the same as the rank of ${}^t\mathbf{A}$ for every matrix $\mathbf{A}$. Use this to prove that the rank of f is the same as the rank of ${}^tf$ for every linear mapping f.

13. Give a more geometric proof of Problem 12 as follows:

(a) Show that the kernel of ${}^tf$ is contained in the orthogonal complement of the image of f; that is, if $\mathbf{v}$ is in the kernel of ${}^tf$ and $\mathbf{w}$ is in the image of f, then $\mathbf{v} \cdot \mathbf{w} = 0$. (We defined "orthogonal complement" in Problem 13, Chapter 5, Section 4.)

(b) Show that the orthogonal complement of the image of f is contained in the kernel of ${}^tf$; that is, if $\mathbf{v}$ is orthogonal to the image of f, then ${}^tf(\mathbf{v}) = \mathbf{0}$. Use Problem 9. The result of parts (a) and (b) together is that the kernel of ${}^tf$ is the same as the orthogonal complement of the image of f; hence, these two subspaces have the same dimension.

(c) Refer back to Chapter 5 to show that the dimension of the kernel of ${}^tf$ is m minus the rank of ${}^tf$ (we are assuming that f maps R^n to R^m, so ${}^tf$ maps R^m to R^n and the kernel of ${}^tf$ is a subspace of R^m).

(d) Refer to Problem 13 in Chapter 5, Section 4 to show that the dimension of the orthogonal complement of the image of f is m minus the rank of f.

(e) Put parts (a)–(d) together to show that the rank of ${}^tf$ = the rank of f.

14. Use 3.10 to show that if a matrix $\mathbf{B}$ is derived from a matrix $\mathbf{A}$ by use of Manipulations 1, 2, and 3 on rows and columns, then $\mathbf{A}$ is invertible if and only if $\mathbf{B}$ is invertible.

15. Show that the determinant of every rotation in R^3 is 1: If f is a rotation around a line through the origin, use 3.11 and the signed volume version of $\det(\mathbf{u}, \mathbf{v}, \mathbf{w})$ (see Section 1, after 1.4) to show $\det f = 1$.

(b) Show that the determinant of every reflection in R^3 is -1: If h is the reflection in a plane through the origin (see Problem 27, Chapter 3, Section 4), pick $\mathbf{u}_1$, $\mathbf{u}_2$, $\mathbf{u}_3$ suitably in 3.11 so you know $h(\mathbf{u}_1)$, $h(\mathbf{u}_2)$, $h(\mathbf{u}_3)$ as well as $\det(\mathbf{u}_1, \mathbf{u}_2, \mathbf{u}_3)$, and show $\det h = -1$.

16. Use 3.11 to give another proof of Problem 9(b), Section 1: the determinant of every projection (on a plane through the origin) is 0.

4. SYMMETRIC MATRICES AND MAPPINGS

4.1 Definition

(a) A square matrix $\mathbf{A} = (a_{pq})$ is *symmetric* if ${}^t\mathbf{A} = \mathbf{A}$, that is, if for all p and q, $a_{pq} = a_{qp}$.

(b) A linear mapping f from R^n to R^n is *symmetric* if ${}^tf = f$.

4.2 Let f be a linear mapping from R^n to R^n and let $\mathbf{A}$ be the corresponding matrix. Then

(a) f is symmetric if and only if $\mathbf{A}$ is;

(b) f is symmetric if and only if $f(\mathbf{v}) \cdot \mathbf{w} = \mathbf{v} \cdot f(\mathbf{w})$ for all $\mathbf{v}$ and $\mathbf{w}$ in R^n.

Proof. (a) follows from 3.2: the matrix of ${}^tf$ is ${}^t\mathbf{A}$; ${}^tf = f$ if and only if their matrices are equal, that is, ${}^t\mathbf{A} = \mathbf{A}$.

(b) follows from 3.3.

Example 1. All diagonal matrices are symmetric.

Example 2. The matrix

$$\begin{pmatrix} 2 & 1 & 3 \\ 1 & 4 & 5 \\ 3 & 5 & 0 \end{pmatrix}$$

is symmetric, but

$$\begin{pmatrix} 2 & 1 & 2 \\ 3 & 1 & 3 \\ 4 & 1 & 4 \end{pmatrix}$$

is not.

Example 3. Perpendicular projection of R^3 on the xy-plane is a symmetric mapping because its matrix is

$$\begin{pmatrix} 1 & 0 & 0 \\ 0 & 1 & 0 \\ 0 & 0 & 0 \end{pmatrix},$$

which is symmetric. We could also check directly that

$$f(\mathbf{v}) \cdot \mathbf{w} = \mathbf{v} \cdot f(\mathbf{w})$$

by writing

$$\mathbf{v} = v_1\mathbf{i} + v_2\mathbf{j} + v_3\mathbf{k}, \qquad \mathbf{w} = w_1\mathbf{i} + w_2\mathbf{j} + w_3\mathbf{k};$$

we find that

$$(v_1\mathbf{i} + v_2\mathbf{j}) \cdot (w_1\mathbf{i} + w_2\mathbf{j} + w_3\mathbf{k}) = (v_1\mathbf{i} + v_2\mathbf{j} + v_3\mathbf{k}) \cdot (w_1\mathbf{i} + w_2\mathbf{j})$$

for all $\mathbf{v}$ and $\mathbf{w}$.

From 3.4–3.6 we get the following results on symmetric matrices and mappings.

4.3 Let $\mathbf{A}$ and $\mathbf{B}$ be symmetric n by n matrices, and let f and g be symmetric linear mappings from R^n to R^n. Then

(a) $a\mathbf{A} + b\mathbf{B}$ is symmetric for all scalars a and b; $af + bg$ is symmetric for all scalars a and b;

(b) $\mathbf{AB}$ is symmetric if $\mathbf{AB} = \mathbf{BA}$; $f \circ g$ is symmetric if $f \circ g = g \circ f$;

(c) if $\mathbf{A}$ is invertible, then $\mathbf{A}^{-1}$ is also symmetric; if f is invertible, then f^{-1} is symmetric.

Proof.

$${}^t(a\mathbf{A} + b\mathbf{B}) = a\,{}^t\mathbf{A} + b\,{}^t\mathbf{B} \qquad \text{(by 3.5)}$$

$$= a\mathbf{A} + b\mathbf{B} \qquad (\text{because } {}^t\mathbf{A} = \mathbf{A} \text{ and } {}^t\mathbf{B} = \mathbf{B}).$$

$${}^t(\mathbf{AB}) = {}^t\mathbf{B}\,{}^t\mathbf{A} \qquad \text{(by 3.4)}$$

$$= \mathbf{BA} = \mathbf{AB} \qquad \text{(by hypothesis)}.$$

$${}^t(\mathbf{A}^{-1}) = ({}^t\mathbf{A})^{-1} \qquad \text{(by 3.6)}$$

$$= \mathbf{A}^{-1} \qquad (\text{because } \mathbf{A} \text{ is assumed symmetric}).$$

The proofs for linear mappings are the same.

There is nothing special to say about the determinant of a symmetric matrix.

There is no particular reason for singling out the symmetric mappings here, but in Chapter 7 we shall see that these mappings are the ones that can be nicely described geometrically as "stretchings" along orthogonal axes (Theorem 4.1). They will also be useful in describing and studying quadratic forms (homogeneous second-degree polynomials), conic sections, quadric surfaces, and so on, in Chapter 8. For the present we study them mainly as an exercise in transposes.

Empirically, many linear mappings that come up in applications of mathematics are symmetric.

PROBLEMS

1. Which of the following matrices are symmetric?

$$\begin{pmatrix} 1 & 2 & 3 \\ 2 & 3 & 4 \\ 3 & 4 & 5 \end{pmatrix}; \quad \begin{pmatrix} 1 & 2 & 1 \\ 3 & 4 & 3 \\ 5 & 6 & 5 \end{pmatrix}; \quad \begin{pmatrix} 1 & 2 & 3 \\ & & \\ 2 & 1 & 3 \end{pmatrix};$$

$$\begin{pmatrix} \cos\theta & -\sin\theta & 0 \\ \sin\theta & \cos\theta & 0 \\ 0 & 0 & 1 \end{pmatrix}.$$

2. For what numbers x, y, z is each of the following matrices symmetric?

$$\begin{pmatrix} x & y & z \\ 2 & 0 & 3 \\ 4 & 3 & 3 \end{pmatrix}; \qquad x\mathbf{I} - \begin{pmatrix} 1 & 0 & 1 \\ 0 & 2 & 3 \\ 1 & 3 & 4 \end{pmatrix}.$$

3. Use 4.2 to give two proofs that perpendicular projection of R^3 onto the x-axis is a symmetric mapping.

4. Use 4.2(a) to argue that most rotations about the origin in R^2 are not symmetric (for each rotation it suffices to produce a single $\mathbf{v}$ and a single $\mathbf{w}$ such that $f(\mathbf{v}) \cdot \mathbf{w} \neq \mathbf{v} \cdot f(\mathbf{w})$). Then use 4.2(b) to get the same result. To be precise, there are two rotations that are symmetric; through what angles?

5. Find an example of two symmetric 2 by 2 matrices whose product is not symmetric.

6. (a) For every square matrix $\mathbf{A}$, show that $({}^t\mathbf{A})\mathbf{A}$ is symmetric.

(b) Suppose $\mathbf{A}$ is invertible. Show $({}^t\mathbf{A})\mathbf{A}^{-1}$ is symmetric if and only if $\mathbf{A}^2 = ({}^t\mathbf{A})^2$.

7. Define a linear mapping f from R^n to R^n to be *skew* (or skew-symmetric) if ${}^tf = -f$. A matrix $\mathbf{A}$ is said to be skew if ${}^t\mathbf{A} = -\mathbf{A}$.

(a) Show that the diagonal entries of a skew matrix are all zero.

(b) Find a, b, c so that this matrix is skew:

$$\begin{pmatrix} 0 & 1 & 2 \\ a & 0 & 3 \\ b & c & 0 \end{pmatrix}$$

(c) Find a formula for all 2 by 2 skew matrices. Show that the nonzero ones have positive determinant.

(d) The determinant of every 3 by 3 skew matrix is zero. In fact, show that the determinant of every n by n skew matrix is zero if n is odd.

(e) f is skew if and only if $f(\mathbf{v}) \cdot \mathbf{w} = -\mathbf{v} \cdot f(\mathbf{w})$ for all vectors $\mathbf{v}$ and $\mathbf{w}$ in R^n.

(f) If f is skew and $\mathbf{v}$ is a vector not turned by f, that is, $f(\mathbf{v})$ is a scalar times $\mathbf{v}$, then show $f(\mathbf{v}) = \mathbf{0}$.

(g) If f is defined on R^3 thus: $f(\mathbf{v}) = \mathbf{r} \times \mathbf{v}$ for a fixed vector $\mathbf{r}$, then f is skew. Do this twice, with and without matrices.

(h) Every skew linear mapping on R^3 is onc of thosc described in (g). Use matrices to prove this.

5. ORTHOGONAL MATRICES AND MAPPINGS

Here is another important kind of linear mapping, whose definition is analogous to that of symmetric mappings:

5.1 Definition

A linear mapping f from R^n to R^n is *orthogonal* if f is invertible and ${}^tf = f^{-1}$.

A square matrix $\mathbf{P}$ is *orthogonal* if ${}^t\mathbf{P} = \mathbf{P}^{-1}$.

We give several equivalent versions of this definition in 5.2 and 5.3

5.2 If f is a linear mapping from R^n to R^n, the following conditions on f are equivalent:

(a) f is orthogonal: ${}^tf = f^{-1}$;

(b) ${}^tf \circ f = I$;

(b′) $f \circ {}^tf = I$;

(c) f preserves dot products; that is, for every two vectors $\mathbf{v}$ and $\mathbf{w}$ in R^n, $f(\mathbf{v}) \cdot f(\mathbf{w}) = \mathbf{v} \cdot \mathbf{w}$;

(d) f preserves lengths, or is a "rigid motion"; that is, for every vector $\mathbf{v}$ in R^n, $\| f(\mathbf{v}) \| = \| \mathbf{v} \|$;

(e) $f(\mathbf{i}_1), f(\mathbf{i}_2), \ldots, f(\mathbf{i}_n)$ form an orthonormal set; that is, each of these vectors has length one and is orthogonal to all the others.

(f) the matrix of f is an orthogonal matrix.

Proof. By the definition of f^{-1} (Chapter 3, 5.1), (a) implies (b) and (b′). Conversely, Problem 5, Chapter 5, Section 4 says that if ${}^tf$ has a one-sided inverse as in (b) or (b′), then f has an inverse which is equal to the one-sided inverse; this is (a).

Next we prove the following implications which will prove that all seven conditions in 5.2 are equivalent:

$$(b) \Rightarrow (c) \Rightarrow (e) \Rightarrow (f) \Rightarrow (a); \text{ and } (c) \Leftrightarrow (d).$$

(b) $\Rightarrow$ (c). By 3.3, $\mathbf{x} \cdot f(\mathbf{w}) = {}^tf(\mathbf{x}) \cdot \mathbf{w}$ for all $\mathbf{x}$ and $\mathbf{w}$. Take $\mathbf{x} = f(\mathbf{v})$ and get $f(\mathbf{v}) \cdot f(\mathbf{w}) = {}^tf(f\mathbf{v})) \cdot \mathbf{w} = I(\mathbf{v}) \cdot \mathbf{w} = \mathbf{v} \cdot \mathbf{w}$.

(c) $\Rightarrow$ (d). Set $\mathbf{v} = \mathbf{w}$ in (c) and take square roots; remember, $\| \mathbf{v} \| = (\mathbf{v} \cdot \mathbf{v})^{1/2}$.

(d) $\Rightarrow$ (c). Recall $\mathbf{v} \cdot \mathbf{w} = \frac{1}{2}(\| \mathbf{v} + \mathbf{w} \|^2 - \| \mathbf{v} \|^2 - \| \mathbf{w} \|^2)$. Problem 15, Chapter 1, Section 6 did this for $\mathbf{v}$ and $\mathbf{w}$ in R^3. The proof in R^n is the same. Hence if f preserves sums and lengths, it must preserve dot products. In symbols,

$$\begin{aligned} f(\mathbf{v}) \cdot f(\mathbf{w}) &= \tfrac{1}{2}(\| f(\mathbf{v}) + f(\mathbf{w}) \|^2 - \| f(\mathbf{v}) \|^2 - \| f(\mathbf{w}) \|^2) \\ &= \tfrac{1}{2}(\| f(\mathbf{v} + \mathbf{w}) \|^2 - \| f(\mathbf{v}) \|^2 - \| f(\mathbf{w}) \|^2) \\ &= \tfrac{1}{2}(\| \mathbf{v} + \mathbf{w} \|^2 - \| \mathbf{v} \|^2 - \| \mathbf{w} \|^2) = \mathbf{v} \cdot \mathbf{w}. \end{aligned}$$

(c) $\Rightarrow$ (e). $f(\mathbf{i}_p) \cdot f(\mathbf{i}_q) = \mathbf{i}_p \cdot \mathbf{i}_q = 0$ if $p \neq q$ and $= 1$ if $p = q$.

(e) $\Rightarrow$ (f). Let $\mathbf{P}$ be the matrix of f. The columns of $\mathbf{P}$ are $f(\mathbf{i}_1), \ldots, f(\mathbf{i}_n)$, which are orthonormal by (e). Now the pq-entry in ${}^t\mathbf{P}\,\mathbf{P}$ is the dot product of the pth row of ${}^t\mathbf{P}$ by the qth column of $\mathbf{P}$, that is, $f(\mathbf{i}_p) \cdot f(\mathbf{i}_q)$, which is 0 or 1 according as $p \neq q$ or $p = q$. Thus ${}^t\mathbf{P}\,\mathbf{P} = \mathbf{I}$, so ${}^t\mathbf{P} = \mathbf{P}^{-1}$.

(f) $\Rightarrow$ (a). We know the matrix of ${}^tf$ is ${}^t\mathbf{P}$ and the matrix of f^{-1} is $\mathbf{P}^{-1}$. So ${}^tf = f^{-1}$ if and only if they have the same matrix, that is, ${}^t\mathbf{P} = \mathbf{P}^{-1}$.

Example 1. Every rotation in R^3 is an orthogonal mapping because it is geometrically clear that (d) holds. Similarly, any rotation in R^2 is an orthogonal mapping.

Example 2. Every reflection in a plane (through the origin) in R^3 is an orthogonal mapping for the same reason. Reflection in a plane through the origin in R^3 may be defined as a linear mapping that (1) leaves all vectors in the plane fixed, i.e., $f(\mathbf{v}) = \mathbf{v}$ for all $\mathbf{v}$ in the plane; (2) carries every vector perpendicular to the plane into its negative, i.e., $f(\mathbf{w}) = -\mathbf{w}$ for every $\mathbf{w}$ perpendicular to this plane; and (3) every other vector in R^3 can be expressed as a sum $\mathbf{v} + \mathbf{w}$ of a vector $\mathbf{v}$ in the plane and a vector $\mathbf{w}$ perpendicular to the plane, and then $f(\mathbf{v} + \mathbf{w}) = f(\mathbf{v}) + f(\mathbf{w}) = \mathbf{v} - \mathbf{w}$.

5.3 The following conditions on a square matrix $\mathbf{P}$ are equivalent:

(a) $\mathbf{P}$ is orthogonal, ${}^t\mathbf{P} = \mathbf{P}^{-1}$;

(b) ${}^t\mathbf{P}\,\mathbf{P} = \mathbf{I}$;

(b′) $\mathbf{P}\,{}^t\mathbf{P} = \mathbf{I}$;

(c) The columns of $\mathbf{P}$ are orthonormal, that is, they are vectors of length one, each orthogonal to all the others;

(d) The rows of $\mathbf{P}$ are orthonormal.

Proof. Let f denote the linear mapping associated with $\mathbf{P}$. By 5.2(f), f is orthogonal if and only if $\mathbf{P}$ is; 5.3(b) is the same as 5.2(b), 5.3(b′) is the same as 5.2(b′); and 5.3(c) is the same as 5.2(e) since the columns of $\mathbf{P}$ are $f(\mathbf{i}_1), \ldots, f(\mathbf{i}_n)$. Since all parts of 5.2 are equivalent, so are 5.3(a)–(c). As for 5.3(d) it asserts that the columns of ${}^t\mathbf{P}$ are orthonormal, that is, as we now know, ${}^t\mathbf{P}$ is an orthogonal matrix. But ${}^t\mathbf{P}$ is orthogonal if and only if $\mathbf{P}$ is, because ${}^t({}^t\mathbf{P}) = \mathbf{P}$, so ${}^t({}^t\mathbf{P}) = ({}^t\mathbf{P})^{-1}$ if and only if $\mathbf{P} = ({}^t\mathbf{P})^{-1}$, which means (taking inverses) $\mathbf{P}^{-1} = {}^t\mathbf{P}$, that is, $\mathbf{P}$ is orthogonal.

Example 3.

$$\mathbf{P} = \begin{pmatrix} \frac{1}{2}\sqrt{2} & -\frac{1}{2}\sqrt{2} \\ \frac{1}{2}\sqrt{2} & \frac{1}{2}\sqrt{2} \end{pmatrix}$$

is an orthogonal matrix because the corresponding mapping is rotation through 45 degrees in R^2 and we may appeal to Example 1 and 5.2(f); or, directly from 5.3(b),

$${}^t\mathbf{P}\,\mathbf{P} = \frac{1}{2}\begin{pmatrix} 1 & 1 \\ -1 & 1 \end{pmatrix}\begin{pmatrix} 1 & -1 \\ 1 & 1 \end{pmatrix} = \frac{1}{2}\begin{pmatrix} 2 & 0 \\ 0 & 2 \end{pmatrix} = \mathbf{I}.$$

It is also instructive to check 5.3(c) and 5.3(d) in this example. The columns are $[\frac{1}{2}\sqrt{2}, \frac{1}{2}\sqrt{2}]$ and $[-\frac{1}{2}\sqrt{2}, \frac{1}{2}\sqrt{2}]$ which are indeed orthonormal (check the three dot products). The rows are $[\frac{1}{2}\sqrt{2}, -\frac{1}{2}\sqrt{2}]$ and $[\frac{1}{2}\sqrt{2}, \frac{1}{2}\sqrt{2}]$, which are also orthonormal.

Example 4. Reflection in the xy-plane has an orthogonal matrix:

$$\begin{pmatrix} 1 & 0 & 0 \\ 0 & 1 & 0 \\ 0 & 0 & -1 \end{pmatrix}.$$

Example 5. Find a, b, c, d to make this matrix orthogonal:

$$\begin{pmatrix} a & 0 & b \\ 0 & 1 & c \\ a & 0 & d \end{pmatrix}.$$

We want $[b, c, d]$ to have length 1 and be orthogonal to $[a, 0, a]$ and to $[0, 1, 0]$. In R^3 the easiest way to do this is to take a suitable scalar multiple of the cross product $[a, 0, a] \times [0, 1, 0]$. This cross product is $[-a, 0, a]$; the only scalar multiples of it with length 1 are $\pm[\frac{1}{2}\sqrt{2}, 0, -\frac{1}{2}\sqrt{2}]$. Now the three columns are mutually orthogonal (check the first two). The third and the second have length 1. To make the first have length 1, we need $a = \pm\frac{1}{2}\sqrt{2}$. Thus there are four answers, $a = \pm\frac{1}{2}\sqrt{2}$, $c = 0$, $b = -d = \pm\frac{1}{2}\sqrt{2}$, where the $\pm$ in a need not be the same as in b. (If a and d are both positive, the matrix corresponds to a 45 degree rotation around the y-axis. When a and d are both negative, it is a rotation of 135 degrees. The other two answers have determinant -1, so cannot be rotations. Each is a composite of a rotation and a reflection.)

You can and should construct more examples of orthogonal mappings and matrices from the following propositions 5.4–5.7.

5.4 Every composite of orthogonal mappings is orthogonal; every product of orthogonal matrices is orthogonal.

Proof. If ${}^tf = f^{-1}$ and ${}^tg = g^{-1}$, then ${}^t(f \circ g) = {}^tg \circ {}^tf$ (by 3.4) $= g^{-1} \circ f^{-1} = (f \circ g)^{-1}$ (by Chapter 3, 5.6).

5.5 The inverse of an orthogonal function is orthogonal; similarly for matrices.

Proof. ${}^t(f^{-1}) = {}^t({}^tf) = f = (f^{-1})^{-1}$.

5.6 The transpose of an orthogonal function is orthogonal; similarly for matrices.

Proof. The transpose is the inverse, which is orthogonal by 5.5.

5.7 If f is orthogonal, af is orthogonal (a a scalar) if and only if $a = \pm 1$; similarly for matrices.

Proof. Left to Problem 11.

5.8 The sum of orthogonal functions is not usually orthogonal. Problem 12 asks you to find examples.

5.9 If f is orthogonal, then $\det f = \pm 1$.

Proof. $\det {}^t f = \det f$, and $\det f^{-1} = 1/\det f$. If ${}^t f = f^{-1}$ then $\det f = 1/\det f$, so $(\det f)^2 = 1$, so $\det f = \pm 1$.

5.10 Every orthogonal mapping in R^2 is either a rotation around the origin (if its determinant is $+1$) or a reflection in a line (if its determinant is -1).

Proof. Let f be the orthogonal mapping. Then $f(\mathbf{i})$ is some vector in R^2 of length 1. Find a rotation, g, that carries $\mathbf{i}$ into $f(\mathbf{i})$. Since $\mathbf{i} \cdot \mathbf{j} = 0$, and f is orthogonal, $f(\mathbf{i}) \cdot f(\mathbf{j}) = 0$, so $f(\mathbf{j})$ is perpendicular to $f(\mathbf{i})$. Furthermore, $f(\mathbf{j})$ has length 1 since $\mathbf{j}$ has. There are only two vectors in R^2 that satisfy both of these restrictions, and they are negatives of each other. One of them is $g(\mathbf{j})$ because g is also orthogonal and is subject to the same analysis as we just applied to f.

Case 1. $f(\mathbf{j})$ is the same as $g(\mathbf{j})$. Since already $f(\mathbf{i}) = g(\mathbf{i})$, $f(a\mathbf{i} + b\mathbf{j}) = af(\mathbf{i}) + bf(\mathbf{j}) = ag(\mathbf{i}) + bg(\mathbf{j}) = g(a\mathbf{i} + b\mathbf{j})$ for all a and b, which says that f and g are the same mapping on R^2; f is the rotation g, and, of course $\det f = \det g = +1$ by Problem 7, Section 2.

Case 2. $f(\mathbf{j}) \neq g(\mathbf{j})$. By the preceding analysis, necessarily $f(\mathbf{j}) = -g(\mathbf{j}) = g(-\mathbf{j})$. Then f is no longer the same as g, but if h is the linear mapping which sends $\mathbf{i}$ to $\mathbf{i}$ and $\mathbf{j}$ to $-\mathbf{j}$, then $f(\mathbf{i}) = (g \circ h)(\mathbf{i})$ and $f(\mathbf{j}) = (g \circ h)(\mathbf{j})$, and the same argument as in Case 1 shows that $f = g \circ h$; that is, f is the composite of reflection in the x-axis (which is what h is) and a rotation, g. You should write the matrix of h and check that $\det h = -1$. Since $\det g = +1$ as before, we know that $\det f = (\det g)(\det h) = -1$ in this case.

We can push Case 2 harder. We shall find a vector $\mathbf{v}$ that is not moved at all by f. Let α be the angle through which g rotates vectors. Take $\mathbf{v}$ to be the result of rotating $\mathbf{i}$ just $\frac{1}{2}\alpha$. Then $h(\mathbf{v})$ is the reflection of $\mathbf{v}$ in the x-axis and it also makes an angle of $\frac{1}{2}\alpha$ with the x-axis, but is on the opposite side of the axis from $\mathbf{v}$. If we rotate $h(\mathbf{v})$ through an angle α, we get back to $\mathbf{v}$ again. Thus $f(\mathbf{v}) = g(h(\mathbf{v})) = \mathbf{v}$.

It follows that any vector along the line of $\mathbf{v}$ is also left fixed by f. If $\mathbf{w}$ is any vector perpendicular to $\mathbf{v}$, then $f(\mathbf{w})$ is perpendicular to $f(\mathbf{v}) = \mathbf{v}$, hence is parallel to $\mathbf{w}$. It has the same length as $\mathbf{w}$, so $f(\mathbf{w}) = \pm\mathbf{w}$. Now $f(\mathbf{w}) = \mathbf{w}$ is impossible because every vector in R^2 is a linear combination of $\mathbf{v}$ and $\mathbf{w}$, and if f leaves $\mathbf{v}$ and $\mathbf{w}$ fixed, it will leave every vector fixed, so $f = I$, $\det f = +1$, contradicting the fact proved above that in Case 2, $\det f = -1$. Therefore, $f(\mathbf{w}) = -\mathbf{w}$, which means that f is reflection in the line of $\mathbf{v}$. This completes Case 2, and so also 5.10.

The proof of 5.10 depended heavily on the fact that we were in R^2, but there is a general theorem. In Chapter 7, 2.6 we shall see a generalization to R^3. In R^n we would have to define rotations and reflections before we could make sense of 5.10. This can be done and the result is that every orthogonal mapping in R^n is either a rotation or the composite of a rotation and a reflection, according as its determinant is $+1$ or -1.

PROBLEMS

1. (a) If f is the linear mapping with matrix

$$\begin{pmatrix} \frac{3}{5} & \frac{4}{5} \\ -\frac{4}{5} & \frac{3}{5} \end{pmatrix}$$

show that f is orthogonal.

(b) Also check by direct computation that $\|f(\mathbf{v})\| = \mathbf{v}$ for every $\mathbf{v} = x\mathbf{i} + y\mathbf{j}$.

(c) Is f a rotation or a reflection? If a rotation, through what angle; if a reflection, about what line?

2. Same as problem 1, for the matrix

$$\begin{pmatrix} \frac{3}{5} & \frac{4}{5} \\ \frac{4}{5} & -\frac{3}{5} \end{pmatrix}$$

3. Verify in three different ways that the following matrix is orthogonal:

$$42^{-1/2}\begin{pmatrix} 5 & -\sqrt{3} & \sqrt{14} \\ -1 & 3\sqrt{3} & \sqrt{14} \\ -4 & -2\sqrt{3} & \sqrt{14} \end{pmatrix}$$

4. (a) Fill out the following to be a 3 by 3 orthogonal matrix. How many correct answers are there?

$$\begin{pmatrix} 3/13 & 4/13 & 12/13 \\ 4/5 & -3/5 & 0 \\ \cdot & \cdot & \cdot \end{pmatrix}.$$

(b) Do the same for the 4 by 4 matrix

$$\frac{1}{5}\begin{pmatrix} 3 & 4 & 0 & 0 \\ 4 & -3 & 0 & 0 \\ \cdot & \cdot & 5 & 0 \\ \cdot & \cdot & \cdot & \cdot \end{pmatrix}.$$

(c) Do the same for the 2 by 2 matrix

$$\begin{pmatrix} a & b \\ \cdot & \cdot \end{pmatrix}.$$

5. Work out Example 5 in the text by studying rows instead of columns.

6. (a) Verify that the following matrix is orthogonal

$$\frac{1}{3}\begin{pmatrix} 1 & 2 & 2 \\ -2 & -1 & 2 \\ 2 & -2 & 1 \end{pmatrix}$$

(b) In fact, this is the matrix of a rotation. Get a clue to this by finding the axis of rotation: a vector $\mathbf{v}$ that satisfies $\mathbf{Av} = \mathbf{v}$. (This should involve three homogeneous linear equations in the components of $\mathbf{v}$.)

(c) Finish showing this is a rotation as follows: (1) Write a formula for all vectors $\mathbf{w}$ perpendicular to $\mathbf{v}$; (2) show that $\mathbf{Aw}$ is also perpendicular to $\mathbf{v}$; (3) show that the angle between $\mathbf{w}$ and $\mathbf{Aw}$ is the same for all $\mathbf{w}$ perpendicular to $\mathbf{v}$.

7. Prove directly by matrix multiplication that 5.3(b) implies 5.3(c) and 5.3(b′) implies 5.3(d).

8. Show that every orthogonal mapping preserves angles: If f is orthogonal and $\mathbf{v}$ and $\mathbf{w}$ are any vectors in R^3, then the angle between $\mathbf{v}$ and $\mathbf{w}$ is the same as the angle between $f(\mathbf{v})$ and $f(\mathbf{w})$.

9. Show that the composite of two rotations in R^2 is a rotation; the composite of two reflections is a rotation; and the composite of a rotation and a reflection is a reflection.

10. If an orthogonal mapping f in R^2 is not a rotation, you may take h to be reflection in any line you please, and write $f = g \circ h$ for some rotation g. (Argue with determinants, or deduce this as a corollary of Problem 9.)

11. Prove 5.7.

12. Prove 5.8.

13. In Example 5, show that, in the case $a = +\frac{1}{2}\sqrt{2}$ and $b = -d = -\frac{1}{2}\sqrt{2}$, the mapping is $g \circ h$ with h a reflection in a plane (try the xy-plane) and g a rotation. About what axis and by how many degrees?

Then write the same mapping as $h \circ g'$ and answer the same questions about the rotation g'.

14. If f is an orthogonal mapping and $\mathbf{v}$ is a nonzero vector such that $f(\mathbf{v}) = a\mathbf{v}$ for some scalar a, then $a = \pm 1$.

CHAPTER

7

Eigenvalues

1. DIAGONALIZABLE TRANSFORMATIONS

Let us return to the spirit of Chapter 3 and take another look at the geometric properties of a linear function whose matrix is diagonal. Such functions occur frequently in nature. Our first question is the usual attempt at relating the two personalities of the function: If the matrix is diagonal (analytic personality), what kind of a function is it (geometric personality)?

Let the matrix of f be

$$\begin{pmatrix} a_{11} & 0 & \cdots & 0 \\ 0 & a_{22} & \cdots & 0 \\ \vdots & \vdots & \cdots & \vdots \\ 0 & 0 & \cdots & a_{nn} \end{pmatrix}.$$

It is difficult to say anything lucid about the geometric relation of $\mathbf{v}$ to $f(\mathbf{v})$ for general vectors $\mathbf{v}$ (see Problem 1), but for the special vectors along the coordinate axes, things are very simple:

$$f(\mathbf{i}_p) = a_{pp}\mathbf{i}_p \qquad \text{for every } p$$

(check this by operating with the matrix of f on the vector $\mathbf{i}_p = [0, \ldots, 0, 1, 0, \ldots, 0]$). In fact, for any vector $\mathbf{v}$ parallel to $\mathbf{i}_p$,

$$f(\mathbf{v}) = a_{pp}\mathbf{v} \tag{1.1}$$

because $\mathbf{v} = x\mathbf{i}_p$, so $f(\mathbf{v}) = f(x\mathbf{i}_p) = xf(\mathbf{i}_p) = xa_{pp}\mathbf{i}_p = a_{pp}x\mathbf{i}_p = a_{pp}\mathbf{v}$. Then Eq. (1.1) says something quite geometric: If $\mathbf{v}$ is a vector along one of the coordinate axes, then $f(\mathbf{v})$ is parallel to $\mathbf{v}$ (we can even say $f(\mathbf{v})$ is in the same direction as $\mathbf{v}$ when $a_{pp} > 0$). Thus we may think of f as a *stretching function*: To operate with f on a vector, multiply the first component of the vector by a_{11}, the second by a_{22}, and so on; or, more geometrically, operating by f stretches the vector by a factor of a_{11} in the $\mathbf{i}_1$-direction, by a factor of a_{22} in the $\mathbf{i}_2$-direction, and so forth.

Of course, as we have remarked (and see Problem 1), such a stretching transformation usually changes both the lengths and directions of most vectors, except those along the coordinate axes. In the special case where $a_{11} = a_{22} = \cdots = a_{nn}$, however, the function is merely scalar multiplication by a_{11}; all vectors are stretched by this same factor; no vectors are turned; $f(\mathbf{v})$ is parallel to $\mathbf{v}$, for all $\mathbf{v}$. See Problem 12 for the proper generalization of this special case.

1.1 Definition

If f is a linear function from R^n to R^n, an *eigenvector* of f is a nonzero vector $\mathbf{v}$ in R^n such that $f(\mathbf{v})$ is a scalar multiple of $\mathbf{v}$.

1.2 Definition

If f is a linear function from R^n to R^n, an *eigenvalue* of f is a scalar λ such that $f(\mathbf{v}) = \lambda\mathbf{v}$ for some nonzero vector $\mathbf{v}$. Of course, such a vector $\mathbf{v}$ is an eigenvector of f and is called an eigenvector *belonging to* the eigenvalue λ, and λ is called the eigenvalue *corresponding to* the eigenvector $\mathbf{v}$.

We can now summarize the geometric properties of a diagonal matrix in this language:

1.3 If f is a linear function from R^n to R^n and its matrix is diagonal with diagonal entries $a_{11}, \ldots, a_{nn}$, then every nonzero scalar multiple of $\mathbf{i}_p$ is an eigenvector belonging to the eigenvalue a_{pp} ($p = 1, 2, \ldots, n$).

"Eigenvector" and "eigenvalue" are hybrid terms, half German (*eigen* = (its) own) and half English. We use them because they seem to be the most common of the numerous synonyms, such as "characteristic value," "proper value," "characteristic root," or "latent value," or "latent root." Similarly, there are "characteristic vectors" and "proper vectors."

We also speak of eigenvalues and eigenvectors of a square matrix. These are just the eigenvalues and the eigenvectors of the corresponding linear function. See Examples 2 and 3. Now if we are to adopt the geometric point of view seriously, we should admit that vectors along the coordinate axes are not particularly better than any other vectors. Of course, nonzero vectors along $\mathbf{i}_1, \mathbf{i}_2, \ldots, \mathbf{i}_n$ will form a basis of R^n and are orthogonal, but otherwise are not particularly noteworthy.

1.4 Definition

An n by n matrix or a linear function from R^n to R^n is called *orthogonally diagonalizable* if there is some orthogonal basis $\mathbf{v}_1, \ldots, \mathbf{v}_n$ of R^n (this means $\mathbf{v}_p \cdot \mathbf{v}_q = 0$ for all $p \neq q$ and the $\mathbf{v}$'s are independent) such that each $\mathbf{v}_p$ is an eigenvector of the matrix or of the function.

In other words, an orthogonally diagonalizable linear function is one whose matrix would have been diagonal if we had been using a different coordinate system, coordinate axes along $\mathbf{v}_1, \ldots, \mathbf{v}_n$ instead of $\mathbf{i}_1, \ldots, \mathbf{i}_n$. It is a stretching function, but the "axes" along which the stretching takes place are the eigenvectors $\mathbf{v}_1, \ldots, \mathbf{v}_n$, not necessarily the x_1-, x_2-, $\ldots$, x_n-axes. As before, the stretching factors are the corresponding eigenvalues.

1.5 Definition

An n by n matrix or a linear function from R^n to R^n is (not necessarily orthogonally) *diagonalizable* if there is a basis (not necessarily orthogonal) consisting of eigenvectors. In this case the function will still be a stretching function, but the axes of the stretching may not be orthogonal.

Every diagonal matrix is diagonalizable (see 1.3), but the converse is not true.

Example 1. It is easy to manufacture diagonalizable functions whose matrices are not diagonal. For example, in R^2 consider the linear function f that stretches $\mathbf{i} + \mathbf{j}$ by a factor of 2 and $\mathbf{i} - \mathbf{j}$ by a factor of 3, that is,

$$f(\mathbf{i}+\mathbf{j}) = 2\mathbf{i}+2\mathbf{j},$$
$$f(\mathbf{i}-\mathbf{j}) = 3\mathbf{i}-3\mathbf{j}. \tag{1.2}$$

By adding the equations, multiplying by $\frac{1}{2}$, and using the linearity of f, we get $f(\mathbf{i}) = (\frac{5}{2})\mathbf{i} - (\frac{1}{2})\mathbf{j}$. By subtracting and multiplying by $\frac{1}{2}$, we get

$$f(\mathbf{j}) = (-\tfrac{1}{2})\mathbf{i} + (\tfrac{5}{2})\mathbf{j}.$$

Thus the matrix of f must be

$$\begin{pmatrix} \frac{5}{2} & -\frac{1}{2} \\ -\frac{1}{2} & \frac{5}{2} \end{pmatrix}.$$

The linear function with this matrix will indeed have the properties (1.2) and so will be orthogonally diagonalizable. Note that the matrix is symmetric. In Section 4 we shall sketch the proof that every symmetric matrix is orthogonally diagonalizable, and conversely.

Example 2. We find all the eigenvectors and all the eigenvalues of the matrix

$$\begin{pmatrix} 1 & -1 \\ -1 & 1 \end{pmatrix}.$$

In Section 2 you will find a routine for this, but a better view of eigenvectors and eigenvalues is to be had by working a few examples with hammer and tongs.

We are to find all $[x, y]$ such that $[x, y] \neq [0, 0]$ and

$$\begin{pmatrix} 1 & -1 \\ -1 & 1 \end{pmatrix}\begin{pmatrix} x \\ y \end{pmatrix} = \lambda\begin{pmatrix} x \\ y \end{pmatrix}.$$

This is a pair of homogeneous linear equations in unknowns x and y with coefficients involving λ. We leave it to the reader to verify that the matrix of coefficients is

$$\begin{pmatrix} 1-\lambda & -1 \\ -1 & 1-\lambda \end{pmatrix}.$$

Reducing toward echelon form we interchange rows, add $(1 - \lambda)$ times the new first row to the second, to get

$$\begin{pmatrix} -1 & 1-\lambda \\ 0 & \lambda^2-2\lambda \end{pmatrix},$$

which is far enough. The linear equations are now

$$\begin{aligned} -x + (1 - \lambda)y &= 0, \\ (\lambda^2 - 2\lambda)y &= 0. \end{aligned} \tag{1.3}$$

The second equation says $y = 0$ or $\lambda = 0$ or 2. If $y = 0$, the first equation gives $x = 0$, so $[x, y] = [0, 0]$, which is not an eigenvector. Hence λ must be 0 or 2. There are no other eigenvalues.

If $\lambda = 0$, then Eqs. (1.3) say y is arbitrary and $x = y$; the eigenvectors belonging to $\lambda = 0$ are all $[x, x]$ except $[0, 0]$, that is, all the nonzero scalar multiples of $[1, 1]$.

If $\lambda = 2$, then again y is arbitrary and $x = -y$; the eigenvectors belonging to $\lambda = 2$ are all nonzero scalar multiples of $[1, -1]$.

Note that the eigenvectors belonging to 0 comprise the kernel of the mapping, except that **0** is in the kernel but is not an eigenvector. See Problem 14.

Example 3. The matrix

$$\begin{pmatrix} 0 & 1 \\ 0 & 0 \end{pmatrix}$$

is not diagonalizable because we can find its eigenvectors as in Example 2 and no collection of these eigenvectors can form a basis of R^2: The equations that must be satisfied if $[x, y]$ is to be an eigenvector are

$$\begin{aligned} 0x + 1y &= \lambda x, \\ 0x + 0y &= \lambda y. \end{aligned}$$

The second implies $\lambda = 0$ or $y = 0$. However, if $\lambda \neq 0$, then not only is $y = 0$ but, from the first equation, $x = 0$ and $[x, y] = \mathbf{0}$, which we reject. Thus $\lambda = 0$ and the system has as solution space all multiples of $[1, 0]$; every two eigenvectors are dependent; there is no basis of R^2 consisting of eigenvectors of this matrix.

Example 4. To get a function that is diagonalizable but not orthogonally diagonalizable we proceed as in Example 1. Let f map R^2 to R^2 and suppose f stretches the vector $\mathbf{i}$ by a factor of 2 and the vector $\mathbf{i} + \mathbf{j}$ (which is not orthogonal to $\mathbf{i}$) by a factor of 3. Then $f(\mathbf{i}) = 2\mathbf{i}$, $f(\mathbf{i} + \mathbf{j}) = 3\mathbf{i} + 3\mathbf{j}$, so $f(\mathbf{i}) + f(\mathbf{j}) = 3\mathbf{i} + 3\mathbf{j}$, $f(\mathbf{j}) = 3\mathbf{i} + 3\mathbf{j} - f(\mathbf{i}) = 3\mathbf{i} + 3\mathbf{j} - 2\mathbf{i} = \mathbf{i} + 3\mathbf{j}$. The matrix of f is

$$\begin{pmatrix} 2 & 1 \\ 0 & 3 \end{pmatrix}.$$

We leave it to the reader to show that this is not orthogonally diagonalizable. The eigenvectors **i** and **i** + **j** are not orthogonal, but you must still show that f does not have other eigenvectors that *are* orthogonal. See Problem 9. Note that this matrix is not symmetric; it cannot be, because of Theorem 4.1 later in this Chapter.

PROBLEMS

1. Consider the linear function f whose matrix is

$$\begin{pmatrix} 1 & 0 & 0 \\ 0 & 2 & 0 \\ 0 & 0 & -3 \end{pmatrix}.$$

(a) Show that for every vector **v** in the xy-plane $f(\mathbf{v})$ will also be in the xy-plane.

(b) Show that, for every vector **v** in a coordinate plane, $f(\mathbf{v})$ is also in that same coordinate plane.

(c) Find an example of a vector in the xz-plane that is rotated 90 degrees by the linear function (that is, **v** and $f(\mathbf{v})$ are to be perpendicular).

(d) If $\mathbf{v} = \mathbf{i} + 2\mathbf{j} + 3\mathbf{k}$, find the angle between **v** and $f(\mathbf{v})$. Find the ratio of the lengths of **v** and $f(\mathbf{v})$. Note that this vector is both turned and stretched by the mapping f.

(e) Repeat (d) for $\mathbf{v} = [-2, 1, 0]$.

(f) Find an example of a vector in the xy-plane that is rotated 90 degrees by f.

(g) What is the maximum angle between a vector **v** in the xy-plane and its image $f(\mathbf{v})$? Small hint: The angle between **v** and $f(\mathbf{v})$ is the same as the angle between $a\mathbf{v}$ and $f(a\mathbf{v})$ (Why?) so without loss of generality you can assume $\|\mathbf{v}\| = 1$. You will need calculus for this.

2. Find *all* the eigenvectors of the matrix

$$\begin{pmatrix} 2 & 0 & 0 \\ 0 & 3 & 0 \\ 0 & 0 & -1 \end{pmatrix}$$

and find the corresponding eigenvalues.

3. Find all the eigenvectors and eigenvalues of the matrix

$$\begin{pmatrix} 2 & 0 & 0 \\ 0 & 2 & 0 \\ 0 & 0 & -1 \end{pmatrix}.$$

4. In 1.3 we assert that if a 3 by 3 matrix is diagonal, then the vectors along the coordinate axes are eigenvectors and the corresponding eigenvalues are the diagonal entries of the matrix. Prove that if these diagonal entries are all distinct, then there are no other eigenvectors.

5. Prove the converse of Problem 4: If two diagonal entries of a diagonal matrix are equal, then there are eigenvectors of the matrix that are not along the coordinate axes. How can the set of all eigenvectors of a diagonal matrix be described?

6. Find *all* eigenvectors of the function whose matrix is

$$\begin{pmatrix} 1 & 1 \\ 0 & 1 \end{pmatrix}.$$

Find the corresponding eigenvalues. Is this matrix orthogonally diagonalizable?

7. Solve Problem 6 with the matrix

$$\begin{pmatrix} 1 & 1 \\ 1 & 1 \end{pmatrix}.$$

8. Solve Problem 6 with the matrix

$$\begin{pmatrix} \frac{\sqrt{2}}{2} & \frac{-\sqrt{2}}{2} \\ \frac{\sqrt{2}}{2} & \frac{\sqrt{2}}{2} \end{pmatrix}.$$

What is this function geometrically? How does this explain your answer?

9. Solve Problem 6 with the matrix

$$\begin{pmatrix} 2 & 1 \\ 0 & 3 \end{pmatrix}.$$

10. Prove that if **v** is an eigenvector of f belonging to λ, then every nonzero scalar multiple of **v** is an eigenvector of f belonging to λ.

11. If **v** and **w** are eigenvectors of f belonging to the same eigenvalue λ, then every linear combination of **v** and **w** (except **0**) is an eigenvector of f and belongs to λ. The set of all eigenvectors of f belonging to one eigenvalue, together with **0**, form a subspace.

12. If f is a diagonalizable linear mapping from R^n to R^n, and if $\mathbf{e}_1, \ldots, \mathbf{e}_n$ is a basis consisting of eigenvectors, then the dimension of the subspace of all eigenvectors belonging to one λ (see Problem 11) equals the number of **e**'s that belong to λ. If f has a diagonal matrix, this is also the number of times λ appears as a diagonal entry in the matrix of f.

13. Deduce Problem 4 from Problem 12.

14. (a) Prove that the kernel of a linear mapping f is the set of eigenvectors of f belonging to 0, together with the zero vector.

(b) Prove that a linear mapping f from R^n to R^n is invertible if and only if 0 is not an eigenvalue of f.

15. The description of a diagonal function as a stretching function (after Eq. (1.1)) says that to operate with f, you must first stretch by a factor a_{11} in the $\mathbf{i}_1$-direction, and so on. In other words, f is being described as a composite of simpler functions. Translate this description into an expression of the diagonal matrix of f as a product of corresponding simpler matrices.

16. Show that the eigenvalues of $\mathbf{A} - \lambda\mathbf{I}$ are $-\lambda +$ the eigenvalues of $\mathbf{A}$.

2. COMPUTATION OF EIGENVALUES AND EIGENVECTORS

In Examples 2 and 3 and in several problems in the preceding section we computed eigenvectors and eigenvalues for some special 2 by 2 matrices. For more complicated matrices we need a better routine. The trick is to compute the eigenvalues first. There are many cases where the eigenvalues are all that we want, however. For example, if we know the function is a stretching function (orthogonally diagonalizable), we may be interested only in knowing the stretching factors and not the axes along which the stretching takes place.

2.1 A number λ is an eigenvalue of a square matrix $\mathbf{A}$ if and only if $\det(\mathbf{A} - \lambda\mathbf{I}) = 0$. Equivalently, λ is an eigenvalue of a linear mapping f if and only if $\det(f - \lambda I) = 0$.

Proof. λ is an eigenvalue of f if and only if $f(\mathbf{v}) = \lambda\mathbf{v}$ for some nonzero $\mathbf{v}$. This is equivalent to $(f - \lambda I)(\mathbf{v}) = \mathbf{0}$ for some nonzero $\mathbf{v}$; that is, $f - \lambda I$ has a nonzero kernel. By Chapter 6, 1.6, this happens if and only if $\det(f - \lambda I) = 0$.

The matrix $\mathbf{A} - \lambda\mathbf{I}$ is the same as $\mathbf{A}$ except on the main diagonal, where $\mathbf{A}$ has the entry a_{pp}, but $\mathbf{A} - \lambda\mathbf{I}$ has $a_{pp} - \lambda$. We shall see in 2.2 that $\det(\mathbf{A} - \lambda\mathbf{I})$ is a polynomial in λ of degree n (n is the number of rows in $\mathbf{A}$). The procedure is to find the roots of this nth-degree polynomial. For each root λ we get a linear homogeneous system of n linear equations in n unknowns $(\mathbf{A} - \lambda\mathbf{I})\mathbf{x} = \mathbf{0}$ that has a nonzero solution space (that is what the condition $\det(\mathbf{A} - \lambda\mathbf{I}) = 0$ amounts to). We find the solution space and have all the eigenvectors belonging to λ. We proceed similarly with each of the eigenvalues in turn.

Example 1. Find the eigenvalues and eigenvectors of

$$\begin{pmatrix} 2 & 1 \\ 1 & 2 \end{pmatrix}.$$

Here $\mathbf{A} - \lambda\mathbf{I}$ is

$$\begin{pmatrix} 2-\lambda & 1 \\ 1 & 2-\lambda \end{pmatrix}$$

and

$$\begin{aligned} \det(\mathbf{A} - \lambda\mathbf{I}) &= (2-\lambda)^2 - 1 \\ &= \lambda^2 - 4\lambda + 3 \\ &= (\lambda - 3)(\lambda - 1). \end{aligned}$$

The roots of this polynomial are $\lambda_1 = 3$ and $\lambda_2 = 1$. These are the only eigenvalues. (If the polynomial had not factored so easily, we would have had to use the quadratic formula.) Then

$$\mathbf{A} - \lambda_1\mathbf{I} = \mathbf{A} - 3\mathbf{I} = \begin{pmatrix} -1 & 1 \\ 1 & -1 \end{pmatrix}$$

and its solution space is spanned by [1, 1] (check this). This is the space of eigenvectors belonging to the eigenvalue 3. Similarly,

$$\mathbf{A} - \lambda_2\mathbf{I} = \begin{pmatrix} 1 & 1 \\ 1 & 1 \end{pmatrix},$$

whose solution space is spanned by [1, −1]; this is the space of eigenvectors belonging to 1. There are no other eigenvectors.

Example 2. Find the eigenvalues and eigenvectors of

$$\begin{pmatrix} 0 & 2 & -2 & 0 \\ 1 & 1 & 0 & -1 \\ -1 & 1 & -2 & 1 \\ -1 & 1 & -2 & 1 \end{pmatrix}.$$

We begin reducing $\mathbf{A} - \lambda\mathbf{I}$ to echelon form, for two reasons. We want to compute $\det(\mathbf{A} - \lambda\mathbf{I})$ to find the eigenvalues, and we want to solve $(\mathbf{A} - \lambda\mathbf{I})\mathbf{x} = \mathbf{0}$ to find the eigenvectors. Interchange the top two rows and then make zeros in the rest of the first column:

$$\begin{pmatrix} 1 & 1-\lambda & 0 & -1 \\ 0 & 2+\lambda-\lambda^2 & -2 & -\lambda \\ 0 & 2-\lambda & -2-\lambda & 0 \\ 0 & 2-\lambda & -2 & -\lambda \end{pmatrix}.$$

Interchange second and fourth rows, subtract the new second row from the third and $(1 + \lambda)$ times the second from the fourth, to get

$$\begin{pmatrix} 1 & 1-\lambda & 0 & -1 \\ 0 & 2-\lambda & -2 & -\lambda \\ 0 & 0 & -\lambda & \lambda \\ 0 & 0 & 2\lambda & \lambda^2 \end{pmatrix}.$$

Add twice the third row to the fourth and we have a triangular matrix, which is close enough to echelon form for our purposes. We get

$$\begin{aligned} \det(\mathbf{A} - \lambda\mathbf{I}) &= \pm 1(2-\lambda)(-\lambda)(2\lambda + \lambda^2) \\ &= \pm\lambda^2(2-\lambda)(2+\lambda), \end{aligned}$$

the $\pm$ sign coming from our failing to keep track of how many times we interchanged rows. Thus the eigenvalues are 0 (twice), 2, and −2. The eigenvectors are found from the equations

$$\begin{aligned} x_1 + (1-\lambda)x_2 \qquad\qquad - x_4 &= 0, \\ (2-\lambda)x_2 - 2x_3 \quad - \lambda x_4 &= 0, \\ -\lambda x_3 \quad + \lambda x_4 &= 0, \\ (2\lambda + \lambda^2)x_4 &= 0. \end{aligned}$$

If $\lambda = 0$, the solutions are $[-x_3 + x_4,\ x_3,\ x_3,\ x_4]$. Thus the eigenvectors belonging to 0 are all the nonzero linear combinations of $[-1, 1, 1, 0]$ and $[1, 0, 0, 1]$. Note again that this is the kernel of **A**.

If $\lambda = 2$, the solutions are $[x_2, x_2, 0, 0]$. Thus the eigenvectors belonging to 2 are all the nonzero scalar multiples of $[1, 1, 0, 0]$.

If $\lambda = -2$, the solutions are $[x_4, 0, x_4, x_4]$. Thus the eigenvectors belonging to -2 are all the nonzero scalar multiples of $[1, 0, 1, 1]$.

Note that the four particular eigenvectors we picked out form a basis of R^4: $[-1, 1, 1, 0]$, $[1, 0, 0, 1]$, $[1, 1, 0, 0]$, $[1, 0, 1, 1]$, but they are not orthogonal; that is, **A** is diagonalizable (but not orthogonally diagonalizable; Why?).

2.2 If **A** is an n by n matrix, then $\det(\mathbf{A} - \lambda\mathbf{I})$ is a polynomial in λ of degree n. The coefficient of λ^n is $(-1)^n$, the coefficient of λ^{n-1} is $(-1)^{n-1} \sum_p a_{pp}$ (we call $\sum_p a_{pp}$ the *trace* of **A**) and the coefficient of λ^0 is det **A**.

As an example we check 2.2 for $n = 3$:

$$\det\begin{pmatrix} a_{11} - \lambda & a_{12} & a_{13} \\ a_{21} & a_{22} - \lambda & a_{23} \\ a_{31} & a_{32} & a_{33} - \lambda \end{pmatrix}$$

$$= (a_{11} - \lambda)(a_{22} - \lambda)(a_{33} - \lambda) + a_{12}a_{23}a_{31} + a_{13}a_{21}a_{32}$$

$$- [a_{21}a_{12}(a_{33} - \lambda) + a_{31}a_{13}(a_{22} - \lambda) + a_{32}a_{23}(a_{11} - \lambda)].$$

If we imagine multiplying out all parentheses, we can see this is a cubic polynomial in λ. Moreover, all the terms of degrees 2 and 3 come from the product

$$(a_{11} - \lambda)(a_{22} - \lambda)(a_{33} - \lambda)$$

and they are as indicated in 2.2. As for the coefficient of λ^0, we can always obtain the constant term of a polynomial by computing the value of the polynomial for $\lambda = 0$. If we do this with $\det(\mathbf{A} - \lambda\mathbf{I})$, we get $\det(\mathbf{A} - 0\mathbf{I}) = \det \mathbf{A}$.

For the case of general n, the quickest argument uses the formula of Chapter 6, 2.2, which expresses a determinant as a sum of products of matrix entries. The argument is much the same as in the case where $n = 3$; since the entries in $\mathbf{A} - \lambda\mathbf{I}$ are all polynomials of degree 0 or 1, every product of n entries is a polynomial of degree $\leq n$, hence any sum of such products is, too, which proves the first assertion in 2.2.

Only one product has a term of degree n, and that product is $(a_{11} - \lambda)(a_{22} - \lambda) \cdots (a_{nn} - \lambda)$; the coefficient of λ^n here is $(-1)^n$, proving the second assertion in 2.2. Moreover, this product is the only one having λ^{n-1} in it, because to get an $(n - 1)$st power of λ you must multiply at least $n - 1$ diagonal entries in $\mathbf{A} - \lambda\mathbf{I}$ (the off-diagonal entries have no λ's in them); the nth factor in the product need not be specified. But according to the formula we are using from Chapter 6, the n entries multiplied together in one product must all come from different rows and from different columns. Once $n - 1$ of the factors are entries from the diagonal, we have no choice but to use as the nth factor the remaining diagonal entry.

To summarize, all terms of degree $n - 1$ in λ come from the product of diagonal entries $(a_{11} - \lambda) \cdots (a_{nn} - \lambda)$. We leave it to the reader to show that the coefficient of λ^{n-1} in this product is $(-1)^{n-1} \sum_{p=1}^{n} a_{pp}$.

2.3 An n by n matrix has at most n eigenvalues, because a polynomial of degree n has at most n roots.

2.4 Definition

The polynomial $\det(\mathbf{A} - \lambda\mathbf{I})$ is called the *characteristic polynomial* of $\mathbf{A}$. This explains the synonym "characteristic root" for eigenvalue, since the eigenvalues of $\mathbf{A}$ are the (real) roots of this characteristic polynomial of $\mathbf{A}$.

In the next section we make some remarks on the imaginary roots of this characteristic polynomial. In particular, we show that if $\mathbf{A}$ is symmetric, then all n of the roots of its characteristic polynomial are real.

We have already noticed some facts about eigenvalues of other special matrices: The eigenvalues of a diagonal matrix are the diagonal entries (Problem 5, Section 1); the eigenvalues of a skew matrix $({}^t\mathbf{A} = -\mathbf{A})$ are zero (Problem 21, Chapter 6, Section 4); the (real) eigenvalues of an orthogonal matrix $({}^t\mathbf{A} = \mathbf{A}^{-1})$ are ± 1 (Problem 14, Chapter 6, Section 5).

2.5 If an n by n matrix has n distinct eigenvalues (that is, the characteristic polynomial has only real roots and no repeated roots) then the matrix is diagonalizable.

Proof. Let $\lambda_1, \ldots, \lambda_n$ be the eigenvalues of the matrix $\mathbf{A}$ and let $\mathbf{v}_1, \ldots, \mathbf{v}_n$ be eigenvectors with $\mathbf{v}_p$ belonging to λ_p for each p. We have only to prove that the v's form a basis of R^n, in other words, that $\{\mathbf{v}_1, \ldots, \mathbf{v}_n\}$ is an independent set. If not, then $x_1\mathbf{v}_1 + \cdots + x_n\mathbf{v}_n = \mathbf{0}$ for some scalars $x_1, \ldots, x_n$, not all zero. From among all such equations choose one with the fewest possible nonzero terms, say

$$y_1\mathbf{v}_1 + \cdots + y_r\mathbf{v}_r = \mathbf{0}, \tag{2.1}$$

when we write only the nonzero terms. Hence, $y_p \neq 0$ for $p = 1, \ldots, r$. (It is conceivable, and even likely, that the equation $\sum y_p\mathbf{v}_p = \mathbf{0}$ with the fewest nonzero terms has its nonzero terms scattered among $y_1\mathbf{v}_1, \ldots, y_n\mathbf{v}_n$ rather than bunched at the beginning, as we have in Eq. (2.1). In this case we merely renumber the eigenvectors and the eigenvalues so that the eigenvectors involved in these nonzero terms are now called $\mathbf{v}_1, \ldots, \mathbf{v}_r$.)

No nontrivial linear combination of $\mathbf{v}_1, \ldots, \mathbf{v}_{r-1}$ can be $\mathbf{0}$, for this would be an equation like (2.1) with fewer nonzero terms. Now apply to Eq. (2.1) the linear mapping f corresponding to the matrix $\mathbf{A}$.

$$\mathbf{0} = f(y_1\mathbf{v}_1 + \cdots + y_r\mathbf{v}_r) = y_1 f(\mathbf{v}_1) + \cdots + y_r f(\mathbf{v}_r),$$

$$\mathbf{0} = y_1\lambda_1\mathbf{v}_1 + \cdots + y_r\lambda_r\mathbf{v}_r.$$

Subtract from this λ_r times (2.1) to get

$$\mathbf{0} = y_1(\lambda_1 - \lambda_r)\mathbf{v}_1 + \cdots + y_{r-1}(\lambda_{r-1} - \lambda_r)\mathbf{v}_{r-1},$$

which is exactly a linear combination of $\mathbf{v}_1, \ldots, \mathbf{v}_{r-1}$ equal to $\mathbf{0}$; therefore, it must be trivial, which means $y_p(\lambda_p - \lambda_r) = 0$ for $p = 1, \ldots, r - 1$. Since $\lambda_p \neq \lambda_r$ by hypothesis, we have $y_p = 0$ for $p = 1, \ldots, r - 1$. This is also impossible, since by our choice of Eq. (2.1), all y_p are nonzero. Hence there must be no p's in the range $p = 1, \ldots, r - 1$; that is, $r = 1$, and Eq. (2.1) becomes $y_1\mathbf{v}_1 = \mathbf{0}$. This forces $y_1 = 0$, since an eigenvector is not $\mathbf{0}$, by definition, which is again a contradiction. Thus the assumptions we made in Eq. (2.1) lead to contradictions, which shows that $\mathbf{v}_1, \ldots, \mathbf{v}_n$ are independent. This completes the proof.

2.6 Every orthogonal mapping from R^3 to R^3 is either a rotation about a line through the origin or a composite $g \circ h$, where h is reflection in a plane through the origin and g is a rotation about a line perpendicular to that plane. We can tell the difference between the two cases by determinants: a rotation has determinant 1, and a composite of a reflection and a rotation has determinant -1.

Proof. Let f denote the orthogonal mapping under consideration. Then f has at least one real eigenvalue because the characteristic polynomial $\det(f - \lambda I)$ is a cubic polynomial, and every cubic polynomial with real coefficients has at least one real root. By Problem 14, Chapter 6, Section 5, this eigenvalue is 1 or -1. If $\mathbf{e}$ is an eigenvector corresponding to this eigenvalue, then

$$f(\mathbf{e}) = \pm\mathbf{e}.$$

Now let V be the subspace consisting of all vectors in R^3 that are orthogonal to $\mathbf{e}$. This is a plane through the origin, and f sends this plane into itself: If $\mathbf{v}$ is in V, then $f(\mathbf{v}) \cdot \mathbf{e} = \mathbf{v} \cdot {}^tf(\mathbf{e}) = \mathbf{v} \cdot f^{-1}(\mathbf{e}) = \mathbf{v} \cdot (\pm\mathbf{e}) = 0$. So $f(\mathbf{v})$ is orthogonal to $\mathbf{e}$, which says $f(\mathbf{v})$ is in V.

We treat V as if it were R^2 and use Chapter 6, 5.10. More accurately, we reread the proof of that theorem and notice that it applies perfectly well to V if we replace $\mathbf{i}$ and $\mathbf{j}$ in that proof by an orthonormal basis $\mathbf{e}_2$ and $\mathbf{e}_3$ of V. Our conclusion is that if $f(\mathbf{v}) \cdot \mathbf{w} = \mathbf{v} \cdot f^{-1}(\mathbf{w})$ for all $\mathbf{v}$ and $\mathbf{w}$ in V (which is true even for all $\mathbf{v}$ and $\mathbf{w}$ in R^3), then the action of f on vectors in V is either to rotate them around the origin or to reflect them in a line in V. Let us consider these cases separately.

Case 1. f rotates vectors in V by an angle α. Let g denote the rotation in R^3 around the line of $\mathbf{e}$ through an angle α. Then for vectors $\mathbf{v}$ in V, $f(\mathbf{v}) = g(\mathbf{v})$.

Case 1a. If, besides, $f(\mathbf{e}) = \mathbf{e}$, then $f(\mathbf{e}) = g(\mathbf{e})$, and so f and g agree on all vectors parallel to $\mathbf{e}$ and all vectors perpendicular to $\mathbf{e}$. But every vector $\mathbf{v}$ in R^3 can be written as a sum of a vector $\mathbf{v}_1$ parallel to $\mathbf{e}$ (the projection of $\mathbf{v}$ on $\mathbf{e}$) and a vector $\mathbf{v}_2$ perpendicular to $\mathbf{e}$. Since f and g agree on both $\mathbf{v}_1$ and $\mathbf{v}_2$, they also agree on $\mathbf{v}$.

$$\begin{aligned} f(\mathbf{v}) &= f((\mathbf{v} \cdot \mathbf{e})\mathbf{e}) + f(\mathbf{v} - (\mathbf{v} \cdot \mathbf{e})\mathbf{e}) \\ &= g((\mathbf{v} \cdot \mathbf{e})\mathbf{e}) + g(\mathbf{v} - (\mathbf{v} \cdot \mathbf{e})\mathbf{e}) \\ &= g(\mathbf{v}). \end{aligned}$$

Hence $f = g$.

Case 1b. If instead, $f(\mathbf{e}) = -\mathbf{e}$, then define h to be the reflection in the plane V, and prove $f = g \circ h$: Just as in Case 1a, it is sufficient to check that f and $g \circ h$ agree on all vectors $\mathbf{v}$ in V and on $\mathbf{e}$. If $\mathbf{v}$ is in V, then $h(\mathbf{v}) = \mathbf{v}$ and $(g \circ h)(\mathbf{v}) = g(\mathbf{v}) = f(\mathbf{v})$. When $\mathbf{v} = \mathbf{e}$, $(g \circ h)(\mathbf{e}) = g(-\mathbf{e}) = -\mathbf{e} = f(\mathbf{e})$. So we conclude $f = g \circ h$.

Case 2. f reflects vectors in V in a line in V. Then choose $\mathbf{e}_2$ to be a unit vector along this line, and $\mathbf{e}_3$ to be a unit vector in V perpendicular to this line. Case 2 says $f(\mathbf{e}_2) = \mathbf{e}_2$ and $f(\mathbf{e}_3) = -\mathbf{e}_3$.

Case 2a. $f(\mathbf{e}) = \mathbf{e}$, and still $f(\mathbf{e}_2) = \mathbf{e}_2$, $f(\mathbf{e}_3) = -\mathbf{e}_3$. Then f is simply the reflection in the plane of $\mathbf{e}$ and $\mathbf{e}_2$ (followed by a rotation through a zero angle) because every linear combination of $\mathbf{e}$ and $\mathbf{e}_2$ is left fixed and vectors perpendicular to the plane are changed in sign.

Case 2b. $f(\mathbf{e}) = -\mathbf{e}$ and still $f(\mathbf{e}_2) = \mathbf{e}_2$, $f(\mathbf{e}_3) = -\mathbf{e}_3$. This time f is a 180-degree rotation around the line of $\mathbf{e}_2$.

By Problem 15, Chapter 6, Section 3 we know the last statement of the theorem: rotations have determinant 1, and reflections have determinant -1, so a composite of a reflection and a rotation has determinant $1 \times -1 = -1$.

PROBLEMS

1. Repeat Problems 6–9 in Section 1 using 2.1 to find the eigenvalues.

2. Find the eigenvalues and the eigenvectors of the following matrices. Tell which matrices are diagonalizable or orthogonally diagonalizable.

(a) $\begin{pmatrix} 2 & 1 \\ 1 & 2 \end{pmatrix}$; (b) $\begin{pmatrix} 1 & 2 \\ 2 & 2 \end{pmatrix}$; (c) $\begin{pmatrix} 1 & 2 & -3 \\ 2 & -2 & 0 \\ -3 & 0 & 3 \end{pmatrix}$;

(d) $\begin{pmatrix} 2 & 0 & -1 \\ 0 & 1 & 0 \\ 0 & 0 & 1 \end{pmatrix}$; (e) $\begin{pmatrix} 1 & 1 & 0 \\ 0 & 1 & 1 \\ 0 & 0 & 1 \end{pmatrix}$;

(f) $\begin{pmatrix} 1 & -1 & -2 & 3 \\ -4 & 4 & 0 & 4 \\ -3 & 3 & 2 & 3 \\ -3 & 3 & -2 & 7 \end{pmatrix}$.

3. Give an example of a matrix satisfying the conclusion of 2.5 but not the hypothesis. In other words, show that the converse of 2.5 is false.

4. Prove that the sum of all the eigenvalues of an n by n matrix $\mathbf{A}$ is the trace of $\mathbf{A}$, and the product of the eigenvalues is det $\mathbf{A}$.

For this to be correct, some eigenvalues have to be used more than once; specifically, each eigenvalue is to be used as many times as it is a root of the characteristic polynomial. Assume also that all the roots of the characteristic polynomial are real. If there are complex roots, the statement is still true if we modify the definitions to allow complex eigenvalues and complex eigenvectors as in Section 3.

3. COMPLEX NUMBERS

So far we have restricted ourselves to real scalars in this text. This gives us a fairly firm geometric intuition, since R^1 is the usual "real line" or "real number line," R^2 can be visualized as a plane and R^3 as ordinary 3-space.

However, there are cases where it is important to have not only real, but complex scalars available. The first has just occurred in the preceding section: The characteristic polynomial $\det(f - \lambda I)$ may have roots that are not real, but are complex numbers. These complex numbers should also have some significance for the linear mapping.

Let us recall that a complex number is a number of the form $a + bi$ with a and b real and with $i = \sqrt{-1}$. Thus the set of all complex numbers can be identified with the set of all pairs, $[a, b]$ of real numbers, but for our purposes we do not do that; we prefer to think of the set C of all complex numbers as a new set of scalars, so as a kind of one-dimensional space. It is not, of course, geometrically much like R^1. If the complex numbers are to be an effective set of scalars, we must be able to add, subtract, multiply and divide them. The formulas are

$$(a + bi) \pm (c + di) = (a \pm c) + (b \pm d)i;$$

$$(a + bi)(c + di) = (ac - bd) + (ad + bc)i$$

This is the obvious product since

$$i^2 = -1.$$

To divide, we rationalize denominators (remember $i = \sqrt{-1}$)

$$(a + bi)/(c + di) = (a + bi)(c - di)/(c^2 + d^2)$$

$$= [(ac + bd)/(c^2 + d^2)] + [(bc - ad)/c^2 + d^2)]i.$$

The real numbers are considered as special cases of complex numbers; the real number a is usually identified with the complex number

$a + 0i$. The identity elements for addition and multiplication of complex numbers are both real, namely 0 and 1.

Then we can define C^n as the set of all n-tuples of complex numbers. They can be added and subtracted, exactly as vectors in R^n can. Similarly, they can be multiplied by a complex scalar by multiplying each complex entry in the n-tuple by the complex scalar. Virtually everything we have done so far in this text can be done for C^n as well as for R^n, with complex scalars replacing real scalars. Specifically, all of Chapters 3–6 carry over verbatim when R^n is changed to C^n and scalars are the complex numbers. This is true because the only properties of the real numbers used in those chapters were the properties of the addition, subtraction, multiplication, and division of real numbers (for example, the commutative, associative and distributive laws) and all these properties are equally true of the addition, subtraction, multiplication, and division of complex numbers.

Then, if $\mathbf{A}$ is an n by n matrix with real (or even with complex entries), it makes sense to talk about $\mathbf{A}$ as describing a linear mapping from C^n to C^n, and to talk about complex eigenvectors of $\mathbf{A}$, and also of complex eigenvalues.

Instead of 2.1 saying that the eigenvalues of $\mathbf{A}$ are the real roots of $\det(\mathbf{A} - \lambda\mathbf{I}) = 0$, it can say, with the same proof, that the (complex) eigenvalues of $\mathbf{A}$ are exactly all the roots of $\det(\mathbf{A} - \lambda\mathbf{I}) = 0$.

As an example of the usefulness of this concept, even for real (as distinct from general, complex) problems, we prove a theorem about real, symmetric matrices:

3.1 If f is a symmetric linear mapping from R^n to R^n, then f has a (real) eigenvector in R^n. In fact, if V is any nonzero subspace of R^n and $f(V) \subset V$, then f has an eigenvector in V.

All the complex eigenvalues of f are real.

Proof. Let the matrix of f be $\mathbf{A}$, which is then a square matrix (with real entries) with ${}^t\mathbf{A} = \mathbf{A}$. Since real numbers are special complex numbers, and every matrix with complex entries acts as a linear transformation from C^n to C^n, we consider $\mathbf{A}$ as describing such a linear transformation. Now ask for nonzero vectors (say n by 1 matrices, column vectors) $\mathbf{X}$ with complex entries and a complex scalar λ such that

$$\mathbf{AX} = \lambda\mathbf{X}.$$

Of course what we want are a real scalar λ and a vector $\mathbf{X}$ with real entries that satisfy this equation, but temporarily we settle for the more general, complex ones, because we know they exist: The characteristic equation of $\mathbf{A}$ is $\det(\mathbf{A} - \lambda\mathbf{I}) = 0$; it is a polynomial equation in λ. By the "Fundamental Theorem of Algebra" every such polynomial equation has a complex solution, λ. With this λ, then, the results of Chapter 6, extended to C^n instead of R^n, say that $\mathbf{A} - \lambda\mathbf{I}$ has rank less than n. The results of Chapter 4 then say that $\mathbf{A} - \lambda\mathbf{I}$ has a nonzero kernel. If $\mathbf{X}$ is a nonzero vector in this kernel, then $(\mathbf{A} - \lambda\mathbf{I})\mathbf{X} = 0$, so $\mathbf{AX} = \lambda\mathbf{IX} = \lambda\mathbf{X}$ as desired.

We now proceed to prove that λ is real.

For this we need the concept of the *conjugate* $\bar{x}$ of a complex number x, and a generalization of this to matrices. If $x = a + bi$ with a and b real, define $\bar{x}$ to be $a - bi$. It is easy to show

$$\overline{x + y} = \bar{x} + \bar{y}, \qquad \overline{xy} = \bar{x}\,\bar{y}$$

for all complex numbers x and y, and similar results for subtraction and division. Moreover, x is real if and only if $x = \bar{x}$.

For a matrix $\mathbf{B}$ with complex entries we define the *conjugate transpose*, $\mathbf{B}^*$, of $\mathbf{B}$ (or the adjoint of $\mathbf{B}$, though it bears no relation to the adjoint in Chapter 6, the matrix of cofactors) to be the matrix whose entries are the conjugates of the entries in ${}^t\mathbf{B}$. (If $\mathbf{B}$ is 1 by 1, you can think of $\mathbf{B}$ as a complex number, and then $\mathbf{B}^*$ becomes just the conjugate of this complex number.) The properties of the conjugate transpose of a matrix are similar to the properties of the conjugate of a scalar. The ones we need here are

$$(\mathbf{BC})^* = \mathbf{C}^*\mathbf{B}^* \qquad \text{and} \qquad (\lambda\mathbf{B})^* = \bar{\lambda}\mathbf{B}^*.$$

They are both easy to check. Now take the equation $\mathbf{AX} = \lambda\mathbf{X}$ and multiply both sides by $\mathbf{X}^*$ on the left:

$$\mathbf{X}^*\mathbf{AX} = \lambda\mathbf{X}^*\mathbf{X}.$$

Since $\mathbf{A} = \mathbf{A}^*$ because $\mathbf{A}$ is symmetric (so ${}^t\mathbf{A} = \mathbf{A}$) and its entries are real (so are all equal to their conjugates),

$$\mathbf{X}^*\mathbf{AX} = \mathbf{X}^*\mathbf{A}^*\mathbf{X} = (\mathbf{AX})^*\mathbf{X} = (\lambda\mathbf{X})^*\mathbf{X} = \bar{\lambda}\mathbf{X}^*\mathbf{X}.$$

Comparing these last two strings of equations, we get

$$\lambda(\mathbf{X}^*\mathbf{X}) = \bar{\lambda}(\mathbf{X}^*\mathbf{X}).$$

Now if the entries in $\mathbf{X}$ are the complex numbers $x_1, \ldots, x_n$, then $\mathbf{X}^*\mathbf{X}$ is the 1 by 1 matrix with entry $\bar{x}_1x_1 + \cdots + \bar{x}_nx_n$. Each $\bar{x}_jx_j$ is a positive real number unless $x_j = 0$. Thus $\mathbf{X}^*\mathbf{X}$ has its entry positive, hence nonzero, unless all x_j's are 0, which they are not, because $\mathbf{X}$

was chosen to be a nonzero vector. Therefore we can divide the last displayed equation by the nonzero number $\mathbf{X}^*\mathbf{X}$ to get $\lambda = \bar{\lambda}$, proving λ is real, and we have found a real eigenvalue. In fact, we have shown that all the roots of the characteristic polynomial of $\mathbf{A}$ are real.

To find a real eigenvector, we just go through the mechanics of Section 2: with λ equal to the real eigenvalue just found, use the matrix $\mathbf{A} - \lambda\mathbf{I}$ as coefficient matrix on homogeneous equations in n unknowns. This will be a matrix with real entries, so when reduced to echelon form the solutions you get will be real vectors, and will be eigenvectors of $\mathbf{A}$ belonging to the eigenvalue λ.

Now suppose V is a subspace of R^n but the mapping f carries every vector in V into some other vector in V. The trick is to treat V just as we did R^n in the preceding part of the proof. You can take an orthonormal basis $\mathbf{e}_1, \ldots, \mathbf{e}_m$ of V to substitute for $\mathbf{i}_1, \ldots, \mathbf{i}_n$. Then for each $p = 1, \ldots, m$, $f(\mathbf{e}_p) = b_{p1}\mathbf{e}_1 + \cdots + b_{pm}\mathbf{e}_m$ and the matrix $\mathbf{B} = (b_{pq})$ is an m by m symmetric matrix (symmetric because $f(\mathbf{e}_p) \cdot \mathbf{e}_q = \mathbf{e}_p \cdot f(\mathbf{e}_q)$ and these two numbers are b_{pq} and b_{qp} respectively from the formula for $f(\mathbf{e}_n)$ and the orthonormality of the $\mathbf{e}$'s). It has a real eigenvalue, μ and a real eigenvector $\mathbf{v} = [x_1, \ldots, x_m]$. That is, $\mathbf{Bv} = \mu\mathbf{v}$. Translated from m-tuples to linear combinations of $\mathbf{e}$'s, we get $f(\mathbf{v}) = \mu\mathbf{v}$ for $v = x_1\mathbf{e}_1 + \cdots + x_m\mathbf{e}_m$, which is in V. This completes the proof of 3.1.

PROBLEMS

1. Solve the following systems of equations by reducing the augmented matrix toward echelon form.

(a)
$$\begin{aligned} 2x + 3y + (8 + 6i)z &= 13, \\ -2ix + (1 - 3i)y + (8 - 6i)z &= 4 - 13i, \\ iy + (-1 + 3i)z &= 1 + 4i. \end{aligned}$$

(b)
$$\begin{aligned} (1 - i)x + (2 - i)y &= 1, \\ x + iy &= 2. \end{aligned}$$

(c)
$$\begin{aligned} (1 - i)x + (-1 + i)y + (-1 - i)z &= 0, \\ ix + y + z &= 0, \\ -x + 2y + iz &= 0. \end{aligned}$$

2. Solve as many of the systems in Problem 1 as possible by Cramer's rule.

3. Find the ranks of the following matrices:

$$\begin{pmatrix} 1 & 1+i & 1-i & 2 \\ 1+i & 2i & 2 & 2+2i \end{pmatrix};$$

$$\begin{pmatrix} 3+i & 7+2i & 1+3i & 3+4i \\ 1 & 2 & i & 1+i \\ i & 1+2i & 1 & i \end{pmatrix}.$$

4. Find $\mathbf{A}^*$ for each of the following matrices $\mathbf{A}$:

$$\begin{pmatrix} 1 & 2 & 3 \\ 4 & 5 & 6 \\ 7 & 8 & 9 \end{pmatrix}; \qquad \begin{pmatrix} 1+i & 2+i & 3+i \\ 4+2i & 5+2i & 6+2i \\ 7+3i & 8+3i & 9+3i \end{pmatrix}.$$

5. Show that $\det \mathbf{A}^*$ is the complex conjugate of $\det \mathbf{A}$ for every square complex matrix $\mathbf{A}$. Check this for the following matrices:

$$\begin{pmatrix} 1 & 2+i \\ 2+i & 3 \end{pmatrix}; \qquad \begin{pmatrix} 1 & 2+i & 3-i \\ i & 0 & -i \\ 1+i & 2 & 3 \end{pmatrix}$$

6. Identify the complex number $x + yi$ with the vector $[x, y]$ in R^2. Now take a fixed complex number $a + bi$ and multiply all $x + yi$'s by it. This gives a mapping f from R^2 to R^2: $f([x, y]) = [x', y']$ where x' and y' are the real and imaginary components of $(x + yi)(a + bi)$.

(a) Show that f is a linear mapping.

(b) Show that the matrix of f is

$$\begin{pmatrix} a & b \\ -b & a \end{pmatrix} \quad \text{or} \quad a\mathbf{I} + b\begin{pmatrix} 0 & 1 \\ -1 & 0 \end{pmatrix}.$$

(c) If $a = \cos\theta$ and $b = \sin\theta$, the matrix of f is the matrix of rotation in R^2 through an angle θ; multiplying a complex number by $\cos\theta + i\sin\theta$ rotates the corresponding vector through an angle θ.

(d) Use (c) to prove De Moivre's Theorem:

$$(\cos\theta + i\sin\theta)^n = \cos n\theta + i\sin n\theta$$

for $n = 1, 2, \ldots$, or in fact for any integer n, positive, negative, or zero. (Show first that multiplication by $(\cos\theta + i\sin\theta)^n$ is the same mapping as multiplication by $\cos n\theta + i\sin n\theta$.)

7. Find the characteristic polynomial of the following symmetric matrix, and show directly that all its roots are real.

$$\begin{pmatrix} 1 & 2 & 3 \\ 2 & 3 & 4 \\ 3 & 4 & 5 \end{pmatrix}.$$

4. PRINCIPAL AXIS THEOREM

4.1 Theorem

Every symmetric linear mapping is orthogonally diagonalizable. Conversely, every orthogonally diagonalizable mapping is symmetric.

Proof. First, the converse: Suppose f is a linear mapping from R^n to R^n and $\mathbf{e}_1, \ldots, \mathbf{e}_n$ is an orthogonal set of eigenvectors of f. We are to show $f(\mathbf{v}) \cdot \mathbf{w} = \mathbf{v} \cdot f(\mathbf{w})$ for all $\mathbf{v}$ and $\mathbf{w}$ in R^n. Since $\{\mathbf{e}_1, \ldots, \mathbf{e}_n\}$ is an independent set (Problem 2, Chapter 5, Section 2 or the proof of Chapter 5, 3.1) of n vectors in n-space, it is a basis of R^n and all vectors in R^n are linear combinations of them. Write $\mathbf{v} = \sum_{p=1}^{n} v_p\mathbf{e}_p$ and $\mathbf{w} = \sum_{p=1}^{n} w_p\mathbf{e}_p$, then compute $f(\mathbf{v}) \cdot \mathbf{w}$. We have

$$f(\mathbf{v}) = f(\sum v_p\mathbf{e}_p) = \sum v_p f(\mathbf{e}_p) = \sum v_p\lambda_p\mathbf{e}_p$$

and so

$$f(\mathbf{v}) \cdot \mathbf{w} = (\sum v_p\lambda_p\mathbf{e}_p) \cdot (\sum w_p\mathbf{e}_p) = \sum v_p\lambda_p w_p\mathbf{e}_p \cdot \mathbf{e}_p$$

because all the cross-product terms involve $\mathbf{e}_p \cdot \mathbf{e}_q$ and so vanish. The computation of $\mathbf{v} \cdot f(\mathbf{w})$ is similar and leads to the same number. Hence, f is symmetric.

By 3.1, f has an eigenvector $\mathbf{e}_1$. We need a whole basis of eigenvectors. To find more, let V be the orthogonal complement of the space spanned by $\mathbf{e}_1$, that is, the subspace of R^n consisting of all vectors orthogonal to $\mathbf{e}_1$. (See Problem 13, Chapter 5, Section 4.) Then for every $\mathbf{v}$ in V, $f(\mathbf{v})$ is also in V because

$$f(\mathbf{v}) \cdot \mathbf{e}_1 = \mathbf{v} \cdot f(\mathbf{e}_1) = \mathbf{v} \cdot \lambda\mathbf{e}_1 = \lambda(\mathbf{v} \cdot \mathbf{e}_1) = 0.$$

In 3.1 we have also sketched a sharper theorem, which asserts that if V has nonzero vectors in it, then f has an eigenvector $\mathbf{e}_2$ in V. Of course, $\mathbf{e}_2$ is orthogonal to $\mathbf{e}_1$ since all vectors in V are orthogonal to $\mathbf{e}_1$. Then take the orthogonal complement V' of the space spanned by $\mathbf{e}_1$ and $\mathbf{e}_2$ and show that f sends V' into itself, so there is an eigenvector in V', and so on. This process stops only when we get vectors $\mathbf{e}_1, \ldots, \mathbf{e}_m$

spanning a space whose orthogonal complement is zero. But this means the space spanned by $\mathbf{e}_1, \ldots, \mathbf{e}_m$ is all R^n (Problem 13, Chapter 5, Section 4). This completes the proof.

As a corollary of 4.1, we can derive a property of two of the eigenvalues of a symmetric mapping, which can be used effectively for machine computation of these eigenvalues.

4.2 The smallest eigenvalue of a symmetric mapping f is the minimum value of $f(\mathbf{v}) \cdot \mathbf{v}$ as $\mathbf{v}$ ranges over vectors of length 1.

Remark. $f(\mathbf{v}) \cdot \mathbf{v}$ is the subject of all of the next chapter.

Proof of 4.2. Let the eigenvalues be arranged in increasing order $\lambda_1 \leqq \lambda_2 \leqq \cdots \leqq \lambda_n$ and let $\mathbf{e}_1, \ldots, \mathbf{e}_n$ be a corresponding basis of eigenvectors as in 4.1. A vector $\mathbf{v} = x_1\mathbf{e}_1 + \cdots + x_n\mathbf{e}_n$ has length 1 if and only if $\mathbf{v} \cdot \mathbf{v} = 1$, which means

$$x_1^2 + x_2^2 + \cdots \cdot x_n^2 = 1.$$

The formula for $f(\mathbf{v}) \cdot \mathbf{v}$ is $(x_1\lambda_1\mathbf{e}_1 + \cdots + x_n\lambda_n\mathbf{e}_n) \cdot \mathbf{v}$ or

$$f(\mathbf{v}) \cdot \mathbf{v} = x_1^2\lambda_1 + x_2^2\lambda_2 + \cdots + x_n^2\lambda_n.$$

Since the x_p^2 are nonnegative numbers adding up to 1, this formula for $f(\mathbf{v}) \cdot \mathbf{v}$ is a weighted average of $\lambda_1, \ldots, \lambda_n$ and so lies somewhere between the smallest one, λ_1, and the largest one, λ_n (or we may add the inequalities $x_i^2\lambda_1 \leq x_i^2\lambda_i \leq x_i^2\lambda_n$ for $i = 1, \ldots, n$, to get $\lambda_1 \leq f(\mathbf{v}) \cdot \mathbf{v} \leq \lambda_n$). Hence the minimum value of $f(\mathbf{v}) \cdot \mathbf{v}$ is at least λ_1. But λ_1 is one possible value of $f(\mathbf{v}) \cdot \mathbf{v}$, namely when $\mathbf{v} = \mathbf{e}_1$ or $x_1 = 1, x_2 = \cdots = x_n = 0$. Hence λ_1 is the minimum as desired.

Similarly, λ_n, the largest eigenvalue, is the maximum value of $f(\mathbf{v}) \cdot \mathbf{v}$ as $\mathbf{v}$ ranges over vectors of length 1.

Example 1. All the eigenvalues of the following matrix are positive:

$$\begin{pmatrix} 1 & 2 & 0 \\ 2 & 5 & 1 \\ 0 & 1 & 2 \end{pmatrix}$$

because if $\mathbf{v} = x\mathbf{i} + y\mathbf{j} + z\mathbf{k}$ then

$$f(\mathbf{v}) \cdot \mathbf{v} = x^2 + 4xy + 2z^2 + 2yz + 5y^2$$

and we can show that this is always positive by completing the squares

$$\begin{aligned} f(\mathbf{v}) \cdot \mathbf{v} &= (x + 2y)^2 - 4y^2 \\ &\quad + 2(z + \tfrac{1}{2}y)^2 - \tfrac{1}{2}y^2 \\ &\quad + 5y^2 \\ &= (x + 2y)^2 + 2(z + \tfrac{1}{2}y)^2 + \tfrac{1}{2}y^2 > 0 \end{aligned}$$

for all x, y, z (even if $\mathbf{v}$ does not have length 1). Hence the minimum value of $f(\mathbf{v})$ is positive, so the minimum eigenvalue of f is positive.

This kind of example comes up in problems of oscillations of physical or biological systems. The system often has associated with it a symmetric matrix, and the system is stable, that is, does not oscillate wildly, if and only if all the eigenvalues of the matrix are positive.

CHAPTER

8

Quadratic Forms and Change of Basis

1. QUADRATIC FORMS

1.1 Definition

A *quadratic form in n variables* is a polynomial function of $x_1, \ldots, x_n$ that is homogeneous and of second degree. In other words, it is a sum of monomials of the form ax_px_q $(p, q = 1, 2, \ldots, n)$. When $p = q$, we get a "square term" of the form ax_p^2; the other terms are "cross-product terms." Every term has degree two.

Instead of thinking of a quadratic form as a function of $x_1, \ldots, x_n$, we shall think of it as a function of the n-tuple $[x_1, \ldots, x_n]$. Thus a quadratic form is a certain kind of (nonlinear) function from R^n to R^1. We shall often use Q for such a function and $Q(\mathbf{x})$ for its value at $\mathbf{x}$.

Example 1. $x^2 + 4xy + 3y^2$ is a quadratic form in x and y. The corresponding function from R^2 to R^1 is

$$Q(\mathbf{r}) = x^2 + 4xy + 3y^2,$$

where $\mathbf{r}$ denotes $x\mathbf{i} + y\mathbf{j}$ or $[x, y]$.

Example 2. $x^2 + y^2 + z^2$ is a quadratic form in three variables. The function from R^3 to R^1 is

$$Q(\mathbf{r}) = x^2 + y^2 + z^2 \qquad \text{or} \qquad Q(\mathbf{r}) = \|\mathbf{r}\|^2.$$

Example 3. $Q(\mathbf{x}) = x_1^2 - 2x_2^2 - 3x_3^2 - 4x_4^2 + 5x_1x_2 + 6x_1x_3 + 7x_1x_4 + 8x_2x_3 + 9x_2x_4 + 10x_3x_4$ is a quadratic form in four variables; written as a function from R^4 to R^1, the function associates to each vector

$$\mathbf{x} = [x_1, x_2, x_3, x_4] = x_1\mathbf{i}_1 + x_2\mathbf{i}_2 + x_3\mathbf{i}_3 + x_4\mathbf{i}_4,$$

the number (in R^1) given by the long formula above.

Nonexample 4. $x^2 + xy + y^2 + 2x + 3y + 1$ is not a quadratic form. It is a second-degree polynomial, but is not homogeneous; the last three terms do not have degree two.

When we study a quadratic form in many variables, one deceptively innocent question that arises is the order in which to write the terms. There are n square terms

$$a_1x_1^2 + a_2x_2^2 + \cdots + a_nx_n^2,$$

together with many (how many?) cross-product terms. The simple but surprisingly fruitful idea is to split each of these cross-product terms bx_px_q in half, writing it as $\frac{1}{2}bx_px_q + \frac{1}{2}bx_qx_p$. Then for each ordered pair of indices (p, q) we have one term in x_px_q, including the square terms where $p = q$, which are not to be split. We get a total of n^2 terms, which strongly suggests a square array.

1.2 Definition

If Q is a quadratic form in n variables $x_1, \ldots, x_n$, the *matrix* of Q is the matrix (a_{pq}) where a_{pp} = the coefficient of x_p^2 in the polynomial, and a_{pq} = half the coefficient of x_px_q in the polynomial.

For example, the matrix of the quadratic form $x^2 + 4xy + 3y^2$ in

Example 1 is

$$\begin{pmatrix} 1 & 2 \\ 2 & 3 \end{pmatrix}$$

since the coefficient of x^2 is 1 and of y^2 is 3 (the diagonal entries are the coefficients of the square terms), and half the coefficient of xy is 2.

The matrices of the quadratic forms in Examples 2 and 3 are, respectively, the 3 by 3 identity matrix and

$$\begin{pmatrix} 1 & \frac{5}{2} & 3 & \frac{7}{2} \\ \frac{5}{2} & -2 & 4 & \frac{9}{2} \\ 3 & 4 & -3 & 5 \\ \frac{7}{2} & \frac{9}{2} & 5 & -4 \end{pmatrix}.$$

The reader should be warned that not all functions from R^n to R^1 have matrices. A linear function does have a matrix, but it is 1 by n. A quadratic form also has one, but it is n by n. Other nonlinear functions have no matrices associated with them at all.

1.3 To each quadratic form Q from R^n to R^1 there corresponds a symmetric linear function f from R^n to R^n. Given Q, f is found as the linear function whose matrix is the same as the matrix of Q. Given f, Q may be retrieved from the formula

$$Q(\mathbf{x}) = f(\mathbf{x}) \cdot \mathbf{x} \qquad \text{for all } \mathbf{x} \text{ in } R^n. \tag{1.1}$$

Proof. Since the matrix of a quadratic form is symmetric by its very construction, the corresponding linear function will be symmetric. To prove (1.1), take any $\mathbf{x} = [x_1, \ldots, x_n]$ in R^n. Then

$$Q(\mathbf{x}) = \sum_{p=1}^{n} \sum_{q=1}^{n} a_{pq} x_p x_q. \tag{1.2}$$

Also

$$f(\mathbf{x}) = \sum y_p \mathbf{i}_p \qquad \text{with} \qquad y_p = \sum_{q=1}^{n} a_{pq} x_q$$

and

$$f(\mathbf{x}) \cdot \mathbf{x} = \sum_{p=1}^{n} y_p x_p = \sum_{p=1}^{n} \sum_{q=1}^{n} a_{pq} x_q x_p = Q(\mathbf{x}).$$

Probably a better way to visualize Eq. (1.1) is with matrices. Let $\mathbf{X}$ be the n by 1 matrix with entries equal to the components of $\mathbf{x}$ and let $\mathbf{A}$ be the matrix of both Q and f. Then

$$f(\mathbf{x}) \cdot \mathbf{x} = \mathbf{x} \cdot f(\mathbf{x}) = {}^t\mathbf{X}(\mathbf{AX})$$

$$= (x_1, \ldots, x_n) \begin{pmatrix} a_{11} & \cdots & a_{1n} \\ a_{21} & \cdots & a_{2n} \\ \vdots & \vdots & \vdots \\ a_{n1} & \cdots & a_{nn} \end{pmatrix} \begin{pmatrix} x_1 \\ x_2 \\ \vdots \\ x_n \end{pmatrix}$$

$$= a_{11}x_1^2 + a_{12}x_1x_2 + a_{13}x_1x_3 + \cdots = Q(\mathbf{x}).$$

What Eq. (1.1) exhibits is a function (or correspondence) associating to each quadratic form a symmetric linear mapping f. (Here is a function whose inputs and outputs are themselves functions!) This correspondence is a one-to-one correspondence, which means that it is invertible in the same sense that linear functions can be invertible: There is a correspondence in the reverse direction, associating to each symmetric linear function f a quadratic form (in fact, this corresponding quadratic form is defined by Eq. (1.1)), and the composite of the two correspondences is the identity correspondence; if a quadratic form Q gives a symmetric function f, which gives a quadratic form defined by Eq. (1.1), this quadratic form will be the same as Q; and, starting with any symmetric linear function f, f determines a quadratic form Q by Eq. (1.1), which in turn determines a symmetric function that will be f again.

We know all about symmetric linear functions from Chapters 6 and 7. They are stretching transformations (that is, orthogonally diagonalizable). For each such f there is a basis $\mathbf{e}_1, \ldots, \mathbf{e}_n$ of R^n that is orthogonal and consists of eigenvectors of f. Let us also arrange that $\| \mathbf{e}_p \| = 1$ for every p. Every vector $\mathbf{x}$ in R^n is a linear combination of the $\mathbf{e}$'s:

$$\mathbf{x} = \xi_1\mathbf{e}_1 + \cdots + \xi_n\mathbf{e}_n.$$

(We do not want to use the letters $x_1, \ldots, x_n$ for the coefficients of this linear combination since we usually reserve them for the components of $\mathbf{x}$ along $\mathbf{i}_1, \ldots, \mathbf{i}_n$:

$$\mathbf{x} = [x_1, \ldots, x_n] = x_1\mathbf{i}_1 + \cdots + x_n\mathbf{i}_n,$$

and these coefficients of $\mathbf{e}_1, \ldots, \mathbf{e}_n$ will probably be different numbers.)

$$f(\mathbf{x}) = \sum_{p=1}^{n} \xi_p f(\mathbf{e}_p) = \sum \xi_p\lambda_p\mathbf{e}_p,$$

$$Q(\mathbf{x}) = f(\mathbf{x}) \cdot \mathbf{x} = (\sum \xi_p\lambda_p\mathbf{e}_p) \cdot (\sum \xi_p\mathbf{e}_p) = \sum \lambda_p\xi_p^2$$

since in multiplying out this dot product, $\mathbf{e}_p \cdot \mathbf{e}_p = \| \mathbf{e}_p \|^2 = 1$ and $\mathbf{e}_p \cdot \mathbf{e}_q = 0$ by the orthogonality. Thus, if our coordinate axes were along $\mathbf{e}_1, \ldots, \mathbf{e}_n$ instead of along $\mathbf{i}_1, \ldots, \mathbf{i}_n$, the quadratic polynomial Q would have had only square terms in it, and no cross-product terms—its matrix would have been diagonal and the diagonal entries would have been the eigenvalues of f. This elimination of the cross products is one way of thinking of the principal axis theorem (Chapter 7, Section 4), and allows us to answer several important questions.

Example 5. What is the graph of the equation $41x^2 - 24xy + 34y^2 = 25$? The matrix of the quadratic form on the left of the equation is

$$\begin{pmatrix} 41 & -12 \\ -12 & 34 \end{pmatrix}$$

whose eigenvectors may be computed as in Chapter 7 to be

$$\mathbf{e}_1 = \tfrac{3}{5}\mathbf{i} + \tfrac{4}{5}\mathbf{j} \qquad \text{and} \qquad \mathbf{e}_2 = \tfrac{4}{5}\mathbf{i} - \tfrac{3}{5}\mathbf{j}$$

belonging to eigenvalues 25 and 50, respectively. If $\mathbf{x} = \xi_1\mathbf{e}_1 + \xi_2\mathbf{e}_2$, then

$$Q(\mathbf{x}) = 25\xi_1^2 + 50\xi_2^2,$$

so the graph of $Q(\mathbf{x}) = 25$ is the set of all (heads of) vectors $\mathbf{x} = \xi_1\mathbf{e}_1 + \xi_2\mathbf{e}_2$ such that

$$25\xi_1^2 + 50\xi_2^2 = 25.$$

We know that the set of all points (x, y) or the heads of all position vectors $x\mathbf{i} + y\mathbf{j}$ with x and y satisfying $25x^2 + 50y^2 = 25$ form an ellipse with semimajor axis (major radius) 1 unit long along the x-axis and with semiminor axis $1/\sqrt{2}$ unit long along the y-axis. We need only replace $\mathbf{i}$ and $\mathbf{j}$ by $\mathbf{e}_1$ and $\mathbf{e}_2$ in this argument, and x and y by ξ_1 and ξ_2, to see that the set of vectors

$$\mathbf{x} \quad (= \xi_1\mathbf{e}_1 + \xi_2\mathbf{e}_2 = x\mathbf{i} + y\mathbf{j})$$

satisfying $Q(\mathbf{x}) = 25$ is an ellipse with semimajor axis 1 unit long along $\mathbf{e}_1$ and semiminor axis $1/\sqrt{2}$ unit long along $\mathbf{e}_2$. (See Fig. 8.1.) Thus the eigenvalue computations from Chapter 7 have answered the question in this example completely. We shall go into this idea of change of basis in R^n more carefully in Section 2. (See especially Examples 1 and 2, Section 2 and the problems in Section 3.)

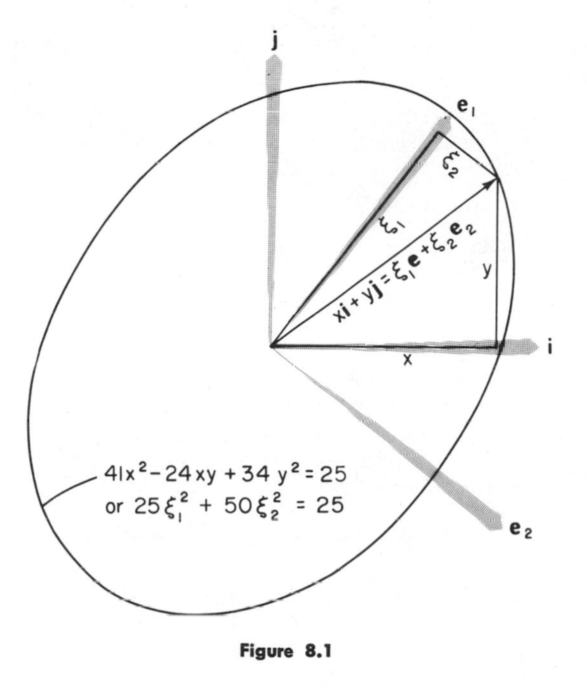

Figure 8.1

PROBLEMS

1. Write the matrix of each of the following quadratic forms. Retrieve each quadratic form from the formula ${}^t\mathbf{XAX}$.

(a) $x^2 + 2xy + y^2$.

(b) $x_1^2 + x_2^2 + x_3^2 + x_4^2$.

(c) $x_1^2 + x_2^2 + x_3^2 + 3x_4^2 + x_1x_2 + 2x_1x_3 + x_1x_4 + 5x_2x_3 + x_2x_4$.

(d) $x^2 + 2xy + y^2$ thought of as a function of x, y, z.

2. (i) Write the quadratic forms corresponding to the following matrices. (ii) Describe geometrically the linear function having the same matrix. Use the principal axis theorem if necessary. (iii) Compute $Q(\mathbf{i})$ and $Q(\mathbf{j})$ from this geometric description (use $Q(\mathbf{x}) = f(\mathbf{x}) \cdot \mathbf{x}$). Check with the results in (i).

(a) $\begin{pmatrix} 1 & 0 & 0 \\ 0 & 1 & 0 \\ 0 & 0 & 1 \end{pmatrix}$; (b) $\begin{pmatrix} 1 & 0 & 0 \\ 0 & 2 & 0 \\ 0 & 0 & 3 \end{pmatrix}$; (c) $\begin{pmatrix} 0 & 1 \\ 1 & 0 \end{pmatrix}$;

(d) $\begin{pmatrix} 1 & 1 \\ 1 & 1 \end{pmatrix}$; (e) $\begin{pmatrix} 1 & 0 & 0 \\ 0 & 0 & -1 \\ 0 & -1 & 0 \end{pmatrix}$.

3. Describe the graphs of the following equations in x and y (shape, size, and location):

(a) $x^2 + 2xy + y^2 = 1$; (b) $x^2 + 2xy + y^2 = 0$;
(c) $x^2 + 3xy + y^2 = 1$; (d) $x^2 + 3xy + y^2 = 0$;
(e) $x^2 + xy + y^2 = 1$; (f) $x^2 + xy + y^2 = 0$;
(g) $x^2 + xy + 2y^2 = 1$; (h) $x^2 + 2xy + y^2 = x + y$.

4. Describe the graphs of the following equations in x, y, z.

(a) $x^2 + y^2 + 2z^2 = 1$. (b) $x^2 + y^2 - 2z^2 = 1$.
(c) $x^2 - y^2 - 2z^2 = 1$. (d) $x^2 - y^2 - 2z^2 = 0$.
(e) $x^2 + 2y^2 = 1$. (f) $x^2 + xy + y^2 + z^2 = 1$.
(g) $8x^2 + 10y^2 + 8z^2 + 2(6)^{1/2}xy - 12xz - 2(6)^{1/2}yz = 1$.
(h) $5x^2 + 17y^2 + 5z^2 - 4xy - 10xz + 4yz = 1$.

5. Let Q be a quadratic form in two variables (thus a function from R^2 to R^1), $Q(\mathbf{x}) = ax_1^2 + bx_1x_2 + cx_2^2$, let its matrix be $\mathbf{A}$, and let the eigenvalues of $\mathbf{A}$ be λ_1 and λ_2; then show the following.

(a) $\det \mathbf{A} = \lambda_1\lambda_2$.

(b) $\det \mathbf{A} = (b^2 - 4ac)/(-4)$.

(c) λ_1 and λ_2 have the same sign if and only if $b^2 - 4ac < 0$.

(d) Trace of $\mathbf{A} = \lambda_1 + \lambda_2 = a + c$ (see Chapter 7, 2.2 and Problem 4, Chapter 7, Section 2).

(e) λ_1 and λ_2 are both positive if and only if $a + c > 0$ and $b^2 - 4ac < 0$. These two conditions are also equivalent to $b^2 - 4ac < 0$ and $a > 0$.

(f) The graph of $ax_1^2 + bx_1x_2 + cx_2^2 = 1$ is an ellipse if $b^2 - 4ac < 0$ and $a + c > 0$ or, equivalently, $b^2 - 4ac < 0$ and $a > 0$.

(g) If the graph of $ax_1^2 + bx_1x_2 + cx_2^2 = 1$ is an ellipse, then $b^2 - 4ac < 0$ and $a + c > 0$.

(h) If $b^2 - 4ac > 0$, then the graph of $ax_1^2 + bx_1x_2 + cx_2^2 = 1$ is a hyperbola.

(i) $b^2 - 4ac = 0$ if and only if one of λ_1 and λ_2 is zero.

(j) If $b^2 - 4ac = 0$, then the graph of $ax_1^2 + bx_1x_2 + cx_2^2 = 1$ is a pair of parallel lines if $a + c > 0$, and is empty if $a + c < 0$.

(k) What is the graph of $ax_1^2 + bx_1x_2 + cx_2^2 = 1$ if $b^2 - 4ac < 0$ and $a + c < 0$?

(l) Use parts (f), (g), (h), (j) to find quickly the shapes of the curves in Problem 3.

6. Let Q be a quadratic form in three variables and let $\mathbf{A}$ be its matrix. Use the technique of Problem 5 to show the following.

(a) If the graph of $Q(\mathbf{x}) = 1$ is an ellipsoid, then $\det \mathbf{A} > 0$ and trace of $\mathbf{A} > 0$.

(b) If the graph of $Q(\mathbf{x}) = 1$ is a hyperboloid of two sheets, then $\det \mathbf{A} > 0$.

(c) If the graph of $Q(\mathbf{x}) = 1$ is a hyperboloid of one sheet, then $\det \mathbf{A} < 0$.

(d) If $\det \mathbf{A} = 0$, then the graph of $Q(\mathbf{x}) = 1$ is a cylinder (in the general sense of a surface swept out by a line moving always parallel to a fixed line) or is empty.

(e) Use parts (a)–(d) to find quickly the shapes of the surfaces in Problem 4(f)–(h).

7. Explain why, if a figure in the xy-plane has the equation $ax^2 + bxy + ay^2 = c$, its principal axes make angles of 45 degrees with the x- and y-axes.

8. Show that the sum of two quadratic forms in three variables is again a quadratic form in three variables, and that a quadratic form in three variables multiplied by a number is also a quadratic form in three variables. Find some quadratic forms that form a basis of the space of all quadratic forms; that is, find forms $Q_1, Q_2, \ldots, Q_k$ such that every quadratic form in three variables is a linear combination of $Q_1, \ldots, Q_k$ and such that $\{Q_1, \ldots, Q_k\}$ is an independent set. What is the number k? (That is, what is the dimension of the space of quadratic forms in three variables?)

9. Answer the questions in Problem 8 for the space of quadratic forms in n variables.

10. "Quadratic forms" can be defined without reference to components at all, thus: First, define a *bilinear form* on R^n to be a function (like the dot product) of two vectors in R^n, with functional values in R^1, that is a linear function of each one of the vectors when the other one is held constant. That is, if B is the bilinear function, we are demanding that:

$$B(\mathbf{u}, \mathbf{v} + \mathbf{v}') = B(\mathbf{u}, \mathbf{v}) + B(\mathbf{u}, \mathbf{v}'),$$
$$B(\mathbf{u}, a\mathbf{v}) = aB(\mathbf{u}, \mathbf{v}),$$
$$B(\mathbf{u} + \mathbf{u}', \mathbf{v}) = B(\mathbf{u}, \mathbf{v}) + B(\mathbf{u}', \mathbf{v}),$$
$$B(a\mathbf{u}, \mathbf{v}) = aB(\mathbf{u}, \mathbf{v})$$

for all $\mathbf{u}$, $\mathbf{u}'$, $\mathbf{v}$, $\mathbf{v}'$ in R^n and all scalars a.

(a) Show that for every quadratic form Q from R^n to R^1, the function B, defined as

$$B(\mathbf{u}, \mathbf{v}) = \tfrac{1}{2}[Q(\mathbf{u} + \mathbf{v}) - Q(\mathbf{u}) - Q(\mathbf{v})], \tag{1.3}$$

is a bilinear form.

(b) If B is any bilinear form let $\mathbf{u}$ be a fixed (but arbitrary) vector in R^n and consider the function B_1 of one variable defined by

$$B_1(\mathbf{v}) = B(\mathbf{u}, \mathbf{v}).$$

Then B_1 is a linear function from R^n to R^1, so necessarily there is a vector $\mathbf{u}'$ in R^n such that $B_1(\mathbf{v}) = \mathbf{u}' \cdot \mathbf{v}$; in fact, there is exactly one such $\mathbf{u}'$. Of course $\mathbf{u}'$ will change if $\mathbf{u}$ changes, that is, $\mathbf{u}'$ is a function of $\mathbf{u}$. Write $\mathbf{u}' = f(\mathbf{u})$ and show that f is linear. This shows that every bilinear form can be written as a combination of a linear function and a dot product:

$$B(\mathbf{u}, \mathbf{v}) = f(\mathbf{u}) \cdot \mathbf{v}. \tag{1.4}$$

(c) If Q is a quadratic form, show

$$Q(a\mathbf{x}) = a^2 Q(\mathbf{x}) \tag{1.5}$$

for every vector $\mathbf{x}$ and every scalar a.

(d) Conversely, show that if Q is any function from R^n to R^1 that satisfies (1.5) and for which the function B in (1.3) is bilinear, then

$$Q(\mathbf{x}) = B(\mathbf{x}, \mathbf{x}) \qquad \text{for all vectors} \quad \mathbf{x}. \tag{1.6}$$

Then use Eqs. (1.4) and (1.6) to show that Q is a quadratic form.

(e) Summary: A function Q from R^n to R^1 is a quadratic form if and only if Eq. (1.5) holds and the function B defined by Eq. (1.3) is a bilinear form.

2. CHANGE OF COORDINATES

In studying diagonalizable mappings and again in studying quadratic forms we encountered orthogonal bases of R^n other than $\{\mathbf{i}_1, \ldots, \mathbf{i}_n\}$ and had the urge to replace the standard coordinate system by one having its axes along those other basis vectors. In fact, any basis $\{\mathbf{e}_1, \ldots, \mathbf{e}_n\}$ of R^n (even a nonorthogonal one) describes a new coordinate system, with coordinate axes along the vectors $\mathbf{e}_1, \ldots, \mathbf{e}_n$, and unit points at the heads of $\mathbf{e}_1, \ldots, \mathbf{e}_n$. In this section we investigate what happens when we change to such a new coordinate system. Specifically we are interested in what happens to the components of a vector, the coordinates of a point, the equations of curves, surfaces, and so on, as well as what happens to the matrix of a linear mapping. The geometric object (vector, point, curve, linear mapping) remains fixed; analytic representations (components, coordinates, equations, matrix) change. For example, an arrow represents a vector in 3-space independent of where the coordinate system is, but the components of this one vector in the standard coordinate system will be different from the components of the same vector in some other coordinate system.

Suppose we have two bases of R^n $\{\mathbf{i}_1, \ldots, \mathbf{i}_n\}$ and $\{\mathbf{e}_1, \ldots, \mathbf{e}_n\}$ and hence two coordinate systems. Suppose we also have a (fixed, but arbitrary†) vector $\mathbf{x}$ in R^n. Then $\mathbf{x}$ is a linear combination of $\mathbf{i}_1, \ldots, \mathbf{i}_n$:

$$\mathbf{x} = x_1\mathbf{i}_1 + \cdots + x_n\mathbf{i}_n = \sum x_p\mathbf{i}_p.$$

But $\mathbf{x}$ is also a linear combination of $\mathbf{e}_1, \ldots, \mathbf{e}_n$; let us call these coefficients $x_1', \ldots, x_n'$:

$$\mathbf{x} = x_1'\mathbf{e}_1 + \cdots + x_n'\mathbf{e}_n = \sum x_p'\mathbf{e}_p.$$

The numbers $x_1, \ldots, x_n$ are the *components of* $\mathbf{x}$ *with respect to the basis* $\mathbf{i}_1, \ldots, \mathbf{i}_n$; they are the components we have used regularly throughout this text. The numbers $x_1', \ldots, x_n'$ are the *components* of this same $\mathbf{x}$ *with respect to* $\{\mathbf{e}_1, \ldots, \mathbf{e}_n\}$. (Notice, however, that each x_p' is the "component of $\mathbf{x}$ along $\mathbf{e}_p$" discussed in Chapter 1 only when the basis $\mathbf{e}_1, \ldots, \mathbf{e}_n$ is orthonormal: $\mathbf{e}_p \cdot \mathbf{e}_q = 1$ if $p = q$ and $= 0$ if $p \neq q$.) Our first task is to find $x_1', \ldots, x_n'$, given $x_1, \ldots, x_n$, and vice versa.

† This self-contradictory phrase is a locution describing a very common pattern of thought in mathematics (and other disciplines): $\mathbf{x}$ will be fixed during this discussion, but since we shall assume no special properties of $\mathbf{x}$ at all, any results we finally arrive at will apply to *all* vectors in R^n.

We do this by defining a linear mapping f associating to each $\mathbf{x}$ in R^n the n-tuple (also in R^n) $[x_1', \ldots, x_n']$. That is, given $\mathbf{x}$, we write $\mathbf{x}$ as a linear combination of $\mathbf{e}_1, \ldots, \mathbf{e}_n$, which can be done in exactly one way since $\{\mathbf{e}_1, \ldots, \mathbf{e}_n\}$ is a basis of R^n; then use the coefficients of this linear combination as components of an n-tuple and call that n-tuple $f(\mathbf{x})$. It is easy to show f is linear: If $\mathbf{x} = \sum x_p'\mathbf{e}_p$ and $\mathbf{y} = \sum y_p'\mathbf{e}_p$, then

$$\mathbf{x} + \mathbf{y} = \sum x_p'\mathbf{e}_p + \sum y_p'\mathbf{e}_p = \sum (x_p' + y_p')\mathbf{e}_p,$$

so

$$\begin{aligned} f(\mathbf{x} + \mathbf{y}) &= [x_1' + y_1', \ldots, x_n' + y_n'] \\ &= [x_1', \ldots, x_n'] + [y_1', \ldots, y_n'] \\ &= f(\mathbf{x}) + f(\mathbf{y}); \end{aligned}$$

in addition

$$a\mathbf{x} = a(\sum x_p'\mathbf{e}_p) = \sum a x_p'\mathbf{e}_p,$$

so

$$f(a\mathbf{x}) = [ax_1', \ldots, ax_n'] = af(\mathbf{x}).$$

Note also that f is invertible, because its image is spanned by $\mathbf{i}_1, \ldots, \mathbf{i}_n$, therefore is R^n; so f has rank n.

Finding the matrix of f will complete our first task since, if we call this matrix $\mathbf{P}$,

$$\begin{pmatrix} x_1' \\ x_2' \\ \vdots \\ x_n' \end{pmatrix} = \mathbf{P} \begin{pmatrix} x_1 \\ x_2 \\ \vdots \\ x_n \end{pmatrix} \qquad \text{or} \qquad \mathbf{X}' = \mathbf{PX}, \tag{2.1}$$

where we use $\mathbf{X}$ for the n by 1 matrix whose entries are the components $x_1, \ldots, x_n$ of $\mathbf{x}$. Similarly, $\mathbf{X}'$ is the matrix whose entries are the new components $x_1', \ldots, x_n'$ of $\mathbf{x}$. Compute $f(\mathbf{e}_1) : \mathbf{e}_1 = 1\mathbf{e}_1 + 0\mathbf{e}_2 + \cdots + 0\mathbf{e}_n$, so $f(\mathbf{e}_1) = [1, 0, \ldots, 0]$. Similarly, $f(\mathbf{e}_p) = [0, \ldots, 0, 1, 0, \ldots, 0] = \mathbf{i}_p$ for every $\mathbf{P}$. In other words, $f^{-1}(\mathbf{i}_p) = \mathbf{e}_p$ so we know the matrix $\mathbf{Q}$ of f^{-1}: its columns are the (standard) components of $\mathbf{e}_1, \ldots, \mathbf{e}_n$. If we call this matrix $\mathbf{Q}$, the matrix of f is then $\mathbf{Q}^{-1}$. To summarize,

2.1 If $\{\mathbf{e}_1, \ldots, \mathbf{e}_n\}$ is a basis of R^n, and if $\mathbf{Q}$ is the matrix whose pth column consists of the components of $\mathbf{e}_p$ ($\mathbf{Q}$ is the matrix of the linear mapping that sends the old basis $\{\mathbf{i}_1, \ldots, \mathbf{i}_n\}$ to the new basis $\{\mathbf{e}_1, \ldots, \mathbf{e}_n\}$), write $\mathbf{P} = \mathbf{Q}^{-1}$. Then the components $x_1', \ldots, x_n'$ of a vector with respect to the basis $\mathbf{e}_1, \ldots, \mathbf{e}_n$ are given by the formula

(2.1) in terms of the components $x_1, \ldots, x_n$ of this same vector with respect to the basis $\mathbf{i}_1, \ldots, \mathbf{i}_n$. Another version of Eq. (2.1) is

$$\begin{pmatrix} x_1 \\ \vdots \\ x_n \end{pmatrix} = \mathbf{Q} \begin{pmatrix} x_1' \\ \vdots \\ x_n' \end{pmatrix} \quad \text{or} \quad \mathbf{X} = \mathbf{QX}'. \tag{2.2}$$

Since the coordinates of a point are the components of its position vector, the same formulas give the new coordinates $(x_1', \ldots, x_n')$ of a point in terms of the original coordinates $(x_1, \ldots, x_n)$ of the same point.

Example 1. Let $\mathbf{e}_1, \mathbf{e}_2$ be the basis of R^2 obtained by rotating the coordinate system 30 degrees counterclockwise. Then $\mathbf{e}_1 = (\frac{1}{2}\sqrt{3})\mathbf{i} + (\frac{1}{2})\mathbf{j}$ (in R^2 we continue to write $\mathbf{i}$ and $\mathbf{j}$ instead of $\mathbf{i}_1$ and $\mathbf{i}_2$) and $\mathbf{e}_2 = (-\frac{1}{2})\mathbf{i} + (\frac{1}{2}\sqrt{3})\mathbf{j}$. Hence

$$\mathbf{Q} = \begin{pmatrix} \frac{1}{2}\sqrt{3} & -\frac{1}{2} \\ \frac{1}{2} & \frac{1}{2}\sqrt{3} \end{pmatrix}$$

and

$$\mathbf{P} = \mathbf{Q}^{-1} = \begin{pmatrix} \frac{1}{2}\sqrt{3} & \frac{1}{2} \\ -\frac{1}{2} & \frac{1}{2}\sqrt{3} \end{pmatrix}.$$

(A quick way to find $\mathbf{Q}^{-1}$: Since $\mathbf{Q}$ is the matrix of a rotation, it is orthogonal (Chapter 6, Section 5), so its inverse is just its transpose!) If the components of a vector are $[x, y]$, then the components with respect to the rotated coordinate system are

$$\begin{pmatrix} \frac{1}{2}\sqrt{3} & \frac{1}{2} \\ -\frac{1}{2} & \frac{1}{2}\sqrt{3} \end{pmatrix} \begin{pmatrix} x \\ y \end{pmatrix} = \begin{pmatrix} \frac{1}{2}\sqrt{3}x + \frac{1}{2}y \\ -\frac{1}{2}x + \frac{1}{2}\sqrt{3}y \end{pmatrix}.$$

In particular, the vector $2\mathbf{i} - \mathbf{j}$ (for which $x = 2, y = -1$) has new components $\sqrt{3} - \frac{1}{2}$ and $-1 - \frac{1}{2}\sqrt{3}$, approximately 1.2 and -1.9. You should draw a picture and check that these results are plausible. Furthermore, we could have obtained the same result from the fact that

$$\mathbf{i} = \tfrac{1}{2}\sqrt{3}\,\mathbf{e}_1 - \tfrac{1}{2}\mathbf{e}_2$$

and

$$\mathbf{j} = \tfrac{1}{2}\mathbf{e}_1 + \tfrac{1}{2}\sqrt{3}\mathbf{e}_2:$$

then

$$\begin{aligned} 2\mathbf{i} - \mathbf{j} &= 2(\tfrac{1}{2}\sqrt{3}\mathbf{e}_1 - \tfrac{1}{2}\mathbf{e}_2) - (\tfrac{1}{2}\mathbf{e}_1 + \tfrac{1}{2}\sqrt{3}\mathbf{e}_2) \\ &= \sqrt{3}\mathbf{e}_1 - \mathbf{e}_2 - \tfrac{1}{2}\mathbf{e}_1 - \tfrac{1}{2}\sqrt{3}\mathbf{e}_2 \\ &= (\sqrt{3} - \tfrac{1}{2})\mathbf{e}_1 + (-1 - \tfrac{1}{2}\sqrt{3})\mathbf{e}_2. \end{aligned}$$

It is now just as easy to say what happens to equations of curves, and so on, when changing coordinates. Suppose a set of points in R^n is described by one or more equations, that is, it is the set of all points in R^n whose coordinates $x_1, \ldots, x_n$ satisfy these equations. Simply substitute for $x_1, \ldots, x_n$ the expressions given by Eq. (2.2) and we get new equations involving the new coordinates $x_1', \ldots, x_n'$ of the same point; the new coordinates of a point satisfy these new equations if and only if the point is in our original set; these new equations, then, also describe the same set of points.

Example 2. Consider the surface in R^3 whose equation is $x^2 + xy + y^2 + z^2 = 1$. Rotate the coordinate axes 45 degrees around the z-axis and find the new equation of the same surface: Here

$$\mathbf{e}_1 = 2^{-1/2}(\mathbf{i} + \mathbf{j}), \qquad \mathbf{e}_2 = 2^{-1/2}(-\mathbf{i} + \mathbf{j}),$$

and

$$\mathbf{e}_3 = \mathbf{k},$$

so

$$\mathbf{Q} = \begin{pmatrix} 2^{-1/2} & -2^{-1/2} & 0 \\ 2^{-1/2} & 2^{-1/2} & 0 \\ 0 & 0 & 1 \end{pmatrix}.$$

If (x', y', z') denote the new coordinates of a point whose old coordinates are (x, y, z), then Eq. (2.2) says

$$\begin{pmatrix} x \\ y \\ z \end{pmatrix} = \mathbf{Q} \begin{pmatrix} x' \\ y' \\ z' \end{pmatrix}$$

or

$$x = 2^{-1/2}(x' - y'),$$

$$y = 2^{-1/2}(x' + y'),$$

$$z = z'.$$

Note that for transforming equations we want Eq. (2.2), not Eq. (2.1); we want the old coordinates in terms of the new. Substituting into the equation $x^2 + xy + y^2 + z^2 = 1$, we get the new equation of the same surface

$$\frac{(x' - y')^2}{2} + \frac{x'^2 - y'^2}{2} + \frac{(x' + y')^2}{2} + z'^2 = 1$$

or

$$\tfrac{3}{2}x'^2 + \tfrac{1}{2}y'^2 + z'^2 = 1.$$

Thus the surface is an *ellipsoid,* an ovoid kind of surface that intersects each (new) coordinate plane in an ellipse.

We have watched changes of coordinates carry old equations into new equations. Sometimes this is better formalized by watching changes of coordinates carry an old formula for a function (from R^n to R^1, say) to a new formula for the same function. The function involved may be one side of such an equation. We are especially interested in quadratic forms. Let Q be a quadratic form on R^n. For example, if $n = 2$,

$$Q(\mathbf{x}) = a_{11}x_1^2 + 2a_{12}x_1x_2 + a_{22}x_2^2$$

is thought of as a function associating to each vector $\mathbf{x} = [x_1, x_2]$ in R^2 a number $Q(\mathbf{x})$. If the vector $\mathbf{x}$ is described by components $[x_1', x_2', \ldots, x_n']$ with respect to a new basis, and if $x_1 \neq x_1'$, $x_2 \neq x_2'$, ..., we do not expect $Q(\mathbf{x})$ to be given by the same polynomial in the x''s as in the x's. What is the formula for this same function, thought of now as a function of the variables $x_1', x_2', \ldots, x_n'$? As we shall see, it is another polynomial, still with all terms of second degree, but with different coefficients. As usual, we arrange these coefficients in a symmetric matrix, which we call the *matrix of Q with respect to the new basis*—or the new matrix of Q. We proceed to compute the new matrix of Q if the old matrix of Q is $\mathbf{A}$. According to Section 1,

$$Q(\mathbf{x}) = {}^t\mathbf{XAX},$$

where ${}^t\mathbf{X}$ is the matrix $(x_1 \cdots x_n)$, and $\mathbf{X}$ is its transpose. Then we use Eq. (2.2) to write $\mathbf{X} = \mathbf{QX}'$ so that

$$Q(\mathbf{x}) = {}^t(\mathbf{QX}')\mathbf{A}(\mathbf{QX}') = {}^t\mathbf{X}'({}^t\mathbf{QAQ})\mathbf{X}',$$

which says that the matrix of Q with respect to the new basis is ${}^t\mathbf{QAQ}$; if $n = 2$,

$$Q(\mathbf{x}) = a_{11}'x_1'^2 + 2a_{12}'x_1'x_2' + a_{22}'x_2'^2,$$

where

$$\begin{pmatrix} a_{11}' & a_{12}' \\ a_{12}' & a_{22}' \end{pmatrix} = {}^t\mathbf{Q} \begin{pmatrix} a_{11} & a_{12} \\ a_{12} & a_{22} \end{pmatrix} \mathbf{Q}.$$

2.2 We have shown that if Q is a quadratic form on R^n with matrix $\mathbf{A}$ then the matrix of Q with respect to a new basis is ${}^t\mathbf{QAQ}$ where $\mathbf{Q}$ is the change-of-basis matrix in Eq. (2.2).

Example 3. Rotate axes 45 degrees around the z-axis and compute the new expression for the quadratic form $x^2 + 2axy + y^2 + 2xz + 2yz + z^2$. Here the columns of $\mathbf{Q}$ are the components of the new basis vectors $\mathbf{e}_1 = 2^{-1/2}(\mathbf{i} + \mathbf{j})$, $\mathbf{e}_2 = 2^{-1/2}(-\mathbf{i} + \mathbf{j})$, $\mathbf{e}_3 = \mathbf{k}$, so

$$\mathbf{Q} = 2^{-1/2}\begin{pmatrix} 1 & -1 & 0 \\ 1 & 1 & 0 \\ 0 & 0 & \sqrt{2} \end{pmatrix}.$$

The matrix of the quadratic form is

$$\mathbf{A} = \begin{pmatrix} 1 & a & 1 \\ a & 1 & 1 \\ 1 & 1 & 1 \end{pmatrix},$$

$${}^t\mathbf{Q}\mathbf{A} = 2^{-1/2}\begin{pmatrix} 1 & 1 & 0 \\ -1 & 1 & 0 \\ 0 & 0 & \sqrt{2} \end{pmatrix}\begin{pmatrix} 1 & a & 1 \\ a & 1 & 1 \\ 1 & 1 & 1 \end{pmatrix} = 2^{-1/2}\begin{pmatrix} a+1 & a+1 & 2 \\ a-1 & 1-a & 0 \\ \sqrt{2} & \sqrt{2} & \sqrt{2} \end{pmatrix},$$

and

$${}^t\mathbf{Q}\mathbf{A}\mathbf{Q} = 2^{-1}\begin{pmatrix} 2a+2 & 0 & 2\sqrt{2} \\ 0 & 2a-2 & 0 \\ 2\sqrt{2} & 0 & 2 \end{pmatrix} = \begin{pmatrix} a+1 & 0 & \sqrt{2} \\ 0 & a-1 & 0 \\ \sqrt{2} & 0 & 1 \end{pmatrix}.$$

(We know ${}^t\mathbf{Q}\mathbf{A}\mathbf{Q}$ must come out symmetric, so the three entries below the diagonal serve as checks on the three entries above the diagonal.) The answer is

$$(a+1)x'^2 + (a-1)y'^2 + z'^2 + 2\sqrt{2}x'z'.$$

3. EFFECT OF CHANGE OF BASIS ON MATRICES OF LINEAR FUNCTIONS

Let f be a linear function from R^n to R^n and let $\mathbf{A}$ be its matrix, that is, the qth column of $\mathbf{A}$ consists of the components of $f(\mathbf{i}_q)$. Now suppose instead of $\{\mathbf{i}_1, \ldots, \mathbf{i}_n\}$ we use a new basis $\{\mathbf{e}_1, \ldots, \mathbf{e}_n\}$, as in Section 2. The *matrix of f with respect to the basis* $\{\mathbf{e}_1, \ldots, \mathbf{e}_n\}$ is the matrix whose qth column consists of the components (with respect to $\{\mathbf{e}_1, \ldots, \mathbf{e}_n\}$!) of $f(\mathbf{e}_q)$. Note that there are two changes rung here: We compute $f(\mathbf{e}_q)$ instead of $f(\mathbf{i}_q)$ and we write the answer as a linear combination of the $\mathbf{e}$'s, not of the $\mathbf{i}$'s. This will require two changes in the matrix, as we shall see. Since the new components of $\mathbf{e}_q$ are $[0, 0, \ldots, 1, \ldots, 0]$, Eq. (2.2) says that the old components of $\mathbf{e}_q$

(with respect to $\{\mathbf{i}_1, \ldots, \mathbf{i}_n\}$) are

$$\mathbf{Q}\begin{pmatrix} 0 \\ \vdots \\ 0 \\ 1 \\ 0 \\ \vdots \\ 0 \end{pmatrix} = \text{the } q\text{th column of } \mathbf{Q}.$$

To find the (old) components of $f(\mathbf{e}_q)$ we multiply this qth column of $\mathbf{Q}$ by the matrix $\mathbf{A}$. Thus the old components of $f(\mathbf{e}_1), \ldots, f(\mathbf{e}_n)$ are the columns of $\mathbf{AQ}$. To get the new components of the $f(\mathbf{e}_q)$'s each column of $\mathbf{AQ}$ must be multiplied by $\mathbf{P}$, according to Eq. (2.1). The net result is

3.1 If the matrix of f with respect to $\{\mathbf{i}_1, \ldots, \mathbf{i}_n\}$ is $\mathbf{A}$ and if $\mathbf{P}$ and $\mathbf{Q}$ are the change-of-basis matrices in Section 2, then the matrix of f with respect to $\{\mathbf{e}_1, \ldots, \mathbf{e}_n\}$ is

$$\mathbf{PAQ} = \mathbf{PAP}^{-1} = \mathbf{Q}^{-1}\mathbf{AQ}.$$

Notice that if $\mathbf{Q}$ is an orthogonal matrix as in Chapter 6, Section 5; that is, if the new basis $\mathbf{e}_1, \ldots, \mathbf{e}_n$ is orthonormal like $\mathbf{i}_1, \ldots, \mathbf{i}_n$ (Chapter 6, 5.2(e) or 5.3(c)), then $\mathbf{Q}^{-1} = {}^t\mathbf{Q}$, so the formula $\mathbf{Q}^{-1}\mathbf{AQ}$ in 3.1 is the same as the formula ${}^t\mathbf{QAQ}$ in 2.2. In other words, if we think of $\mathbf{A}$ as the matrix of both a quadratic form Q and of a linear mapping f, then as long as we change only to other orthonormal bases (for example, by rotating axes), the matrices of Q and f will remain equal. (But if we change to another basis by a nonorthogonal matrix $\mathbf{Q}$ then the new matrices of f and of Q will probably no longer coincide.)

Example 1. Suppose f is diagonalizable—that is, there is a basis $\{\mathbf{e}_1, \ldots, \mathbf{e}_n\}$ of R^n such that $f(\mathbf{e}_q) = \lambda_q\mathbf{e}_q$. Then 3.1 says that the matrix of f with respect to $\{\mathbf{e}_1, \ldots, \mathbf{e}_n\}$ is the diagonal matrix

$$\begin{pmatrix} \lambda_1 & 0 & \cdots & 0 \\ 0 & \lambda_2 & \cdots & 0 \\ \cdot & \cdot & \cdots & \cdot \\ 0 & 0 & \cdots & \lambda_n \end{pmatrix}.$$

On the other hand, if $\mathbf{Q}$ is the matrix whose columns are the (old) components of the eigenvectors $\mathbf{e}_1, \ldots, \mathbf{e}_n$, and $\mathbf{A}$ is the original matrix of f, then

$$\mathbf{Q}^{-1}\mathbf{AQ} = \begin{pmatrix} \lambda_1 & 0 & \cdots & 0 \\ 0 & \lambda_2 & \cdots & 0 \\ \cdot & \cdot & \cdots & \cdot \\ 0 & 0 & \cdots & \lambda_n \end{pmatrix}.$$

This gives another way of stating the Principal Axis Theorem (Chapter 7, 4.1).

3.2 Principal Axis Theorem, Matrix Form

If $\mathbf{A}$ is a symmetric matrix, then there is an orthogonal matrix $\mathbf{Q}$ (that is, $\mathbf{Q}^{-1} = {}^t\mathbf{Q}$) such that $\mathbf{Q}^{-1}\mathbf{AQ}$ is a diagonal matrix. The diagonal entries of $\mathbf{Q}^{-1}\mathbf{AQ}$ are the eigenvalues of $\mathbf{A}$. The qth column of $\mathbf{Q}$ is an eigenvector belonging to the eigenvalue that is the q, q-entry in the diagonal matrix $\mathbf{Q}^{-1}\mathbf{AQ}$.

Proof. Chapter 7, 4.1 says that $\mathbf{A}$ is orthogonally diagonalizable. Let $\{\mathbf{e}_1, \ldots, \mathbf{e}_n\}$ be an orthogonal basis consisting of eigenvectors. Multiply each by a nonzero scalar if necessary to get an orthogonal basis consisting of eigenvectors 1 unit long, and let $\mathbf{Q}$ be the corresponding change-of-basis matrix as described in 3.2 or 2.1. Then Example 1 shows that $\mathbf{Q}^{-1}\mathbf{AQ}$ is a diagonal matrix and all statements in 2 are proved except for the orthogonality of $\mathbf{Q}$. This is Chapter 6, 5.3(c).

3.3 Corollary

If Q is a quadratic form on R^n then there is an orthonormal basis $\mathbf{e}_1, \ldots, \mathbf{e}_n$ of R^n with respect to which the matrix of Q is diagonal; that is, $Q(x_1'\mathbf{e}_1 + \cdots + x_n'\mathbf{e}_n) = \lambda_1 x_1'^2 + \cdots + \lambda_n x_n'^2$; that is, when Q is expressed as a polynomial in the new coordinates $x_1', \ldots, x_n'$, there are no cross-product terms.

Proof. The matrix of Q is a symmetric matrix $\mathbf{A}$. If we change the basis by an orthogonal matrix $\mathbf{Q}$, the new matrix of the quadratic form is ${}^t\mathbf{QAQ}$ by 2.2, which is $\mathbf{Q}^{-1}\mathbf{AQ}$ because $\mathbf{Q}$ is orthogonal. By 3.2, this is diagonal for suitable $\mathbf{Q}$. We leave it to the reader (Problem 14) to translate the fact that the matrix of Q is now diagonal to the other assertions in 3.3.

PROBLEMS

1. Rotate the coordinate axes in R^2 counterclockwise 45 degrees. Write the matrices $\mathbf{P}$ and $\mathbf{Q}$ that describe this rotation as in Section 2, and use them to find:

(a) The new coordinates of the points whose old coordinates are $(0, 1)$, $(1, 0)$, $(1, 1)$, $(1, -2)$; check with a sketch;

(b) The old coordinates of the points 1 unit from the origin on the positive ends of the new axes;

(c) Equations expressing the new coordinates (x', y') of an arbitrary point as functions of the old coordinates (x, y) of the same point;

(d) Equations expressing the old coordinates (x, y) as functions of the new coordinates (x', y'); do this in two ways: by solving your equations in part (c) for x and y, then by using Eq. (2.1);

(e) the new equation (in x' and y') of the curves $x^2 + y^2 = a^2$, $xy = a$, $x^2 + xy + y^2 = 1$, $x^2 + xy = 1$;

(f) the new matrix of the linear functions whose old matrices are

$$\begin{pmatrix} 1 & 2 \\ 2 & 1 \end{pmatrix}; \quad \begin{pmatrix} 1 & 0 \\ 0 & 1 \end{pmatrix}; \quad \begin{pmatrix} 0 & 1 \\ 1 & 0 \end{pmatrix}; \quad \begin{pmatrix} a & b \\ b & a \end{pmatrix}$$

(deduce this result from the preceding two);

$$\begin{pmatrix} 1 & 1 \\ 1 & 2 \end{pmatrix}.$$

2. Solve Problem 1 with a rotation through an angle θ.

3. What angle of rotation of axes will give the curve $ax^2 + bxy + cy^2 = 1$ a new equation with no $x'y'$ term? Phrased differently, what rotation of axes will give a new basis of R^2 with respect to which the matrix of the quadratic form $ax^2 + bxy + cy^2$ becomes diagonal?

4. Find a rotation in R^2 that carries $\mathbf{i}$ and $\mathbf{j}$ into a new basis with respect to which the quadratic form $x^2 + 4xy - 2y^2$ will have a diagonal matrix. Use this method: We want ${}^t\mathbf{QAQ}$ to be diagonal and (since $\mathbf{Q}$ is to be a rotation) ${}^t\mathbf{Q} = \mathbf{Q}^{-1}$, so we want $\mathbf{Q}^{-1}\mathbf{AQ}$ to be diagonal. By the Principal Axis Theorem, this can be done by taking the columns of $\mathbf{Q}$ (the new basis vectors) to be orthogonal eigenvectors of $\mathbf{A}$. The angle of rotation is then the angle between $\mathbf{i}$ and the first of these eigenvectors. Compare with the results found by applying Problem 3.

5. Solve Problem 4 with the quadratic form $ax^2 + bxy + ay^2$. Again compare with Problem 3.

6. Solve Problem 4 with the quadratic form $ax^2 + bxy + cy^2$. Compare with Problem 3.

7. Is it possible to find $\mathbf{Q}$ such that $\mathbf{Q}^{-1}\mathbf{A}\mathbf{Q}$ is a diagonal matrix if

$$\mathbf{A} = \begin{pmatrix} 1 & 2 \\ 2 & 1 \end{pmatrix}; \quad \begin{pmatrix} 1 & 1 \\ 0 & 1 \end{pmatrix}; \quad \begin{pmatrix} 0 & 1 \\ 1 & 0 \end{pmatrix}; \quad \begin{pmatrix} 0 & 1 \\ 2 & 0 \end{pmatrix}?$$

Why? In all cases where it is possible, find $\mathbf{Q}$ and the diagonal matrix $\mathbf{Q}^{-1}\mathbf{A}\mathbf{Q}$ (you should be able to find this diagonal matrix without computing $\mathbf{Q}^{-1}$ and multiplying matrices; then check by computing $\mathbf{Q}^{-1}\mathbf{A}\mathbf{Q}$ by multiplying matrices). Note that if we know 3.1, this is essentially similar to Problem 2, Chapter 7, Section 2.

8. Let f be a linear mapping from R^n to R^n and suppose its matrix (with respect to the ordinary basis) is $\mathbf{A}$ and its matrix with respect to some other basis is $\mathbf{B}$. Prove $\det \mathbf{A} = \det \mathbf{B}$. Thus $\det f$ can be defined as the determinant of the matrix of f (with respect to any basis). The determinant is then a geometric property of f, independent of the choice of basis. See Chapter 6, 3.11.

9. Show that if $\mathbf{v}$ is an eigenvector of $\mathbf{A}$ belonging to an eigenvalue λ, then $\mathbf{Q}^{-1}\mathbf{v}$ is an eigenvector of $\mathbf{Q}^{-1}\mathbf{A}\mathbf{Q}$ belonging to λ. Explain this geometrically in view of 3.1.

10. Show that $\mathbf{Q}^{-1}\mathbf{A}^r\mathbf{Q} = (\mathbf{Q}^{-1}\mathbf{A}\mathbf{Q})^r$, $\mathbf{Q}^{-1}(\mathbf{A} + \mathbf{B})\mathbf{Q} = \mathbf{Q}^{-1}\mathbf{A}\mathbf{Q} + \mathbf{Q}^{-1}\mathbf{B}\mathbf{Q}$, $\mathbf{Q}^{-1}\mathbf{I}\mathbf{Q} = \mathbf{I}$, and $\det \mathbf{Q}^{-1}\mathbf{A}\mathbf{Q} = \det \mathbf{A}$. How many of these can you explain by using 3.1?

11. Show that ${}^t(\mathbf{Q}^{-1}\mathbf{A}\mathbf{Q}) = \mathbf{Q}^{-1}({}^t\mathbf{A})\mathbf{Q}$ for all $\mathbf{A}$ if $\mathbf{Q}$ is orthogonal.

12. Use Problem 10 to show that the characteristic polynomial det $(\mathbf{A} - \lambda\mathbf{I})$ of $\mathbf{A}$ is the same as the characteristic polynomial of $\mathbf{Q}^{-1}\mathbf{A}\mathbf{Q}$. Hence the eigenvalues of $\mathbf{A}$ are the same as the eigenvalues of $\mathbf{Q}^{-1}\mathbf{A}\mathbf{Q}$. Compare with Problem 9.

13. Problem 12, Chapter 7, Section 1, introduced two concepts of "multiplicity" of an eigenvalue of a diagonalizable mapping: The number of times λ appears on the diagonal of the matrix of f with respect to a basis of eigenvectors; and the dimension of the space of all eigenvectors belonging to λ. Now show that both of these equal the multiplicity implicit in the preceding problem: the multiplicity of λ as a root of the characteristic polynomial of (the matrix of) f.

14. Finish the required translation in the proof of 3.3.

APPENDIX

A Smattering of Logic

In this course we are compelled to manipulate not only mathematical formulas but logical statements. These manipulations have their own rules, which are simple enough but must be learned more or less explicitly.

We consider statements (sentences, clauses): for example, "$\{\mathbf{u}_1, \ldots, \mathbf{u}_r\}$ is an independent set"; "$\mathbf{u}_1 = \mathbf{0}$"; or "$\mathbf{u}_1 = a\mathbf{u}_2$ for some scalar a."

Every statement has a *negation* (our first manipulation). If A denotes the statement, then not$A$ will denote the negation; not$A$ is the statement "A fails" or "A is not true." For example, "not($\{\mathbf{u}_1, \ldots, \mathbf{u}_r\}$ is independent)" is the same as "$\{\mathbf{u}_1, \ldots, \mathbf{u}_r\}$ is dependent," and "not($\mathbf{u}_1 = \mathbf{0}$)" means the same as "$\mathbf{u}_1 \neq \mathbf{0}$."

The second important manipulation makes a new statement out of two old ones. Given statements A and B, we are often interested in the statement "A *implies* B." This has many English equivalents, such as: "if A, then B"; "A only if B"; "whenever A, then B"; and "B if A." For example, if A is "$\{\mathbf{u}_1, \ldots, \mathbf{u}_r\}$ is an independent set" and B is "$\mathbf{u}_1 = 0$," then the assertion in Chapter 5, 2.9 is "A implies not$B$."

What is the negation of "A implies B"? In English we can express it by "A does not imply B." But how can "A implies B" fail? Only if A is true and B is not. This is our first manipulative rule:

A.1 not(A implies B) = (A and not$B$).

For example, "$\{\mathbf{u}_1, \ldots, \mathbf{u}_r\}$ is an independent set" is itself an implication "C implies D" where C is "$\sum a_p\mathbf{u}_p = \mathbf{0}$" and D is "all $a_p = 0$." Using A.1 we see that the negation, "$\{\mathbf{u}_1, \ldots, \mathbf{u}_r\}$ is dependent," is "$\sum a_i\mathbf{u}_i = \mathbf{0}$ and not all $a_i = 0$." (This is not very good English because we have hidden another logical term, which we shall display in a moment.)

Another very useful connection between negations and implications is that every implication is the same as its *contrapositive*. If the implication is "A implies B," its contrapositive is "not$B$ implies not$A$" or "if B fails, then A fails":

A.2 (A implies B) = (not$B$ implies not$A$).

For example, consider the statement "if $x = 1$, then $x^2 + x - 2 = 0$." The contrapositive of this statement is "if $x^2 + x - 2 \neq 0$, then $x \neq 1$." The first statement is true and so is its contrapositive. Note that the contrapositive is *not* the converse (B implies A); in this example the converse is false. The reader should convince himself of A.2 by other examples.

Note that "A if and only if B" means "A if B and A only if B," that is, "B implies A and A implies B." So "A if and only if B" is the combination assertion "A implies B and conversely." We also say "A and B are equivalent."

The third and fourth logical manipulations apply only to statements about variables, such as $\mathbf{u} \cdot \mathbf{u} = \|\mathbf{u}\|^2$, in which $\mathbf{u}$ denotes any vector in R^n. Such statements can be true for all values of the variable, or for some or no values of the variable; the example just given happens to be true for all $\mathbf{u}$. The third logical manipulation consists in making a new statement by prefixing "*for all*..." to the given statement, thus: for all $\mathbf{u}$, $\mathbf{u} \cdot \mathbf{u} = \|\mathbf{u}\|^2$. This new statement happens to be a true one. As another example, let A denote "$\{\mathbf{u}, \mathbf{i}\}$ is a dependent set"; after prefixing, we have "for all $\mathbf{u}$, $\{\mathbf{u}, \mathbf{i}\}$ is a dependent set." This happens to be false. Note that the English renditions of this can also vary: "$\{\mathbf{u}, \mathbf{i}\}$ is a dependent set for all $\mathbf{u}$" "... for every $\mathbf{u}$," "... for each $\mathbf{u}$," or even "$\{\mathbf{u}, \mathbf{i}\}$ is always a dependent set."

If it is unpleasant to spend so much time on a false statement, let us pass to its negation, which will then be a true statement. Why is

"$\{\mathbf{u}, \mathbf{i}\}$ is dependent for all $\mathbf{u}$" false? Because $\{\mathbf{u}, \mathbf{i}\}$ is independent *for some vectors* $\mathbf{u}$. Our fourth logical manipulation is to prefix "for some $\mathbf{u}$" to a statement about a variable $\mathbf{u}$. Other English equivalents are "there exists a $\mathbf{u}$ such that...," "there is some $\mathbf{u}$ such that...," "there are (one or more) vectors $\mathbf{u}$ such that...," and "there exist $\mathbf{u}$ with...." We have the following rules.

A.3 not(for all $\mathbf{u}$, A) = for some $\mathbf{u}$, not$A$.

A.4 not(for some $\mathbf{u}$, A) = for all $\mathbf{u}$, not$A$.

A.5 not(not$A$) = A.

This is as deep as we intend to go. The examples of these rules are sufficiently complicated to include the applications we need. For example, "$\{\mathbf{u}_1, \ldots, \mathbf{u}_n\}$ is independent" really means "for all $a_1, \ldots, a_n$, ($\sum a_p\mathbf{u}_p = \mathbf{0}$ implies ($a_p = 0$ for all p))". By A.3, the negation is

for some $a_1, \ldots, a_n$, not($\sum a_p\mathbf{u}_p = \mathbf{0}$ implies ($a_p = 0$ for all p)),

which, by A.1 is

for some $a_1, \ldots, a_n$, ($\sum a_p\mathbf{u}_p = \mathbf{0}$ and not($a_p = 0$ for all p)),

which again by A.3 is

for some $a_1, \ldots, a_n$ ($\sum a_p\mathbf{u}_p = \mathbf{0}$ and $a_p \neq 0$ for some p).

If we had even more use for this logic than we do, we should insist on symbols for these four manipulations, to save space. Standard symbols (but we do not insist) are:

not$A$:	$\sim A$	(negation);
A implies B:	$A \Rightarrow B$	(implication);
for all $\mathbf{u}$, A:	$\forall \mathbf{u}, A$	(universal quantification);
for some $\mathbf{u}$, A:	$\exists\, \mathbf{u}, A$	(existential quantification).

The principal difficulties that arise in using this logic stem from the numerous English paraphrases. Here are some examples, together with their translations and their translations into symbols.

Example 1.

No real number x satisfies $x^2 + 1 = 0$
= It is not true that there is a real number x with $x^2 + 1 = 0$
= *Not(for some x, x is a real number and $x^2 + 1 = 0$)*
= $\sim(\exists\, x$, x is a real number and $x^2 + 1 = 0)$.

Of course, by A.4 this is also the same as

$$= \forall x, \sim(x \text{ is a real number and } x^2 + 1 = 0)$$

or, using A.1 and A.5, $\sim(A \text{ and not} B) = A$ implies B,

$$\forall x(x \text{ is a real number} \Rightarrow x^2 + 1 \neq 0);$$

that is,

$$x^2 + 1 \neq 0 \text{ for all real } x.$$

Example 2.

No minors are served at this bar
= Minors are not served at this bar
= *For all (minors) x, x is not served at this bar*
= $\forall x, \sim(x$ is served at this bar)

(where we understand x denotes any minor). By A.4 read backward, this is the same as

$$\sim\exists\, x, x \text{ is served at this bar}$$

or

There is no minor who is served at this bar

or just

No minor is served at this bar.

Example 3.

All men who love their country enlist
= For every (man) x, x loves his country implies x enlists.

What is the negation?

For some x, not(x loves his country implies x enlists)
= For some x, x loves his country and x does not enlist
= There are men who love their country but do not enlist.

In the course of the text, the reader will find numerous exercises in such translation. Occasional consciousness of this problem may help.

Answers to Problems

Three dots in some of the answers indicate that the answer given is not quite complete, but has been cut for brevity.

Chapter 1, Section 1

2. xz–plane; $y = 0$
3. A plane parallel to z-axis; $x + y = 1$
4. $z = 1$
5. All are planes parallel to coordinate planes
6. (a) Sphere...; (b) and (c) cylinders...; (d) and (e) planes...; (f) circular paraboloid (bowl-shaped); (g) hyperbolic paraboloid (saddle-surface); (h) z-axis
7. (a) All points in front of the yz-plane; (b) all points behind (and to the left of) the plane $x + y = 1$; (c) all points in a solid ball (sphere)...
8. "Cylinder" parallel to z-axis
9. (a) y-axis; (b) and (c) lines in xy-plane...; (d) and (e) lines...; (f) ellipse; (g) an infinite, solid wedge...

Chapter 1, Section 3

4. $(a + d, b + e, c + f)$
5. (a) $z = -5 \pm \sqrt{10}$; impossible; (b) $5 \pm \sqrt{27}$, $-2 \pm \sqrt{57}$, $51/14$
11. $(x - a)^2 + (y - b)^2 + (z - c)^2 = r^2$
12. $\sqrt{a^2 + b^2 + c^2}$; $\sqrt{30}$
13. slope of $[a, b]$ is b/a; $\cos \alpha = a/\sqrt{a^2 + b^2}$, $\cos \beta = b/\sqrt{a^2 + b^2}$

Chapter 1, Section 4

2. By 4.4, b, d, and h are parallel, and a, c, e, and g are parallel
4. (b) $(a/2, b/2, c/2)$
6. (d) Components of MN: $[\frac{1}{2}x' - \frac{1}{2}x, \frac{1}{2}y' - \frac{1}{2}y, \frac{1}{2}z' - \frac{1}{2}z]$; components of PQ: $[x' - x, y' - y, z' - z]$.
7. $\cos\alpha = a/\sqrt{a^2 + b^2 + c^2}\ldots$
9. $2/\sqrt{14}$, $1/\sqrt{14}$, $3/\sqrt{14}$ are direction cosines of first vector; $\cos^{-1} 2/\sqrt{14} =$ approx 58 degrees
10. (a) Using Problem 8, $\|[\lambda, \mu, \nu]\| = 1$
(c) Here $\alpha + \beta = \pi/2$, so $\cos\beta = \sin\alpha$; $\cos\gamma = 0$, $\cos^2\alpha + \sin^2\alpha = 1$
12. Since $[1, 2, 0]$ is horizontal and the vector joining the heads is vertical, the angle is $\cos^{-1}\|[1, 2, 0]\|/\|[1, 2, 1]\| = \cos^{-1}\sqrt{\frac{5}{6}} = \sin^{-1}(1/\sqrt{6})$
13. $\sin^{-1}(|c|/\sqrt{a^2 + b^2 + c^2}) = \cos^{-1}\sqrt{(a^2 + b^2)/(a^2 + b^2 + c^2)}$

Chapter 1, Section 5

1. $[5, 0, 0]$ or $5\mathbf{i}$; $3\mathbf{i}$; $\mathbf{i}$; $[1, 1, 0]$ or $\mathbf{i} + \mathbf{j}$
2. (a) $2\mathbf{i} + \mathbf{j} + 3\mathbf{k}\cdots$; (b) $\cdots(b_1 - a_1)\mathbf{i} + (b_2 - a_2)\mathbf{j} + (b_3 - a_3)\mathbf{k}$
3. $\sqrt{14}\ldots$
4. (a) $\frac{1}{3}, \frac{1}{3}$; (c) $(a + b)/3$, $(a - 2b)/3$
5. 1st coefficient = 2nd coefficient = -2 times 3rd coefficient; coefficients = $s[2, 2, -1]$ for all scalars s
6. (a) $x = -1$; (b) impossible
7. $1, 2, -1$
8. (a) $1, -4$; (b) $\frac{1}{6}, \frac{1}{6}$; (c) $0, 0$; (d) $\frac{1}{2}, -\frac{3}{2}$; (e) impossible; (f) $-\frac{1}{5}, \frac{2}{5}$
9. $M = (\frac{3}{2}, \frac{5}{2}, \frac{3}{2})$
11. Minimum of $((1 + t)^2 + (4t - 1)^2 + (2 - t)^2)^{1/2}$ is at $t = 5/18$
The minimum is $18^{-1}(23^2 + 2^2 + 31^2)^{1/2}$ at the point $(23/18, 2/18, 31/18)$
13. $OT = (\frac{1}{3})OP + (\frac{1}{3})OQ + (\frac{1}{3})OR$
15. $\sqrt{3}\,\|\mathbf{u}\|$
16. $2\mathbf{i} + 3\mathbf{j} + 5\mathbf{k}$; $x\mathbf{i} + y\mathbf{j} + (x + y)\mathbf{k}$; 1st, 2nd, 4th
19. All vectors; all vectors in that plane, unless $\mathbf{v}_1, \mathbf{v}_2, \mathbf{v}_3$ lie on a line...
21. $tOP + (1 - t)OQ = OQ + t(OP - OQ) = \mathbf{u} + t\mathbf{v}$
23. All linear combinations of $[1, 2, 1]$, $[1, 1, 1]$ and $[1, 2, 0]$
24. $\mathbf{j}$, $\mathbf{i} - \mathbf{j}$, and $-\mathbf{i} + \mathbf{k}$
25. For every a, b, c you can solve $a\mathbf{i} + b\mathbf{j} + c\mathbf{k} = (y - z)\mathbf{i} + (x - y)\mathbf{j} + z\mathbf{k}$ for $x, y, z\ldots$
27. If $\sum a_p\mathbf{v}_p = \sum b_p\mathbf{v}_p$ with the a_p's not all equal to the b_p's then $\sum (a_p - b_p)\mathbf{v}_p = \mathbf{0}$ with not all $a_p - b_p$ zero. If $a_1 - b_1 \neq 0$, for example, then $\mathbf{v}_1 = -(a_1 - b_1)^{-1}\sum_{p\neq 1}(a_p - b_p)\mathbf{v}_p$

Chapter 1, Section 6

1. $-4, 12, 5\sqrt{2}, -1, -1$
2. $\cos^{-1}(-4/3\sqrt{74}), \ldots, \cos^{-1}(-1/\sqrt{10})$
3. $2\mathbf{i} - \mathbf{j} \perp$ all but $\mathbf{i}$; $\mathbf{i} \perp \mathbf{k}$; $\mathbf{i} + 2\mathbf{j} \perp \mathbf{k}$
4. $\cos^{-1}(1/\sqrt{3}) = 34$ degrees
5. $\mathbf{u}$ is perpendicular to the plane containing $\mathbf{v}$ and $\mathbf{w}$ and every linear combination of $\mathbf{v}$ and $\mathbf{u}$ is in this plane; $\mathbf{u}\cdot(a\mathbf{v} + b\mathbf{w}) = a(\mathbf{u}\cdot\mathbf{v}) + b(\mathbf{u}\cdot\mathbf{w}) = 0 + 0 = 0$

6. $\mathbf{i} + \mathbf{j} - (\frac{5}{4})\mathbf{k}$
9. (a) Six of them are 0; three are 1; $\mathbf{v}_4 \cdot \mathbf{v}_1 = \frac{1}{6}$, $\mathbf{v}_4 \cdot \mathbf{v}_2 = \frac{3}{2}^{3/2}$; $\mathbf{v}_4 \cdot \mathbf{v}_3 = \frac{5}{6}\sqrt{2}$
10. $\mathbf{v} \cdot \mathbf{i} = \mathbf{i} \cdot \mathbf{v} =$ component of $\mathbf{v}$ along $\mathbf{i}$ by 6.5
11. (a) The vector is equivalent to a vector OP with tail at the origin, head at P. Through P draw a line parallel to $\mathbf{v}_3$, cutting the plane of $\mathbf{v}_2$ and $\mathbf{v}_1$ at Q. Then QP is equivalent to $a_3\mathbf{v}_3$ for some scalar a_3 and $OP = OQ +$ "QP." Then through Q draw a line parallel to $\mathbf{v}_2$, cutting the line of $\mathbf{v}_1$ at R. $OQ = OR +$ "RQ" and $OR = a_1\mathbf{v}_1$ and "RQ" $= a_2\mathbf{v}_2$.
(b) $(a_1\mathbf{v}_1 + a_2\mathbf{v}_2 + a_3\mathbf{v}_3) \cdot \mathbf{v}_1 = a_1(\mathbf{v}_1 \cdot \mathbf{v}_1) + a_2(\mathbf{v}_2 \cdot \mathbf{v}_1) + a_3(\mathbf{v}_3 \cdot \mathbf{v}_1) = a_1 + 0 + 0$
(c) component $= \mathbf{v} \cdot \mathbf{v}_p/\|\mathbf{v}_p\| = \mathbf{v} \cdot \mathbf{v}_p = a_p$
14. If $\mathbf{u}$ and $\mathbf{v}$ are vectors along two edges, the diagonals are $\mathbf{u} + \mathbf{v}$ and $\mathbf{u} - \mathbf{v}$; $(\mathbf{u} + \mathbf{v}) \cdot (\mathbf{u} - \mathbf{v}) = \mathbf{u} \cdot \mathbf{u} - \mathbf{v} \cdot \mathbf{v} = \|\mathbf{u}\|^2 - \|\mathbf{v}\|^2 = 0$ since $\|\mathbf{u}\| = \|\mathbf{v}\|$
15. Write $\|\mathbf{v} + \mathbf{w}\|^2 = (\mathbf{v} + \mathbf{w}) \cdot (\mathbf{v} + \mathbf{w}), \ldots$

Chapter 1, Section 7

1. $-\mathbf{j} + \mathbf{k}$, $-2\mathbf{i} + 5\mathbf{j} + 9\mathbf{k}$, $[-1, -3, 2]$, $-a_2\mathbf{i} + a_1\mathbf{j}$
4. $(\mathbf{i} + \mathbf{j}) \times (2\mathbf{i} - \mathbf{j} - \mathbf{k}) = -\mathbf{i} + \mathbf{j} - 3\mathbf{k}$
5. $\mathbf{i} - 5\mathbf{j} + 3\mathbf{k}$
6. $[7, 5, 1]$
7. $(1/\sqrt{6})\mathbf{i} + (1/\sqrt{6})\mathbf{j} - (2/\sqrt{6})\mathbf{k}$
8. $= \mathbf{u} \times \mathbf{v} + a\mathbf{v} \times \mathbf{v}$ and $a\mathbf{v} \times \mathbf{v} = 0$ by (7.5) and (7.7)

Chapter 1, Section 8

1. 2, $\sqrt{n}$
2. All vectors with the fourth component 0
3. $\mathbf{i}_3 - \mathbf{i}_4$ and all its scalar multiples
4. All scalar multiples of $\mathbf{i}_3 + 2\mathbf{i}_4$
5. See Problem 5, Section 6
7. $\|\mathbf{u} - \mathbf{v}\| = \|\mathbf{u} + (-\mathbf{v})\| \leq \|\mathbf{u}\| + \|-\mathbf{v}\| = \|\mathbf{u}\| + \|\mathbf{v}\|$
8. 3, -2; impossible

Chapter 1, Section 9

2. $\mathbf{v}_1 = -2\mathbf{i} + \mathbf{k}$, $\mathbf{v}_2 = \mathbf{j}$
3. $4\mathbf{i} + \mathbf{j} - 2\mathbf{k}$
5. The associative, commutative, and distributive laws hold for addition and scalar multiplications of all functions.... The main trick is to check closure: the sum of continuous (resp. differentiable) functions is continuous (resp. differentiable). These are standard theorems in calculus whose proofs you might review.
7. If $f' - f = 0$ and $g' - g = 0$, then $(f + g)' - (f + g) = f' + g' - f - g = 0 - 0 = 0$, proving $(+0)$.... All solutions: $f = ae^x$

Chapter 2, Section 1

1. $2x - y + z = 0$
2. 2nd, 4th

3. $(2\mathbf{i} + \mathbf{j} - \mathbf{k}) \perp (\mathbf{i} - \mathbf{j} + \mathbf{k})$ because their dot product is 0. (The angle between two planes equals the angle between the perpendicular vectors.)
4. $\cos^{-1}(\frac{4}{9})$. See answer to Problem 3
5. Parallel
6. $\frac{1}{3}$
7. $|d|/\sqrt{a^2 + b^2 + c^2}$
8. $|ax_1 + by_1 + cz_1 - d|/\sqrt{a^2 + b^2 + c^2}$
9. $x + 32y + 26z = 0$
10. $x + 3y + 2z = 6$
11. $x + y = 3$
12. (a) $d = 0$; (b) $b = c = 0$; (c) $a = c = 0$; (d) $a = 0$; (e) $b = 0$; (f) $a = b = 0$; (g) $[a, b, c] = s[1, 2, 3]$ for some nonzero s; $a/1 = b/2 = c/3$

Chapter 2, Section 3

1. $x = at, y = bt, z = ct$
2. $x = 2 + t, y = 1, z = 2$
3. $x = 1 + 3t, y = -1 + 4t, z = 2 - t$
4. $x = 2t, y = -t, z = t$
5. $x = 2, y = 6 + t, z = -1 - t$
6. $x = 2 + 2t, y = 1 - t, z = 3 + 2t$
7. $x = t, y = -2t, z = 0$
8. ($t = -\frac{8}{9}$ so) $x = \frac{2}{9}, y = \frac{17}{9}, z = \frac{11}{9}$
9. ($t = 1$ so) $x = 1, y = 2, z = 1$
10. $\sqrt{47/11}$
11. $1/\sqrt{3}$
12. 3 (point of intersection is (3, 7, 3))
13. If $\|\mathbf{u}\| \neq \|\mathbf{v}\|$, we get an ellipse. For example, if $\mathbf{u} = a\mathbf{i}$, $\mathbf{v} = b\mathbf{j}$, parametric equations are $x = a \cos t$, $y = b \sin t$. Eliminate t to get $x^2/a^2 + y^2/b^2 = 1$.
14. $x = t, y = t^2$ and eliminate t to get $y = x^2$
15. Oscillation on a straight line
16. $-\frac{5}{2}$
17. 1
18. Impossible
19. ± 1
20. $\frac{2}{3}, -\frac{10}{3}$

Chapter 2, Section 4

1. (a) behind the yz-plane; (b) behind the plane $x = 1$; (c) above the plane $z = 2$; (d) behind the vertical plane $x + 2y = 1$; (e) below the plane $x - y + z = 1$; (f) above the plane $x - y - z = 1$
2. (a) triangular prism; corners (0, 0, 0), (0, 2, 0), (0, 1, 2) and (2, 0, 0), (2, 2, 0), (2, 1, 2); (b) triangle with interior; vertices (1, 0, 0), (0, 1, 0), (0, 0, 1); (c) trapezoid with interior; vertices (0, 1, 0), (0, 0, 1) and $(\frac{1}{2}, \frac{1}{2}, 0)$, $(0, \frac{1}{2}, \frac{1}{2})$; (d) line segment; endpoints $(\frac{1}{3}, \frac{1}{3}, 0)$ and $(\frac{1}{4}, \frac{1}{4}, \frac{1}{2})$. Note $x \geq 0$ and $y \geq 0$ are superfluous; (e) triangle with interior; the three inequalities describe an infinite triangular prism; vertices (0, 0, 1), (0, 0, 0), $(\frac{1}{4}, \frac{1}{4}, \frac{1}{2})$

3. (a) (2, 0, 0); (b) (1, 0, 0); (c) (0, 0, 1); (d) $(\frac{1}{4}, \frac{1}{4}, \frac{1}{2})$; (e) (0, 0, 1)
4. (a) 57 tons low-sulfur, 43 tons high-sulfur; (b) $57c_1 + 43c_2 < 100c_1$ if $c_1 < c_2$; (c) use all low-sulfur coal
5. (a) no oranges; 181 grams grapefruit, 42 grams carrots; (b) no grapefruit; 145 grams oranges, 40 grams carrots

Chapter 3, Section 1

3. (c) $\mathbf{u} \cdot (\mathbf{v} + \mathbf{w}) = \mathbf{u} \cdot \mathbf{v} + \mathbf{u} \cdot \mathbf{w}$ and $\mathbf{u} \cdot (a\mathbf{v}) = a(\mathbf{u} \cdot \mathbf{v})$ by (I.6.4b) and (I.6.7)
4. (b) See (d); (d) only if $\varphi(\mathbf{0}) = \varphi(0\mathbf{i}) = d$ but $0\varphi(\mathbf{i}) = \mathbf{0}$; (g) $F(x\mathbf{i} + y\mathbf{j} + z\mathbf{k}) = xF(\mathbf{i}) + yF(\mathbf{j}) + zF(\mathbf{k})$ and take $a = F(\mathbf{i})$, $b = F(\mathbf{j})$, $c = F(\mathbf{k})$; (h) take $\mathbf{u} = [a, b, c]$
5. $c(\mathbf{v} + \mathbf{w}) = c\mathbf{v} + c\mathbf{w}$ and $c(a\mathbf{v}) = a(c\mathbf{v})$
6. No...
7. $f(\mathbf{0}) = f(0\mathbf{v}) = 0f(\mathbf{v}) = \mathbf{0}$ for any $\mathbf{v}$
8. Use Problem 7
9. No; $\|\mathbf{i} + \mathbf{j}\| = \sqrt{2}$, $\|\mathbf{i}\| + \|\mathbf{j}\| = 2$
12. Chapter 1, 7.4 and 7.5
13. Best to do it geometrically
15. If f is linear then $f(a\mathbf{v} + b\mathbf{w}) = f(a\mathbf{v}) + f(b\mathbf{w}) = af(\mathbf{v}) + bf(\mathbf{w})$. Conversely, if $f(a\mathbf{v} + b\mathbf{w}) = af(\mathbf{v}) + bf(\mathbf{w})$ for all a and b, then it is true when $a = b = 1$ and when $a = 1$, $b = 0$; the two resulting statements are 1.1.
16. $h(\mathbf{v} + \mathbf{w}) = g(f(\mathbf{v} + \mathbf{w})) = g(f(\mathbf{v}) + f(\mathbf{w})) = g(f(\mathbf{v})) + g(f(\mathbf{w}))\ldots$
17. -3, -4, $x + 2y + 3z$; $(2y + 3z)\mathbf{i} + y\mathbf{j} + z\mathbf{k}$ for all y and z = all linear combinations of $2\mathbf{i} + \mathbf{j}$ and $3\mathbf{i} + \mathbf{k}$
18. $\frac{1}{2}, \frac{1}{2}, \frac{1}{2}$
19. $2\mathbf{j} - \mathbf{k}$, $-4\mathbf{j} - 3\mathbf{k}$, $(x + y + z)\mathbf{i} + (x - y - z)\mathbf{j} + z\mathbf{k}$; $\mathbf{0}$
20. Let $\mathbf{u} = [2, 3, 1]$. The projection of $\mathbf{v}$ on $\mathbf{u}$ is $(\mathbf{v} \cdot \mathbf{u}/\|\mathbf{u}\|)(\mathbf{u}/\|\mathbf{u}\|) = (\mathbf{v} \cdot \mathbf{u}/\mathbf{u} \cdot \mathbf{u})\mathbf{u}$. Projection of $\mathbf{v}$ on the plane is $\mathbf{v}$ minus this. Ans: $(1/14)[10x - 6y - 2z, -6x + 5y - 3z, -2x - 3y + 13z]$
21. $\mathbf{j}, -\mathbf{k}, -\mathbf{i}$; $\mathbf{i} + \mathbf{j} - \mathbf{k}$
22. (a) $\mathbf{k}, \mathbf{j}, -\mathbf{i}$; (c) $f(x\mathbf{i} + y\mathbf{j} + z\mathbf{k}) = xf(\mathbf{i}) + yf(\mathbf{j}) + zf(\mathbf{k}) = x\eta(\mathbf{i}) + y\eta(\mathbf{j}) + z\eta(\mathbf{k}) = \eta(x\mathbf{i} + y\mathbf{j} + z\mathbf{k})$ for all x, y, z

Chapter 3, Section 2

1. b, a, b by a
3. (a) $2\mathbf{i} - \mathbf{k}$, $\mathbf{i} + \mathbf{j}$, $3\mathbf{i} + 4\mathbf{j} + \mathbf{k}$, $10\mathbf{i} + 7\mathbf{j} - \mathbf{k}$, $\mathbf{i} - 3\mathbf{j} - 2\mathbf{k}$; (b) [1, 4, 1], [0, 10, 2], [−1, 4, −1]
4. (a) [3, 9], [−2, 7], [1, 2], [3, 0]; (b) $\mathbf{i}$, $2\mathbf{j}$, $\mathbf{i}$, $4\mathbf{i} + 4\mathbf{j}$
5. (a) 2 and 4, [2, 3, 4, 5], [1, 2, 3, 4], [7, 12, 17, 22]; (b) [1, 0, 1, −1], [0, 1, 1, 1], [1, −1, 0, −2], [2, 7, 9, 5]

6. $\begin{pmatrix} 1 & 0 & 0 \\ 0 & 1 & 0 \\ 0 & 0 & 0 \end{pmatrix}$, $\mathbf{i} + 2\mathbf{j}$, $\mathbf{i} - 2\mathbf{j}$, $2\mathbf{i}$

7. $\begin{pmatrix} 2 & 0 & 0 \\ 0 & 0 & -2 \\ 0 & 2 & 0 \end{pmatrix}$, $2\mathbf{i} - 6\mathbf{j} + 4\mathbf{k}$, $2\mathbf{i} - 6\mathbf{j} - 4\mathbf{k}$, $4\mathbf{i} + 2\mathbf{j}$

8. Diagonal entries are 1; all others are 0

9. All entries 0

10. Diagonal entries are c; all others are 0

12. If $f(\mathbf{i}) = a\mathbf{i}$, $f(\mathbf{j}) = b\mathbf{j}$, $f(\mathbf{k}) = c\mathbf{k}$, then the matrix of f is $\begin{pmatrix} a & 0 & 0 \\ 0 & b & 0 \\ 0 & 0 & c \end{pmatrix}$.

Conversely, use row by column multiplication

14. $\begin{pmatrix} 1 & 0 & 0 \\ 0 & 1 & 0 \end{pmatrix}$, $\begin{pmatrix} 2 & -3 & 0 \\ 3 & 0 & -2 \end{pmatrix}$, $\begin{pmatrix} a_{11} & a_{12} & a_{13} \\ a_{21} & a_{22} & a_{23} \end{pmatrix}$

15. $\begin{pmatrix} 1 & 0 & 0 \\ 0 & 0 & -1 \\ 0 & 1 & 0 \end{pmatrix} \begin{pmatrix} 0 & 0 & -1 \\ 0 & 1 & 0 \\ 1 & 0 & 0 \end{pmatrix} \begin{pmatrix} 0 & -1 & 0 \\ 1 & 0 & 0 \\ 0 & 0 & 1 \end{pmatrix}$

16. $(\frac{1}{2} \;\; \frac{1}{2} \;\; \frac{1}{2})$

17. Matrix of f is $(0 \;\; 1)$; all $[x, 0]$

18. $\begin{pmatrix} 2 \\ 1 \\ 1 \end{pmatrix}$; all $x(2\mathbf{i} + \mathbf{j} + \mathbf{k})$; 0

19. $\begin{pmatrix} 2/\pi \\ 1/\pi \\ 1/\pi \end{pmatrix}$; all $x(2\mathbf{i} + \mathbf{j} + \mathbf{k})$; 0

20. $\begin{pmatrix} 0 & 0 & 0 \\ 2 & 0 & 0 \\ 0 & 1 & 0 \end{pmatrix}$

21. $(\frac{1}{3} \;\; \frac{1}{2} \;\; 1)$

Chapter 3, Section 3

1. $f + g = I$, $-g$ projects on z-axis then reverses, $f - g$ is reflection in the xy plane, $2f + 2g$ stretches the input vector by a factor of 2, $2f + g$ leaves the head at the same height but moves it twice as far from the z-axis.

2. Projection into xy-plane, then rotation of 135 degrees, then stretching by a factor $\sqrt{2}$; $\mathbf{j} - \mathbf{i}$; $\begin{pmatrix} -1 & -1 & 0 \\ 1 & -1 & 0 \\ 0 & 0 & 0 \end{pmatrix}$

3. $(af)(\mathbf{v} + \mathbf{w}) =_{def} a(f(\mathbf{v} + \mathbf{w})) = a(f(\mathbf{v}) + f(\mathbf{w}))$ (because f is linear) $= a(f(\mathbf{v})) + a(f(\mathbf{w}))$ by (I.5.2d) $=_{def} (af)(\mathbf{v}) + (af)(\mathbf{w}) \cdots$

4. The qth column of $a\mathbf{A}$ is the vector $(af)(\mathbf{i}_q) = a(f(\mathbf{i}_q)) = a$ times the qth column of $\mathbf{A}$. Hence for all p and q, the pq-entry in $a\mathbf{A}$ is a times the pq entry in $\mathbf{A}$.

5. $(a_{pq}) = \sum_{p=1}^{m} \sum_{q=1}^{n} a_{pq} E_{pq}$

Chapter 3, Section 4

3. Define g by $g(\mathbf{v}) = a\mathbf{v}$ as in Problem 2.8. Then $af = g \circ f$

4. $\begin{pmatrix} 3 & 5 \\ -3 & -5 \end{pmatrix}$; $\begin{pmatrix} 30 & 36 & 42 \\ 66 & 81 & 96 \\ 102 & 126 & 150 \end{pmatrix}$; $\begin{pmatrix} 3 & -1 \\ 4 & 0 \\ 7 & -1 \end{pmatrix}$; $\begin{pmatrix} -10 & 0 \\ 0 & -10 \end{pmatrix}$; $\begin{pmatrix} ad - bc & 0 \\ 0 & ad - bc \end{pmatrix}$

5. $(f \circ I)(\mathbf{v}) = f(I(\mathbf{v})) = f(\mathbf{v})$, so $f \circ I = f$ for every function f from R^n to R^n. Therefore $\mathbf{AI} = \mathbf{A}$ for every n by n matrix $\mathbf{A}$

8. $(ad - bc)^{-1}\begin{pmatrix} d & -b \\ -c & a \end{pmatrix}$; see Problem 4. If $ad - bc = 0$, $\mathbf{A}$ has no inverse because if $\mathbf{I} = \mathbf{BA}$ then operating on $\begin{pmatrix} d \\ -c \end{pmatrix}$ we get

$$\begin{pmatrix} d \\ -c \end{pmatrix} = \mathbf{B}\begin{pmatrix} 0 \\ 0 \end{pmatrix} = \begin{pmatrix} 0 \\ 0 \end{pmatrix},$$

so $d = c = 0$; but then both rows of $\mathbf{BA}$ are scalar multiples of $(a\ b)$, one is a scalar multiple of the other, which is not true of the rows of $\mathbf{I}$.

9. $\begin{pmatrix} 2 & 0 & 0 \\ 0 & 2 & 0 \\ 0 & 0 & 2 \end{pmatrix}\begin{pmatrix} 1 & 0 & 0 \\ 0 & 1 & 0 \\ 0 & 0 & 0 \end{pmatrix}$

10. See Problem 19

11. $(f \circ g)(\mathrm{i}) = \mathbf{k}$, $(g \circ f)(\mathrm{i}) = \mathbf{j}$, so $f \circ g \neq g \circ f$

12. The pth row of a diagonal matrix is $a_{pp}\mathbf{i}_p$; the qth column is $a_{qq}\mathbf{i}_q$. Hence the pq-entry of $(a_{pq})(b_{pq})$, if both matrices are diagonal, is $(a_{pp}\mathbf{i}_p) \cdot (b_{qq}\mathbf{i}_q) = a_{pp}b_{pp}$ if $p = q$ and 0 if $p \neq q$. The pq-entry of $(b_{pq})(a_{pq})$ is $(b_{pp}\mathbf{i}_p) \cdot (a_{qq}\mathbf{i}_q)$ which is the same.

14. $\begin{pmatrix} 0 & -1 & 0 \\ 1 & 0 & 0 \\ 0 & 0 & 1 \end{pmatrix}, \begin{pmatrix} 2^{1/2} & -2^{1/2} & 0 \\ -2^{1/2} & 2^{1/2} & 0 \\ 0 & 0 & 1 \end{pmatrix}$, $-\mathbf{I}$, rotations about the z-axis through $\pi/2$, $3\pi/4$, π radians, ...

16. Rotation through $\pi/2$; $\begin{pmatrix} 0 & -1 \\ 1 & 0 \end{pmatrix}$

17. $f(\mathbf{k}) = \mathbf{k}$; $f(\mathbf{k}) = 0$; $f(\mathbf{v})$ is in the xy-plane if $\mathbf{v}$ is, and is on the z-axis if $\mathbf{v}$ is; $f(\mathbf{v})$ is in the x_1x_2-plane if $\mathbf{v}$ is and is in the x_3x_4-plane if $\mathbf{v}$ is

18. The uv-entry in $E_{pq}E_{rs}$ is the dot product of the uth row of E_{pq} (which is $\mathbf{0}$ if $u \neq p$, and $\mathbf{i}_q$ if $u = p$) by the vth column of E_{rs} (which is $\mathbf{0}$ if $v = s$, and $\mathbf{i}_r$ if $v = s$), hence is 0 unless $u = p$ and $v = s$. If $u = p$ and $v = s$, the uv-entry is $\mathbf{i}_q \cdot \mathbf{i}_r$ which is 0 if $q \neq r$ and 1 if $q = r$. Thus $E_{pq}E_{qs} = E_{ps}$, $E_{pq}E_{rs} = 0$ if $q \neq r$

26. (a) Projection of the projection is still the same projection. (b) $\begin{pmatrix} 1 & 0 & 0 \\ 0 & 0 & 0 \\ 0 & 0 & 1 \end{pmatrix}$;

(c) For method, compare Problem 20, Section 1. $\frac{1}{6}\begin{pmatrix} 5 & -1 & -2 \\ -1 & 5 & -2 \\ -2 & -2 & 2 \end{pmatrix}$

27. (a) $g(g(\mathbf{v} + \mathbf{w})) = g(\mathbf{v} - \mathbf{w}) = \mathbf{v} - (-\mathbf{w}) = \mathbf{v} + \mathbf{w}$

(b) See Problem 26(b)

(c) For all $\mathbf{x}$, $\mathbf{w} = (\mathbf{x} \cdot [1, 1, 2]/\|[1, 1, 2]\|^2)[1, 1, 2]$ (see Problem 20, Section 1) and $\mathbf{v}$ is $\mathbf{x} - \mathbf{w}$. Then $g(\mathbf{x}) = \mathbf{v} - \mathbf{w} = \mathbf{x} - 2\mathbf{w}$. Matrix of g is

$$\frac{1}{3}\begin{pmatrix} 2 & -1 & -2 \\ -1 & 2 & -2 \\ -2 & -2 & -1 \end{pmatrix}$$

28. (b) $\begin{pmatrix} 0 & 0 & 0 \\ 0 & 0 & -1 \\ 0 & 1 & 0 \end{pmatrix}$

29. $t_{pq}m_{qr}$ = tons of material r imported by country p for product q; $\sum_q t_{pq}m_{qr}$ = total tons of material r imported by country p for all products

30. $a_{pr}a_{rq}$ = number of flights from p to q stopping at r; $\sum_r a_{pr}a_{rq}$ = total number of flights from p to q with one stop, at some r

31. (b) $\begin{pmatrix} \frac{1}{2}(1+3^{-n}) & \frac{1}{2}(1-3^{-n}) \\ \frac{1}{2}(1-3^{-n}) & \frac{1}{2}(1+3^{-n}) \end{pmatrix}$; $\begin{pmatrix} \frac{1}{2} & \frac{1}{2} \\ \frac{1}{2} & \frac{1}{2} \end{pmatrix}$; after a long time, there is about a 50 percent chance of finding the bean in either slot

Chapter 3, Section 5

1. $\begin{pmatrix} -5 & 2 \\ 3 & -1 \end{pmatrix}$, $\begin{pmatrix} 1 & 0 & 1 \\ -2 & -\frac{1}{2} & -4 \\ 1 & \frac{1}{2} & 2 \end{pmatrix}$

$\frac{1}{11}\begin{pmatrix} -4 & 5 & 5 & -1 \\ 7 & -6 & -17 & -1 \\ -6 & 2 & 2 & -7 \\ 9 & -3 & -3 & 5 \end{pmatrix}$, $\begin{pmatrix} \sin\theta & -\cos\theta & 0 \\ \cos\theta & \sin\theta & 0 \\ 0 & 0 & 1 \end{pmatrix}$

2. (a) Rotation through $-\alpha$; (b) multiplication by $1/a$ (unless $a = 0$, when the mapping is not invertible); (c) not invertible; (d) reflection in the xy-plane

4. (a) $x_1 \neq 0$ and $x_2 \neq 0$, $\begin{pmatrix} x_1^{-1} & 0 \\ 0 & x_2^{-1} \end{pmatrix}$;

(b) $x_1 \neq 0$ and $x_2 \neq 0$, $\begin{pmatrix} 1/x_1 & 0 \\ -y/x_1x_2 & 1/x_2 \end{pmatrix}$;

(c) all diagonal entries nonzero $\begin{pmatrix} x_1 & \cdots & 0 \\ \vdots & \ddots & \vdots \\ 0 & \cdots & x_n \end{pmatrix}^{-1} = \begin{pmatrix} x_1^{-1} & \cdots & 0 \\ \vdots & \ddots & \vdots \\ 0 & \cdots & x_n^{-1} \end{pmatrix}$;

(d) all diagonal entries nonzero; the pth diagonal entry in $\mathbf{A}^{-1}$ is the reciprocal of the pth diagonal entry in $\mathbf{A}$

5. See Problem 8, Section 4

6. See Chapter 6, 2.8 and 2.1.

Chapter 3, Section 6

1. (a) All vectors perpendicular to $2\mathbf{i} + \mathbf{j}$, all numbers; (b) all vectors perpendicular to $\mathbf{u}$, all numbers

2. (a) 0, all scalar multiples of $\mathbf{i} + 3\mathbf{j} - \mathbf{k}$; (b) 0, all scalar multiples of $\mathbf{u}$

3. (a) All scalar multiples of $\mathbf{i}$; (b) all vectors in R^3; (c) all $a\mathbf{j}$; (d) all $a\mathbf{i}$; (e) all vectors in R^2; (f) all $a[1, 2]$; (g) all $a[2, 3, 4] + b[1, 2, 1]$

4. (a) All linear combinations of $\mathbf{j}$ and $\mathbf{k}$; (b) $\mathbf{0}$; (c) same as (a); (d) all linear combinations of $\mathbf{i}$ and $\mathbf{k}$; (e) $\mathbf{0}$; (f) all multiples of $[1, -2]$; (g) all multiples of $\mathbf{k}$

5. $f(\mathbf{u}) = f(\mathbf{u}')$ if and only if $f(\mathbf{u} - \mathbf{u}') = \mathbf{0}$, if and only if $\mathbf{u} - \mathbf{u}'$ = a vector in the kernel of f...

6. The set of $[x, y, z]$ satisfying $(a_{11} \quad a_{12} \quad a_{13})\begin{pmatrix} x \\ y \\ z \end{pmatrix} = (0)$, (that is, $a_{11}x + a_{12}y + a_{13}z = 0$) is the set of position vectors of the points on a plane through the origin unless $a_{11} = a_{12} = a_{13} = 0$...

7. $Ax + By + Cz = 0$ is the same equation as

$$(A \quad B \quad C)\begin{pmatrix} x \\ y \\ z \end{pmatrix} = 0 \quad \text{or} \quad f(\mathbf{v}) = 0$$

with $\mathbf{v} = [x, y, z]$ and f having matrix $(A \quad B \quad C)$

8. The set of $[x, y, z]$ satisfying $a_{11}x + a_{12}y + a_{13}z = 0$ and $a_{21}x + a_{22}y + a_{23}z = 0$ is a line unless $[a_{11}, a_{12}, a_{13}]$ and $[a_{21}, a_{22}, a_{23}]$ are parallel

Chapter 4, Section 1

1. Multiplication by 0 yields a new equation $0 = 0$, which is satisfied by all vectors, even those which do not satisfy the original...

2. (a) $\begin{pmatrix} 1 & 0 & \frac{7}{5} \\ 0 & 1 & -\frac{3}{5} \end{pmatrix}$, $[\frac{7}{5}, -\frac{3}{5}]$; (b) $\begin{pmatrix} 1 & 0 & -1 & 0 \\ 0 & 1 & 2 & 0 \\ 0 & 0 & 0 & 0 \end{pmatrix}$, $x_3[1, -2, 1]$;

(c) $\begin{pmatrix} 1 & 0 & -1 & 0 \\ 0 & 1 & 2 & 0 \\ 0 & 0 & 0 & 1 \end{pmatrix}$ no solutions;

(d) $(a_{11}a_{22} - a_{12}a_{21})^{-1}[b_1a_{22} - b_2a_{12}, a_{11}b_2 - a_{21}b_1]$;

(e) $\frac{1}{3}[1, 1, 2, -1, -4, 0] + (x_6/3)[1, 7, -4, -7, 23, 3]$

3. (a) $\mathbf{I}_3$, $\mathbf{0}$;

(b) $\begin{pmatrix} 1 & 0 & -1 \\ 0 & 1 & 1 \end{pmatrix}$, $x_3[1, -1, 1]$;

(c) $\begin{pmatrix} 1 & 0 & 0 & \frac{1}{2} \\ 0 & 1 & 0 & \frac{1}{2} \\ 0 & 0 & 1 & -1 \end{pmatrix}$, $x_4[-\frac{1}{2}, -\frac{1}{2}, 1, 1]$;

(d) $\begin{pmatrix} 1 & 0 & \frac{3}{2} & \frac{3}{4} \\ 0 & 1 & -2 & -\frac{3}{2} \\ 0 & 0 & 0 & 0 \end{pmatrix}$, $x_3[-\frac{3}{2}, 2, 1, 0] + x_4[-\frac{3}{4}, \frac{3}{2}, 0, 1]$

4. (a) $\begin{pmatrix} -0.2 & 0.3 \\ 0.4 & -0.1 \end{pmatrix}$; (b) $\begin{pmatrix} a^{-1} & 0 \\ -b/ac & c^{-1} \end{pmatrix}$; (c) none;

(d) $\dfrac{1}{12}\begin{pmatrix} 1 & 4 & -1 \\ -13 & 8 & 1 \\ 11 & -4 & 1 \end{pmatrix}$; (e) $\dfrac{1}{a_{11}a_{22} - a_{12}a_{21}}\begin{pmatrix} a_{22} & -a_{12} \\ -a_{21} & a_{11} \end{pmatrix}$;

(f) $\dfrac{1}{2}\begin{pmatrix} -2 & 1 & 0 & 1 \\ -2 & 3 & 2 & -3 \\ 0 & -1 & -4 & 3 \\ 2 & -1 & 2 & -1 \end{pmatrix}$;

(g) $\begin{pmatrix} 1 & 0 & 0 & 0 \\ -a & 1 & 0 & 0 \\ -b+ad & -d & 1 & 0 \\ X & -e+fd & -f & 1 \end{pmatrix}$, where $X = -c + ae + bf - fad$

5. $P_1\colon x - 3y + z = 0$, $P_2\colon x + y + z = 0$, $P_3\colon x + y = 2$, $P_4\colon x + y = 1$; do not intersect. P_1, P_2, and P_3 intersect at $(2, 0, -2)$. P_1 and P_2 intersect in the set of all $(-t, 0, t)$, the line $x = -t$, $y = 0$, $z = t$.

6. (a) $(-1, 1, 0)$; (b) none; (c) $(1, 0, 2)$ only if $a = 1\ldots$

7. $x + 7y + 3z = 18$

8. $3x_1 + 3x_2 - 2x_3 + 2x_4 = 2$

9. $-\mathbf{i} + \mathbf{j}$ + all scalars times $(\mathbf{i} - 2\mathbf{j} + \mathbf{k})$; impossible

10. See 2.2

13. Yes, if n columns in echelon form of coefficient matrix are distinguished...

14. All scalar multiples of $\mathbf{i}$

15. $\lambda = 1$, all scalar multiples of $\mathbf{i}$

17. $\lambda = -1$, all $t(\mathbf{i} - \mathbf{j})$; $\lambda = 5$, all $t(\mathbf{i} + \mathbf{j})$

Chapter 4, Section 2

2. If the rank is 0, the echelon form is the zero matrix and the original matrix can be retrieved by performing (reverse) manipulations on the echelon form....

3. 1, 2, 1, 2, 3, 3 if a, b, c are distinct, 3, 1 or 0, 2

4. (a), (d), and (e) nonunique solutions (both ranks $= 2$); (b) no solutions (rank augmented matrix $= 3$); (c) nonunique solutions if $2a - b - c = 0$; no solutions if $2a - b - c \neq 0$; (f) nonunique solutions (both ranks $= 1$); (g) nonunique solutions if $a = 2b$ (both ranks $= 1$), no solutions if $a \neq 2b$ (rank augmented matrix $= 2$); (h) unique solution (both ranks $= 2$); (i) no solutions (rank augmented matrix $= 3$); (j) unique solution if $a = 2$, no solutions if $a \neq 2$

5. β, α, α, δ, γ, α (Note first row irrelevant, no answer is (e))

6. Problem 4(h), 4(j) with $a = 2$

8. $\lambda \neq 1, 0, 1, 1, -1, 5$.

9. $t[1, -1]$; $t[1, 0]$; $t[1, 0, 0]$; $t[1, 0, 0]$, $t[0, -1, 1]$, $t[0, 1, 1]$

10. By 2.7, $\mathbf{A}$ can be carried into $\mathbf{I}$ by manipulations of rows. By 2.9, this means $\mathbf{PA} = \mathbf{I}$ where $\mathbf{P}$ is a product of elementary matrices. Then $\mathbf{P}(\mathbf{AB}) = (\mathbf{PA})\mathbf{B} = \mathbf{B}$, so that these same manipulations carry $\mathbf{AB}$ into $\mathbf{B}$. Carry $\mathbf{B}$ into echelon form by further manipulations; $\mathbf{AB}$ can be carried into this same echelon form by both sets of manipulations

11. $\mathbf{PA} = \mathbf{I}$ so $\mathbf{A} = \mathbf{P}^{-1} = (\mathbf{I}_k' \cdots \mathbf{I}_1')^{-1} = \mathbf{I}_1'^{-1} \cdots \mathbf{I}_k'^{-1}$ and each $\mathbf{I}_p'^{-1}$ is elementary

13. (a) $(0, 2, 0)$; (b) $[1, -2, 1]$; (c) $x = t$, $y = 2 - 2t$, $z = t$; (d) echelon form $\begin{pmatrix} 1 & 0 & -1 & 0 \\ 0 & 1 & 2 & 2 \end{pmatrix}$, solutions $[0, 2, 0] + z[1, -2, 1]$; (e) $x = t$, $y = -2t$, $z = t$ or $[x, y, z] = t[1, -2, 1]\ldots$

Chapter 5, Section 1

2. $\{\mathbf{i} - \mathbf{j}, \mathbf{i} - \mathbf{k}\}$ but there are many other possible spanning sets; all consist of two vectors (or more)

3. No, because $\mathbf{0}$ is in every subspace, but is not in the set of solutions.

4. If $b_1 = \cdots = b_m = 0$, the solutions form a subspace of R^n (compare Example 8).
5. $\{(2\mathbf{i} + 3\mathbf{j} + 4\mathbf{k}) \times (\mathbf{i} - \mathbf{j} + \mathbf{k})\}$ or $\{[7, 2, -5]\}$
6. $\{[2, 0, -3, 5]\}$
7. $\{\mathbf{0}\}$
8. $\{\mathbf{j}, \mathbf{k}\}$
9. If $\mathbf{v} = a_1\mathbf{u}_1 + \cdots + a_r\mathbf{u}_r + b\mathbf{0}$, then $\mathbf{v} = a_1\mathbf{u}_1 + \cdots + a_r\mathbf{u}_r$ and conversely...
10. If $\mathbf{A}$ is the original matrix and $\mathbf{B}$ is obtained from $\mathbf{A}$ by one manipulation, then each row of $\mathbf{A}$ is a linear combination of the rows of $\mathbf{B}$ and each row of $\mathbf{B}$ is a linear combination of the rows of $\mathbf{A}$. Hence (Problem 11) the space spanned by the rows of $\mathbf{A}$ is the same as the space spanned by the rows of $\mathbf{B}$...
11. This is essentially Problem 28, Chapter I, Section 5.
12. R^3; the space of all vectors in the xy-plane; ditto; R^3
13. Yes, by the first two
14. R^3
15–18. The set of all vectors (15) in the plane containing $\mathbf{u}$, $\mathbf{v}$, and $\mathbf{w}$; (16) in the plane containing $\mathbf{u}$ and $\mathbf{v}$; (17) on the line containing $\mathbf{u}$ and $\mathbf{v}$; (18) on the line containing $\mathbf{u}$; $\mathbf{0}$ (one vector)
19. $[-1, 1, -3, -3]$ and $[0, 3, -5, -7]$; if these were linear combinations of one vector, they would be "parallel"—their components would be proportional.
21. Yes, any two of these
22. No; the space spanned by two vectors is at most a plane; this space is R^3
23. Yes, any two
24. If the subspaces are the same, then $\mathbf{u}_1$, being in the first, is in the second, so $\mathbf{u}_1$ is a linear combination of $\mathbf{u}_2, \ldots, \mathbf{u}_r$. The converse is a special case of Problem 11

Chapter 5, Section 2

2. If $a_1\mathbf{u}_1 + \cdots + a_n\mathbf{u}_n = 0$, dot multiply both sides by $\mathbf{u}_p$ to get $a_p(\mathbf{u}_p \cdot \mathbf{u}_p) = 0$; since $\mathbf{u}_p \neq 0$, $\mathbf{u}_p \cdot \mathbf{u}_p = \|\mathbf{u}_p\|^2 \neq 0$; therefore $a_p = 0$, and this argument applies to every p. Counterexample: $\{\mathbf{i}, \mathbf{i} + \mathbf{j}\}$ are independent but not orthogonal
3. If $\sum a_p\mathbf{u}_p = \mathbf{0}$, then $\mathbf{0} = f(\mathbf{0}) = f(\sum a_p\mathbf{u}_p) = \sum a_p f(\mathbf{u}_p)$. This implies all a's are 0, by (2.3c) applied to $f(\mathbf{u}_1), \ldots, f(\mathbf{u}_n)$. Counterexample: Let f be projection on the xy-plane in R^3. Then $\mathbf{i}, \mathbf{j}, \mathbf{k}$ are independent but $f(\mathbf{i}) = \mathbf{i}$, $f(\mathbf{j}) = \mathbf{j}$, $f(\mathbf{k}) = \mathbf{0}$ are not.
4. (a) 2; (b) 3, yes; (c) 3; (d) 4, yes
5. (a) Any two; (b) all; (c) any three; (d) all
6. 2, 2
7. $a_1[1, 0, 0, *, *, *] + a_2[0, 1, 0, *, *, *] + a_3[0, 0, 1, *, *, *] = [a_1, a_2, a_3 \cdot, \cdot, \cdot]$ which is $\mathbf{0}$ only if $a_1 = a_2 = a_3 = 0$
10. (a) Yes, all; (b) yes, all; (c) no, any two; (d) yes, all; (e) no, first and third; (f) no, first two
11. Many answers, for example $[1, 0, 0, 0]$
12. $a = 2$
13. If $\mathbf{u}_1, \ldots, \mathbf{u}_n$ is a basis and if a vector is written in two ways: $\sum a_p\mathbf{u}_p = \sum b_p\mathbf{u}_p$ then $\sum (a_p - b_p)\mathbf{u}_p = \mathbf{0}$ and $a_p - b_p = 0$ for all p; thus the two ways are really one way. Conversely, if every vector can be written in only one way, this applies to the vector $\mathbf{0}$ which can be written as $0\mathbf{u}_1 + \cdots + 0\mathbf{u}_n$; hence $\sum a_p\mathbf{u}_p = \mathbf{0}$ implies $a_p = 0$ for all p.

Chapter 5, Section 3

1. [1, 0, 1, 0], [0, 1, 0, 1], [−1, 0, 1, 0]
2. [1, 1, 0, 0], $\frac{1}{2}$[−3, 3, 2, 0], $\frac{1}{11}$[1, −1, 3, 0], or omit the $\frac{1}{2}$ and $\frac{1}{11}$; [1, 0, 0, 0], [0, 1, 0, 0], [0, 0, 1, 0]
3. [−2, 1, 0], [−3, −6, 10]
4. [3, −3, 2, 0], [−17, −5, 18, 22]
5. $\mathbf{i} + 2\mathbf{j} + \mathbf{k}$, $\frac{1}{2}(\mathbf{i} - \mathbf{k})$, $\mathbf{0}$
6. Yes. "If": $\mathbf{u}_p = 0$ implies $\mathbf{v}_p$ is a linear combination of $\mathbf{u}_1, \ldots, \mathbf{u}_{p-1}$, which in turn are linear combinations of $\mathbf{v}_1, \ldots, \mathbf{v}_{p-1}$. Hence the **v**'s are dependent. "Only if": If no $\mathbf{u}_p$ is $\mathbf{0}$, then the **u**'s form a basis of the space they span (because they are orthogonal, hence independent). The space is then r-dimensional and cannot be spanned by fewer than r vectors. Hence it takes all of the **v**'s to span the space, which means no **v** can be a linear combination of the others; the **v**'s are independent.
7. No; change the order of the original set of vectors
8. Show that if $\mathbf{u}_2$ is any linear combination of $\mathbf{v}_1$ and $\mathbf{v}_2$ that is orthogonal to $\mathbf{v}_1 = \mathbf{u}_1$, then by multiplying by a scalar you can arrange $\mathbf{u}_2 = \mathbf{v}_2 + a\mathbf{u}_1$. Then write $\mathbf{u}_2 \cdot \mathbf{u}_1 = 0$ and solve for a. Similarly $\mathbf{u}_p = \mathbf{v}_p + a_1\mathbf{u}_1 + \cdots + a_{p-1}\mathbf{u}_{p-1}$ and $\mathbf{u}_p \cdot \mathbf{u}_1 = 0$ solves for a_1, etc.

Chapter 5, Section 4

2. Rank $\mathbf{AB} \leq \mathbf{A} \leq$ number of columns in $\mathbf{A}$; conversely, imitate Example 5. The row space is spanned by two of the rows; these two rows form a matrix **B**. . .
4. (c) If a system of three equations in three unknowns has coefficient matrix of rank 2 (resp. 3) then the solutions of the corresponding homogeneous system are all scalars times one nonzero vector (resp. are all [0, 0, 0]). This is true because the echelon form has two (resp. three distinguished columns).
6. Refer to Chapter 4, 2.7. Problem 5(a) is 2.7(d), (b) is 2.7(f), (c) is approximately 2.7(d), (d) is 2.7(e).
7. $(1 \quad 0)\begin{pmatrix}1\\0\end{pmatrix} = (1)$ but 2 by 3 times 3 by 2 is more interesting
10. In Problem 9, if $s = m$ then $\{\mathbf{u}_1, \ldots, \mathbf{u}_s\}$ is a basis of T (review the relevant part of the picking argument which shows that every vector of T is a linear combination of $\mathbf{u}_1, \ldots, \mathbf{u}_s$).
11. A basis of S will be a basis of T by Problem 10
12. (c) If **v** is in S, then **v** is a linear combination of $\{\mathbf{u}_1, \ldots, \mathbf{u}_s\}$. If **w** is in S', then **w** is a linear combination of $\{\mathbf{u}_1, \ldots, \mathbf{u}_r, \mathbf{u}'_{r+1}, \ldots, \mathbf{u}_t'\}$. Hence $\mathbf{v} + \mathbf{w} = (a_1\mathbf{u}_1 + \cdots + a_r\mathbf{u}_r + a_{r+1}\mathbf{u}_{r+1} + \cdots + a_s\mathbf{u}_s) + (b_1\mathbf{u}_1 + \cdots + b_r\mathbf{u}_r + b_{r+1}\mathbf{u}'_{r+1} + \cdots + b_t\mathbf{u}_t') = (a_1 + b_1)\mathbf{u}_1 + \cdots + (a_r + b_r)\mathbf{u}_r + a_{r+1}\mathbf{u}_{r+1} + \cdots + a_s\mathbf{u}_s + b_{r+1}\mathbf{u}'_{r+1} + \cdots + b_t\mathbf{u}_t'$, so the set in question is a spanning set. If $\sum a_p\mathbf{u}_p + \sum_{q=r+1}^{t} b_q\mathbf{u}_q' = \mathbf{0}$, then $\sum a_p\mathbf{u}_p = -\sum b_q\mathbf{u}_q'$. The left side is in S, the right side is in S' so the vector (on both sides) is in $S \cap S'$, and $\sum b_q\mathbf{u}_q'$ must be a linear combination of $\mathbf{u}_1, \ldots, \mathbf{u}_r$. Since $\mathbf{u}_1, \ldots, \mathbf{u}_r, \mathbf{u}'_{r+1}, \ldots, \mathbf{u}_t'$ are independent, this says $b_{r+1} = \cdots = b_t = 0$. Then $\sum a_p\mathbf{u}_p = \mathbf{0}$, so $a_1 = \cdots = a_s = 0$.

13. (a) $\mathbf{v}$ is in $S^\perp$ provided $\mathbf{v} \cdot \mathbf{x} = 0$ for all $\mathbf{x}$ in S. If $\mathbf{v} \cdot \mathbf{x} = 0$ and $\mathbf{w} \cdot \mathbf{x} = 0$ then $(\mathbf{v} + \mathbf{w}) \cdot \mathbf{x} = \mathbf{v} \cdot \mathbf{x} + \mathbf{w} \cdot \mathbf{x} = 0$ and $(a\mathbf{v}) \cdot \mathbf{x} = a(\mathbf{v} \cdot \mathbf{x}) = 0 \ldots$ (d) $\dim(S + S^\perp) = r + (n - r) - 0 = n = \dim R^n$. By Problem 11, $S + S^\perp = R^n$; (e) $\dim S^{\perp\perp} = n - \dim S^\perp = n - (n - r) = r$, and $S^{\perp\perp}$ contains S (if $\mathbf{x}$ is in S then for every $\mathbf{v}$ in $S^\perp$, $\mathbf{x} \cdot \mathbf{v} = 0$). By Problem 11, $S^{\perp\perp} = S$.

14. $S + T = R^n$ means every vector in R^n is expressible as a sum of a vector in S and a vector in T. Uniqueness: if a vector in R^n is both $\mathbf{s} + \mathbf{t}$ and $\mathbf{s}' + \mathbf{t}'$ with $\mathbf{s}, \mathbf{s}'$ in S, $\mathbf{t}, \mathbf{t}'$ in T, then $\mathbf{s} + \mathbf{t} = \mathbf{s}' + \mathbf{t}'$, $\mathbf{s} - \mathbf{s}' = \mathbf{t}' - \mathbf{t}$; but $\mathbf{s} - \mathbf{s}'$ is in S, and equals $\mathbf{t} - \mathbf{t}'$ which is in T so $\mathbf{s} - \mathbf{s}'$ is in $S \cap T = 0$, $\mathbf{s} - \mathbf{s}' = 0$, $\mathbf{s} = \mathbf{s}'$, $\mathbf{t} = \mathbf{t}'$.

Chapter 6, Section 1

1. (a) 1; (b) 6; (c) and (d) 0, not invertible; (e) -2; (f) 17; (g) and (h) $ps - qr$, not invertible if and only if $ps = qr$

4. Second set is dependent, the other two are independent

5. $1, -\frac{1}{2}$

7. (a) Reducing $\mathbf{U}$ toward echelon form automatically reduces $\mathbf{A}$ to echelon form; if this form is not $\mathbf{I}$, then both $\det \mathbf{U}$ and $\det \mathbf{A}$ are 0 and 7(a) is true. If this form is $\mathbf{I}$, then $\det \mathbf{A} = s \det \mathbf{I}$ with s determined by keeping track of the manipulations used, and

$$\det \mathbf{U} = s \det\begin{pmatrix} \mathbf{I} & \mathbf{O} \\ \mathbf{O} & \mathbf{B} \end{pmatrix}, \qquad \text{so} \qquad \det \mathbf{U} = \det \mathbf{A} \det\begin{pmatrix} \mathbf{I} & \mathbf{O} \\ \mathbf{O} & \mathbf{B} \end{pmatrix}.$$

Continue, reducing $\mathbf{B}$ and hence $\mathbf{U}$ to echelon form...

(b) Use Manipulation 1 on the rows of $\mathbf{V}$ to reduce to (a)

8. $(\det \mathbf{A})(\det \mathbf{A}^{-1}) = \det \mathbf{A}\mathbf{A}^{-1} = \det \mathbf{I} = 1$

9. (a) Multiply $\mathbf{A}^2 = \mathbf{A}$ by $\mathbf{A}^{-1}$ if it exists

(b) See Problem 26, Chapter 3, Section 4 or: the image is the plane, which has dimension $2 = \text{rank } \mathbf{A}$

11. By 1.3, $\det(a\mathbf{u}_1, \ldots, a\mathbf{u}_n) = a^n \det(\mathbf{u}_1, \ldots, \mathbf{u}_n)$

12. $\begin{pmatrix} 1 & 0 \\ 0 & 0 \end{pmatrix} + \begin{pmatrix} 0 & 0 \\ 0 & 1 \end{pmatrix} = \mathbf{I}$ but $0 + 0 \neq 1$

Chapter 6, Section 2

2. 2; 6; 24; 3,628,800;...

6. $\cdots$(e) $\begin{pmatrix} 3.5 & -2 & 0.5 \\ -2 & 1 & 0 \\ -2.5 & 2 & -0.5 \end{pmatrix}$ (f) almost impractical

(g) $\begin{pmatrix} \mathbf{I} & \mathbf{O} \\ -\mathbf{BA} & \mathbf{B} \end{pmatrix}$ with $\mathbf{B} = \dfrac{1}{ps - qr}\begin{pmatrix} s & -q \\ -r & p \end{pmatrix}$, $\mathbf{A} = \begin{pmatrix} a & b \\ c & d \end{pmatrix}$

(h) $\begin{pmatrix} \mathbf{I} & -\mathbf{AB} \\ \mathbf{O} & \mathbf{B} \end{pmatrix}$ with $\mathbf{A}$ and $\mathbf{B}$ as in (g)

7. See Problem 22, Chapter 3, Section 4 for the matrix; and $\cos^2 \alpha + \sin^2 \alpha = 1$

8. (a) $[\frac{7}{8}, -\frac{1}{4}, \frac{3}{8}]$ (b) $[1, -1, -\frac{1}{3}, \frac{2}{3}]$ (c) $[2, -3, -4, 8, -2]$ (d) $\frac{1}{9}[2a - 2b + c, a + 2b + 2c, -2a - b + 2c]$

Chapter 6, Section 3

2. $((a^tf + b^tg)(\mathbf{v})) \cdot \mathbf{w} = (a^tf(\mathbf{v}) + b^tg(\mathbf{v})) \cdot \mathbf{w} = a^tf(\mathbf{v}) \cdot \mathbf{w} + b^tg(\mathbf{v}) \cdot \mathbf{w} = a\mathbf{v} \cdot f(\mathbf{w}) + b\mathbf{v} \cdot g(\mathbf{w}) = \mathbf{v} \cdot (af(\mathbf{w}) + bg(\mathbf{w})) = \mathbf{v} \cdot (af + bg)(\mathbf{w}) \ldots$

12. The row space of $\mathbf{A}$ is the column space of ${}^t\mathbf{A}\ldots$

13. (a) If ${}^tf(\mathbf{v}') = \mathbf{0}$ and if $\mathbf{w} = f(\mathbf{v})$, then $\mathbf{w} \cdot \mathbf{v}' = \mathbf{v} \cdot {}^tf(\mathbf{v}') = 0$; (b) if $\mathbf{w} \cdot \mathbf{v}' = 0$ whenever $\mathbf{w} = f(\mathbf{v})$, then $\mathbf{v} \cdot {}^tf(\mathbf{v}') = 0$ for all $\mathbf{v}$ and ${}^tf(\mathbf{v}') = \mathbf{0}$

14. $\det \mathbf{B} =$ (nonzero constant) $\det \mathbf{A}$ so one determinant is zero if and only if the other is

15. (b) Pick $\mathbf{u}_1$ and $\mathbf{u}_2$ in the plane, and $\mathbf{u}_3$ perpendicular to the plane. Then $h(\mathbf{u}_1) = \mathbf{u}_1$, $h(\mathbf{u}_2) = \mathbf{u}_2$, $h(\mathbf{u}_3) = -\mathbf{u}_3$, $\det(h(\mathbf{u}_1), h(\mathbf{u}_2), h(\mathbf{u}_3)) = \det(\mathbf{u}_1, \mathbf{u}_2, -\mathbf{u}_3) = -\det(\mathbf{u}_1, \mathbf{u}_2, \mathbf{u}_3)$, and use 3.12.

16. $\det(f(\mathbf{u}_1), f(\mathbf{u}_2), f(\mathbf{u}_3)) = 0$ because $f(\mathbf{u}_1), f(\mathbf{u}_2), f(\mathbf{u}_3)$ are dependent

Chapter 6, Section 4

1. Only the first

2. $y = 2$, $z = 4$; all x

3. $f(\mathbf{v})$ is along the x-axis, magnitude $\|\mathbf{v}\| \cos\alpha$; $f(\mathbf{v}) \cdot \mathbf{w} = \|f(\mathbf{v})\|\,\|\mathbf{w}\| \cos\alpha' = \|\mathbf{v}\|\,\|\mathbf{w}\| \cos\alpha \cos\alpha'$, where α (resp. α') is the angle between $\mathbf{v}$ (resp. $\mathbf{w}$) and the x-axis. But $\mathbf{v} \cdot f(\mathbf{w})$ is the same; matrix $\begin{pmatrix} 1 & 0 & 0 \\ 0 & 0 & 0 \\ 0 & 0 & 0 \end{pmatrix}$

4. For a rotation f, $\|f(\mathbf{v})\| = \|\mathbf{v}\|$ for all $\mathbf{v}$. Hence f is symmetric if and only if the angle between $\mathbf{v}$ and $f(\mathbf{w})$ equals the angle between $f(\mathbf{v})$ and $\mathbf{w}$. For example, if $\mathbf{w} = f(\mathbf{v})$ and both are perpendicular to the axis of rotation and if the angle of rotation is α then the two angles above are 2α and 0, respectively. Hence the only symmetric rotations are through angles of 0 and π, and these are symmetric.

5. Any two that do not commute

6. (a) ${}^t({}^t\mathbf{A}\mathbf{A}) = {}^t\mathbf{A}\,{}^t({}^t\mathbf{A}) = {}^t\mathbf{A}\mathbf{A}$; (b) ${}^t({}^t\mathbf{A}\mathbf{A}^{-1}) = {}^t\mathbf{A}^{-1}\,{}^t\mathbf{A} = {}^t\mathbf{A}\,{}^t\mathbf{A}$. This is equal to ${}^t\mathbf{A}\mathbf{A}^{-1}$ if and only if ${}^t\mathbf{A}^{-1}\mathbf{A}\mathbf{A} = {}^t\mathbf{A}$, $\mathbf{A}^2 = ({}^t\mathbf{A})^2$.

7. (a) $a_{pq} = -a_{qp}$, so if $p = q$, $a_{pp} = -a_{pp}$, $a_{pp} = 0$; (b) $-1, -2, -3$; (c) $\det\begin{pmatrix} 0 & a \\ -a & 0 \end{pmatrix} = a^2$; (d) $\det \mathbf{A} = \det {}^t\mathbf{A} = \det(-\mathbf{A}) = (-1)^n \det \mathbf{A} = -\det \mathbf{A}$ if n is odd; (e) Use 3.3; (f) If $f(\mathbf{v}) = a\mathbf{v}$ with $\mathbf{v} \neq \mathbf{0}$, use (e) with $\mathbf{v} = \mathbf{w}$ to get $a(\mathbf{v} \cdot \mathbf{v}) = 0$ but $\mathbf{v} \cdot \mathbf{v} = \|\mathbf{v}\|^2 \neq 0$, so $a = 0$ and $a\mathbf{v} = \mathbf{0}$; (g) If $\mathbf{r} = [a, b, c]$, the matrix of f is $\begin{pmatrix} 0 & -c & b \\ c & 0 & -a \\ -b & a & 0 \end{pmatrix}$ which is the most general 3 by 3 skew matrix. Without matrices, use (e) and Problem 12, Chapter 1, Section 7.

Chapter 6, Section 5

1. Use 5.3(b), (b′), (c), or (d); (b) $\|f(\mathbf{v})\| = \|\frac{1}{5}[3x + 4y, -4x + 3y]\| = \frac{1}{5}(25x^2 + 0xy + 25y^2)^{1/2}$; (c) rotation through $\cos^{-1}\frac{3}{5}$

2. (c) Reflection in the line of the vector $[2, 1]$

3. Use 5.3
4. (a) $\pm(\frac{1}{65})[36, 48, -25]$; (b) third row $[0, 0, 5, 0]$, fourth row $[0, 0, 0, \pm 5]$; (c) Impossible unless $a^2 + b^2 = 1$. Then second row is $\pm[-b, a]$
5. From 5.3. Orthogonality of rows demands $bc = 0$, $cd = 0$ and $a^2 + bd = 0$. Length 1 demands $a^2 + b^2 = a^2 + d^2 = 1 + c^2 = 1$. Therefore $c = 0$, $b = \pm d$ (but $+$ is impossible because $a^2 + bd = 0$ and $a \neq 0$)...
6. (b) any scalar times $[1, 0, 1]$; (c) $\mathbf{w} = [x, y, -x]$, $\mathbf{Aw} = \frac{1}{3}[-x + 2y, -4x - y, x - 2y]$, angle $= \cos^{-1}(-\frac{1}{3})$
8. The dot product formulas for the cosines of the two angles are equal.
9. $\det \mathbf{A} = 1$ and $\det \mathbf{B} = 1$ imply $\det \mathbf{AB} = 1$, etc.
10. g must be $f \circ h^{-1}$. Then $\det g = \det f\ (\det h)^{-1} = (-1)(-1)^{-1} = +1$
11. ${}^t(af) = a\,{}^tf$, $(af)^{-1} = a^{-1}f^{-1} = a^{-1}\,{}^tf$; they are equal if and only if $a = a^{-1}$
12. $\mathbf{I} + (-\mathbf{I}) = \mathbf{O}$ and others
13. 45 degrees about the y-axis; -45 degrees about the y-axis

Chapter 7, Section 1

1. (a) $f(\mathbf{i}) = \mathbf{i}$, $f(\mathbf{j}) = 2\mathbf{j}$ so $f(x\mathbf{i} + y\mathbf{j}) = x\mathbf{i} + y(2\mathbf{j})$; (c) $\pm\sqrt{3}\mathbf{i} + \mathbf{k}$ or any scalar multiple of these; (d) $f(\mathbf{v}) = \mathbf{i} + 4\mathbf{j} - 9\mathbf{k}$, $\cos\theta = -18/\sqrt{14}\sqrt{98}$, $\theta = 119°\cdots$; (e) $\cos^{-1} 3/\sqrt{10} = 18°34'$; (f) impossible, unless $\mathbf{v} = \mathbf{0}$; (g) minimize $(x^2 + 2y^2)/(x^2 + y^2)^{-1/2}(x^2 + 4y^2)^{-1/2}$ when $x^2 + y^2 = 1$. *Ans.* $x = \pm\sqrt{2/3}$, $y = \pm\sqrt{1/3}$; angle $= \cos^{-1}(2^{3/2}/3) = 19.5°$
2. All scalars times $\mathbf{i}$ belong to 2; multiples of $\mathbf{j}$ belong to 3; multiples of $\mathbf{k}$ belong to -1.
3. All linear combinations of $\mathbf{i}$ and $\mathbf{j}$ are eigenvectors belonging to 2; multiples of $\mathbf{k}$ belong to -1.
4. $f(\sum x_p\mathbf{i}_p) = \sum x_p a_{pp}\mathbf{i}_p$. If this $= \lambda\sum x_p\mathbf{i}_p$, then $x_p a_{pp} = \lambda x_p$ for all p; for each p, $x_p = 0$ or $\lambda = a_{pp}$. Since λ is independent of p, the latter happens for at most one p; hence the former happens for all but one p. That is, $\sum x_p\mathbf{i}_p$ has only one nonzero component.
5. If $a_{11} = a_{22}$ then $f(x_1\mathbf{i}_1 + x_2\mathbf{i}_2) = a_{11}(x_1\mathbf{i}_1 + x_2\mathbf{i}_2)$ for all x_1, x_2 (neither need be 0)...
6. All scalar multiples of $\mathbf{i}_1$ belong to 1; the matrix is not diagonalizable at all.
7. All scalar multiples of $\mathbf{i}_1 + \mathbf{i}_2$ belong to 2; all scalar multiples of $\mathbf{i}_1 - \mathbf{i}_2$ belong to 0; orthogonally diagonalizable
8. None; the function is rotation through 45 degrees and leaves only the zero vector unturned.
9. Multiples of $\mathbf{i}$ belong to 2; of $\mathbf{i} + \mathbf{j}$ belong to 3; diagonalizable, but not orthogonally
10. If $f(\mathbf{v}) = \lambda\mathbf{v}$, then $f(s\mathbf{v}) = sf(\mathbf{v}) = s\lambda\mathbf{v} = \lambda(s\mathbf{v})$
11. $f(a\mathbf{v} + b\mathbf{w}) = af(\mathbf{v}) + bf(\mathbf{w}) = a(\lambda\mathbf{v}) + b(\lambda\mathbf{w}) = \lambda(a\mathbf{v} + b\mathbf{w})$
12. If $\sum x_p\mathbf{e}_p$ is an eigenvector belonging to λ, then $\lambda\sum x_p\mathbf{e}_p = f(\sum x_p\mathbf{e}_p) = \sum x_p\lambda_p\mathbf{e}_p$ so $\lambda x_p = x_p\lambda_p$ for all p; $(\lambda - \lambda_p)x_p = 0$; if $x_p \neq 0$, then $\lambda_p = \lambda$; $\sum x_p\mathbf{e}_p$ is a linear combination of those $\mathbf{e}_p$'s which belong to λ; the space of all eigenvectors belonging to λ is spanned by those $\mathbf{e}_p$'s which belong to λ....
13. Eigenvectors belonging to first eigenvalue form a one-dimensional space containing $\mathbf{i}$...

15. $\begin{pmatrix} a_{11} & 0 & \cdots & 0 \\ 0 & a_{22} & \cdots & 0 \\ \vdots & \vdots & & \vdots \\ 0 & 0 & \cdots & a_{nn} \end{pmatrix}$

$$= \begin{pmatrix} 1 & 0 & \cdots & 0 \\ 0 & 1 & \cdots & 0 \\ \vdots & \vdots & \vdots & \vdots \\ 0 & 0 & \cdots & a_{nn} \end{pmatrix} \cdots \begin{pmatrix} 1 & 0 & \cdots & 0 \\ 0 & a_{22} & \cdots & 0 \\ \vdots & \vdots & & \vdots \\ 0 & 0 & \cdots & 1 \end{pmatrix} \begin{pmatrix} a_{11} & 0 & \cdots & 0 \\ 0 & 1 & \cdots & 0 \\ \vdots & \vdots & \vdots & \vdots \\ 0 & 0 & \cdots & 1 \end{pmatrix}$$

16. $(f - \lambda \mathbf{I}\mathbf{v}) = \mu\mathbf{v}$ means $f(\mathbf{v}) - \lambda\mathbf{v} = \mu\mathbf{v}$, $f(\mathbf{v}) = (\lambda + \mu)\mathbf{v}$, $\lambda + \mu$ is an eigenvalue of f, $\mu = -\lambda +$ an eigenvalue of f

Chapter 7, Section 2

2. (a) Orthogonally diagonalizable; $[1, -1]$ (and of course all its scalar multiples) belongs to 1; $[1, 1]$ belongs to 3; (b) orthogonally diagonalizable $[-1 \pm \sqrt{17}, 4]$ belong to $(3 \pm \sqrt{17})/2$, respectively; (c) orthogonally diagonalizable; $[1, 1, 1]$ belongs to 0, $[(1 + (\lambda/2))(-2 + (\lambda/3)), (-2 + (\lambda/3)), 1] = \frac{1}{3}[2 \mp \sqrt{19}, -5 \pm \sqrt{19}, 3]$ belongs to λ where $\lambda = 1 \pm \sqrt{19}$; (d) diagonalizable; all linear combinations of $[1, 0, 1]$ and $[0, 1, 0]$ belong to 1; $[1, 0, 0]$ belongs to 2; (e) not diagonalizable; all linear combinations of $\mathbf{i}$ belong to 1; (f) diagonalizable, eigenvalues 0, 2, 4, 8, eigenvectors (respectively) $[1, 1, 0, 0]$, $[1, 0, 1, 1]$, $[1, 0, 0, 1]$, $[0, 1, 1, 1]$

3. $\mathbf{I}$

4. By 2.2, if $\det(\mathbf{A} - \lambda\mathbf{I}) = a_o\lambda^n + a_1\lambda^{-1} + \cdots + a_n$ then $a_1/a_o = -\operatorname{tr}\mathbf{A}$ and $a_n/a_o = (-1)^n \det \mathbf{A}$. If $a_o\lambda^n + \cdots + a_n = a_o(\lambda - \lambda_1)(\lambda - \lambda_2)\cdots(\lambda - \lambda_n)$, where $\lambda_1, \ldots, \lambda_n$ are the roots of this polynomial, then $a_1/a_o = -\sum \lambda_p$ and $a_n/a_o = (-1)^n \prod \lambda_p$.

Chapter 7, Section 3

1. (a) $[\frac{1}{2}i, 2, \frac{1}{2}(1 - i)]$; (b) $[2 + i, -1]$; (c) $z[i, 0, 1]$

2. Just (a) and (b)

3. 1, 2

4. $\begin{pmatrix} 1 & 4 & 7 \\ 2 & 5 & 8 \\ 3 & 6 & 9 \end{pmatrix}; \begin{pmatrix} 1 - i & 4 - 2i & 7 - 3i \\ 2 - i & 5 - 2i & 8 - 3i \\ 3 - i & 6 - 2i & 9 - 3i \end{pmatrix}$

5. $\overline{-4i} = 4i$, $\overline{8 + i} = 8 - i$

6. (d) Both mappings are rotation through $n\theta$. Then apply them to $[1, 0]$, that is multiply $1 + 0i$ by the two complex numbers. You get the two complex numbers, and you get the same vector.

7. $0, \frac{1}{2}(9 \pm \sqrt{105})$

Chapter 8, Section 1

1. (a) $\begin{pmatrix} 1 & 1 \\ 1 & 1 \end{pmatrix}$; (b) $\begin{pmatrix} 1 & 0 & 0 & 0 \\ 0 & 1 & 0 & 0 \\ 0 & 0 & 1 & 0 \\ 0 & 0 & 0 & 1 \end{pmatrix}$; (c) $\begin{pmatrix} 1 & \frac{1}{2} & 1 & \frac{1}{2} \\ \frac{1}{2} & 1 & \frac{5}{2} & \frac{1}{2} \\ 1 & \frac{5}{2} & 1 & 0 \\ \frac{1}{2} & \frac{1}{2} & 0 & 3 \end{pmatrix}$;

(d) $\begin{pmatrix} 1 & 1 & 0 \\ 1 & 1 & 0 \\ 0 & 0 & 0 \end{pmatrix}$

2. (i). (a) $x^2 + y^2 + z^2$, (b) $x^2 + 2y^2 + 3z^2$, (c) $2xy$, (d) $x^2 + 2xy + y^2$, (e) $x^2 - 2yz$ (ii). (a) Identity, (b) stretching, (c) reflection in line $x = y$, (d) stretching by factor of 2 along $\mathbf{i} + \mathbf{j}$ and 0 along $\mathbf{i} - \mathbf{j}$, (e) reflection in plane $y = -z$ or stretching by factor 1 along $\mathbf{i}$ and $\mathbf{j} - \mathbf{k}$ and by factor -1 along $\mathbf{j} + \mathbf{k}$

3. (a) Two lines $y = -x - 1$, $y = -x + 1$; (b) the line $y = -x$; (c) $\frac{5}{2}\xi_1^2 - \frac{1}{2}\xi_2^2 = 1$, hyperbola, vertices at $\pm[5^{-1/2}, 5^{-1/2}]$, axis along line $y = x$; (d) two lines (the asymptotes to (c)), $y = \frac{1}{2}(-3 \pm \sqrt{5})x$; (e) $\frac{3}{2}\xi_1^2 + \frac{1}{2}\xi_2^2 = 1$, ellipse, vertices at $\pm[3^{-1/2}, 3^{-1/2}]$ and $\pm[1, -1]$; (f) the origin; (g) $(3 + \sqrt{2})\xi_1^2 + (3 - \sqrt{2})\xi_2^2 = 2$, ellipse, principal radii $\sqrt{(2/(3 \pm \sqrt{2}))}$; (h) two lines, $y = -x$, $y = -x + 1$

4. (a) Ellipsoid; (b) hyperboloid of one sheet (revolve hyperbola around its conjugate axis); (c) hyperboloid of two sheets (revolve hyperbola around its transverse axis); (d) elliptical cone; (e) elliptical cylinder; (f) $2\xi^2 + z^2 = 1$, cylinder, axis on line $z = 0$, $y = x$; (g) $2\xi_1^2 + 8\xi_2^2 + 16\xi_3^2 = 1$, ellipsoid, axes in directions $\mathbf{i} + \mathbf{k}$, $\mathbf{i} - \sqrt{6}\mathbf{j} - \mathbf{k}$, $\sqrt{6}\mathbf{i} + 2\mathbf{j} - \sqrt{6}\mathbf{k}$; (h) $9\xi_1^2 + 18\xi_2^2 = 1$, elliptic cylinder, axis along $\mathbf{i} + \mathbf{k}$; axes of elliptical cross section: $2\mathbf{i} + \mathbf{j} - 2\mathbf{k}$, $\mathbf{i} - 4\mathbf{j} - \mathbf{k}$

5. (a) $\det(\mathbf{A} - \lambda\mathbf{I}) = (\lambda_1 - \lambda)(\lambda_2 - \lambda)$; set $\lambda = 0$;
(b) $\mathbf{A} = \begin{pmatrix} a & b/2 \\ b/2 & c \end{pmatrix}$, $\det \mathbf{A} = ac - b^2/4$; (c) $-4\lambda_1\lambda_2 = b^2 - 4ac$ by (a) and (b); (d) coefficient of λ in (a) is $-(\lambda_1 + \lambda_2)$; (e) use (c) and (d); (f), (g), (h), (k) $\lambda_1\xi_1^2 + \lambda_2\xi_2^2 = 1$ is an ellipse if λ_1 and λ_2 are both positive, a hyperbola if λ_1 and λ_2 have opposite signs, and empty if λ_1 and λ_2 are both negative; (i) use (c); (j) If $\lambda_1 = 0$ we get $\lambda_2\xi_2^2 = 1$, a pair of lines if $\lambda_2 > 0$ (but $\lambda_2 = \lambda_1 + \lambda_2$), empty if $\lambda_2 < 0$

6. $\det \mathbf{A} = \lambda_1\lambda_2\lambda_3$ and graph is like $\lambda_1\xi_1^2 + \lambda_2\xi_2^2 + \lambda_3\xi_3^2 = 1$; an ellipsoid if all λ's are >0; a hyperboloid of one sheet if one is >0, two are <0; a hyperboloid of two sheets if one is <0, two are >0; a cylinder if one is 0 and two are ≥ 0

7. Eigenvectors of $\begin{pmatrix} a & b/2 \\ b/2 & a \end{pmatrix}$ are $[1, 1]$, $[1, -1]$.

9. Dimension $= \frac{1}{2}n(n + 1)$

Chapter 8, Section 3

1. $\mathbf{Q} = 2^{-1/2}\begin{pmatrix} 1 & -1 \\ 1 & 1 \end{pmatrix}$, $\mathbf{P} = 2^{-1/2}\begin{pmatrix} 1 & 1 \\ -1 & 1 \end{pmatrix}$; (a) $2^{-1/2}[1, 1]$, $2^{-1/2}[1, -1]$, $[2^{1/2}, 0]$, $-2^{-1/2}[1, 3]$; (b) $2^{-1/2}[1, 1]$, $2^{-1/2}[-1, 1]$; (c, d) $x' = 2^{-1/2}x + 2^{-1/2}y$, $y' = -2^{-1/2}x + 2^{-1/2}y$; $x = 2^{-1/2}x' - 2^{-1/2}y'$, $y = 2^{-1/2}x' + 2^{-1/2}y'$; (e) $x'^2 + y'^2 = a^2$, $x'^2 - y'^2 = 2a$, $3x'^2 + y'^2 = 2$, $x'^2 - x'y' = 1$;
(f) $\begin{pmatrix} 3 & 0 \\ 0 & -1 \end{pmatrix}$, $\begin{pmatrix} 1 & 0 \\ 0 & 1 \end{pmatrix}$, $\begin{pmatrix} 1 & 0 \\ 0 & -1 \end{pmatrix}$, $\begin{pmatrix} a+b & 0 \\ 0 & a-b \end{pmatrix}$, $\begin{pmatrix} 5 & 1 \\ 1 & 1 \end{pmatrix}$

2. $\mathbf{Q} = \begin{pmatrix} \cos\theta & -\sin\theta \\ \sin\theta & \cos\theta \end{pmatrix}$, $\mathbf{P} = \begin{pmatrix} \cos\theta & \sin\theta \\ -\sin\theta & \cos\theta \end{pmatrix}$, . . .

3. $\frac{1}{2}\tan^{-1}(b/(a - c))$

4. $\lambda = 2$ or -3, $\mathbf{Q} = 5^{-1/2}\begin{pmatrix} 2 & 1 \\ 1 & -2 \end{pmatrix}$, $\theta = \cos^{-1}(2/\sqrt{5})$

5. $\lambda = a \pm b/2$, $\theta = \pi/4$

6. $\lambda = \frac{1}{2}(c + a \pm ((c - a)^2 + b^2)^{1/2})$, $\tan\theta = b^{-1}(c - a \pm ((c - a)^2 + b^2)^{1/2})$

7. First and third are symmetric; can even find orthogonal $\mathbf{Q}$

$\mathbf{Q} = \begin{pmatrix} 1 & 1 \\ 1 & -1 \end{pmatrix}$; impossible (only one independent eigenvector);

$\begin{pmatrix} 1 & 1 \\ 1 & -1 \end{pmatrix}$; $\begin{pmatrix} 1 & 1 \\ \sqrt{2} & -\sqrt{2} \end{pmatrix}$

8. $\det \mathbf{Q}^{-1}\mathbf{A}\mathbf{Q} = (\det \mathbf{Q})^{-1}(\det \mathbf{A})(\det \mathbf{Q}) = \det \mathbf{A}$ because multiplication of numbers is commutative

9. $(\mathbf{Q}^{-1}\mathbf{A}\mathbf{Q})(\mathbf{Q}^{-1}\mathbf{v}) = \mathbf{Q}^{-1}\mathbf{A}\mathbf{v} = \mathbf{Q}^{-1}\mathbf{v} = (\mathbf{Q}^{-1}\mathbf{v})\ldots$

10. $(\mathbf{Q}^{-1}\mathbf{A}\mathbf{Q})^r = \mathbf{Q}^{-1}\mathbf{A}\mathbf{Q}\mathbf{Q}^{-1}\mathbf{A}\mathbf{Q}\cdots\mathbf{Q}^{-1}\mathbf{A}\mathbf{Q}$. The determinant statement is the same as Problem 8

11. ${}^t(\mathbf{Q}^{-1}\mathbf{A}\mathbf{Q}) = {}^t\mathbf{Q}{}^t\mathbf{A}{}^t\mathbf{Q}^{-1} = \mathbf{Q}^{-1}\mathbf{A}\mathbf{Q}$ if ${}^t\mathbf{Q} = \mathbf{Q}^{-1}$

12. $\mathbf{Q}^{-1}(\mathbf{A} - \lambda\mathbf{I})\mathbf{Q} = \mathbf{Q}^{-1}\mathbf{A}\mathbf{Q} - \lambda\mathbf{Q}^{-1}\mathbf{Q} = \mathbf{Q}^{-1}\mathbf{A}\mathbf{Q} - \lambda\mathbf{I}$. Hence the characteristic polynomial of $\mathbf{Q}^{-1}\mathbf{A}\mathbf{Q}$ equals $\det(\mathbf{Q}^{-1}(\mathbf{A} - \lambda\mathbf{I})\mathbf{Q}) = \det(\mathbf{A} - \lambda\mathbf{I})$ by Problem 10.

13. By Problem 12, the multiplicity of λ_p as a root of $\det(\mathbf{A} - \lambda\mathbf{I})$ is the same as its multiplicity as a root of $\det(\mathbf{B} - \lambda\mathbf{I})$ where $\mathbf{B}$ is the result of diagonalizing $\mathbf{A}$. Since $\mathbf{B}$ is diagonal, the second multiplicity can be computed explicitly (Problem 12, Chapter 7, Section 1) and is the number of times λ_p occurs on the diagonal.

INDEX

Each reference in this index refers to a page number and, where possible, to the specific item on that page. This item is set in italics: the first digit is the chapter number; the next number designates the section; and the third, if any, is the number of the proposition, problem (Prob.), or example (Ex.).

M

N

O

P

Q

R

S

T

U

V

Z

D 9
E 0
F 1
G 2
H 3
I 4
J 5